Disruptive and Emerging Technology Trends Across Education and the Workplace

Julie Anne Delello
The University of Texas at Tyler, USA

Rochell R. McWhorter
The University of Texas at Tyler, USA

A volume in the Advances in Business Strategy and Competitive Advantage (ABSCA) Book Series

IGI Global
DISSEMINATOR OF KNOWLEDGE

Published in the United States of America by
IGI Global
Business Science Reference (an imprint of IGI Global)
701 E. Chocolate Avenue
Hershey PA, USA 17033
Tel: 717-533-8845
Fax: 717-533-8661
E-mail: cust@igi-global.com
Web site: http://www.igi-global.com

Library of Congress Cataloging-in-Publication Data

Names: Delello, Julie Anne, 1970- editor. | McWhorter, Rochell R., editor.
Title: Disruptive and emerging technology trends across education and the workplace / Julie Anne Delello and Rochell R. McWhorter, editors.
Description: Hershey, PA : Business Science Reference, [2020] | Includes bibliographical references and index. | Summary: "This book examines the latest instructive methods being utilized in classrooms and organizations as well as the benefits and challenges of adopting these technologies"--Provided by publisher"
Identifiers: LCCN 2019045661 (print) | LCCN 2019045662 (ebook) | ISBN 9781799829140 (hardcover) | ISBN 9781799829157 (paperback) | ISBN 9781799829164 (ebook)
Subjects: LCSH: Technological innovations--Social aspects. | Technological innovations--Economic aspects. | Disruptive technologies. | Educational technology.
Classification: LCC HM846 .D57 2020 (print) | LCC HM846 (ebook) | DDC 303.48/3--dc23
LC record available at https://lccn.loc.gov/2019045661
LC ebook record available at https://lccn.loc.gov/2019045662

This book is published in the IGI Global book series Advances in Business Strategy and Competitive Advantage (ABSCA) (ISSN: 2327-3429; eISSN: 2327-3437)

British Cataloguing in Publication Data
A Cataloguing in Publication record for this book is available from the British Library.

All work contributed to this book is new, previously-unpublished material.
The views expressed in this book are those of the authors, but not necessarily of the publisher.

For electronic access to this publication, please contact: eresources@igi-global.com.

Advances in Business Strategy and Competitive Advantage (ABSCA) Book Series

ISSN:2327-3429
EISSN:2327-3437

Editor-in-Chief: Patricia Ordóñez de Pablos, Universidad de Oviedo, Spain

MISSION

Business entities are constantly seeking new ways through which to gain advantage over their competitors and strengthen their position within the business environment. With competition at an all-time high due to technological advancements allowing for competition on a global scale, firms continue to seek new ways through which to improve and strengthen their business processes, procedures, and profitability.

The **Advances in Business Strategy and Competitive Advantage (ABSCA) Book Series** is a timely series responding to the high demand for state-of-the-art research on how business strategies are created, implemented and re-designed to meet the demands of globalized competitive markets. With a focus on local and global challenges, business opportunities and the needs of society, the **ABSCA** encourages scientific discourse on doing business and managing information technologies for the creation of sustainable competitive advantage.

COVERAGE

- Co-operative Strategies
- Outsourcing
- Value Chain
- Innovation Strategy
- Adaptive Enterprise
- Tacit Knowledge
- Value Creation
- Core Competencies
- Ethics and Business Strategy
- Customer-Orientation Strategy

IGI Global is currently accepting manuscripts for publication within this series. To submit a proposal for a volume in this series, please contact our Acquisition Editors at Acquisitions@igi-global.com or visit: http://www.igi-global.com/publish/.

The Advances in Business Strategy and Competitive Advantage (ABSCA) Book Series (ISSN 2327-3429) is published by IGI Global, 701 E. Chocolate Avenue, Hershey, PA 17033-1240, USA, www.igi-global.com. This series is composed of titles available for purchase individually; each title is edited to be contextually exclusive from any other title within the series. For pricing and ordering information please visit http://www.igi-global.com/book-series/advances-business-strategy-competitive-advantage/73672. Postmaster: Send all address changes to above address.

Titles in this Series

For a list of additional titles in this series, please visit:
http://www.igi-global.com/book-series/advances-business-strategy-competitive-advantage/73672

Reshaping Entrepreneurship Education With Strategy and Innovation
Ayansola Olatunji Ayandibu (University of Zululand, South Africa)
Business Science Reference • © 2020 • 300pp • H/C (ISBN: 9781799831716) • US $205.00

Implementing Knowledge in Organizations for Improved Performance Outcomes Emerging Research and Opportunities
Michael J. Gravier (Bryant University, USA) and Ted Farris (University of North Texas, USA)
Business Science Reference • © 2020 • 260pp • H/C (ISBN: 9781799824879) • US $195.00

CSR and Socially Responsible Investing Strategies in Transitioning and Emerging Economies
Anetta Kuna-Marszałek (University of Łódź, Poland) and Agnieszka Kłysik-Uryszek (University of Łódź, Poland)
Business Science Reference • © 2020 • 346pp • H/C (ISBN: 9781799821939) • US $235.00

Handbook of Research on the Strategic Management of Family Businesses
Jesús Manuel Palma-Ruiz (Universidad Autónoma de Chihuahua, Mexico) Ismael Barros-Contreras (Universidad Austral de Chile, Chile) and Luca Gnan (Università degli Studi di Roma "Tor Vergata", Italy)
Business Science Reference • © 2020 • 539pp • H/C (ISBN: 9781799822691) • US $275.00

Utilizing Gamification in Servicescapes for Improved Consumer Engagement
Miralem Helmefalk (Linnaeus University, Sweden) and Leif Marcusson (Linnaeus University, Sweden)
Business Science Reference • © 2020 • 319pp • H/C (ISBN: 9781799819707) • US $225.00

Cases on Strategic Management Issues in Contemporary Organizations
Mona Rashidirad (University of Brighton, UK)
Business Science Reference • © 2020 • 300pp • H/C (ISBN: 9781799816300) • US $195.00

For an entire list of titles in this series, please visit:
http://www.igi-global.com/book-series/advances-business-strategy-competitive-advantage/73672

IGI Global
DISSEMINATOR OF KNOWLEDGE

701 East Chocolate Avenue, Hershey, PA 17033, USA
Tel: 717-533-8845 x100 • Fax: 717-533-8661
E-Mail: cust@igi-global.com • www.igi-global.com

Editorial Advisory Board

Table of Contents

Detailed Table of Contents

Section 1
Disrupting and Emerging Technology for Teaching and Learning

The first section of this book specifically addresses the application of emerging and disruptive technologies in K-12 and higher education classrooms. This section contains empirical research, reviews of literature, and case studies that describe the benefits and the challenges of such technologies for educational use.

Every time that a student uses the internet on a school-issued computer, a trail of data about their online learning experiences is created. Until recently this data has largely gone unnoticed for educational purposes. This chapter will explore how research into the use of student online behavior data can be used to improve teaching and learning in digital environments. Techniques for the collection and analysis of targeted online learner analytics (TOLA) will be presented along with the findings for how TOLA impacted the teaching and learning process of real classroom teachers. The powerful and new form of data has the potential to disrupt the status quo of education, and this chapter will take a research-based look at that future.

The robot and its technology imploding in the field of education created excitement in the minds of students with renewed, revitalized interest, and curiosity stretching across the areas of science, technology, engineering, and mathematics (STEM). Robotics education along with computational thinking skills are primary keys for unlocking the unlimited creative and innovative potential with engaging hands-on activities. Robots, a fascinating learning instrument, rejuvenate, animate, and revive 21st century skills in tech-savvy language familiar with today's students.

The service-learning pedagogy has flourished in popularity and student outcomes, and with the online student population expanding, institutions of higher learning must develop programs with similar content across both face-to-face and distance-learning applications. Institutions are using eService-Learning (eS-L) programs to address this divide. Commonly reported service-learning outcomes include student's sense of belonging and community engagement, and this chapter aims to present a case with similar outcomes for graduate-level students enrolled in a distance learning course with a substantive service-learning component.

Digital learning management systems (DLMSs) are an important and necessary educational tool for learning activities. Universities are using the Digital LMS tools that are unique to their environment. Different universities have different DLMS tools that are software driven and allow the users (management, instructors, and students) to use them for regular tasks. This research discusses the qualitative data collected using a case study methodology. In this research, a case study design was selected for the qualitative methodology and semi-structured interviews were employed as the data collection method. This case study is based on a deanship of a university implementing a digital learning management system in the country of Jordan. The research provides an analysis of the interviews to gain insights into instructors' perspectives regarding the factors influencing the implementation of the digital learning management system in a higher education sector.

Section 2
Disruptive and Emerging Technology: Social and Ethical Considerations

The second section of the book reminds us of the importance that emerging and disruptive technologies have on individual lives in education and the workplace. Social and ethical considerations such as issues of privacy, invasiveness, disruptive innovations, and the impact of technology on loneliness are addressed.

Over the past 30 years, the internet has evolved from being the web of content to being the web of thoughts and the web of things in business, communication, entertainment, and education. To stay competitive, higher education institutions have had to train students on the wide range of skills and experiences and to move to digital platforms to better meet the needs of students, employees, and organizations. This chapter provides an overview of the development of online education, the 1.0 to 5.0 phases of web development, and how the field of education has adapted to these phases. Particular emphasis is placed on the use of mobile learning such as MOOCs, course collaboration software, and how smartphones can be used in courses to interact with peers and faculty. This chapter then presents a case study illustrating how online courses can successfully integrate Web 4.0 and 5.0 technology. It concludes by discussing the benefits and challenges of adopting some disruptive technologies and on how educational institutions can meet the needs of the next generation of students.

There is a growing concern regarding student use of digital devices such as smartphones during classroom instruction. This pilot study used a mixed-method survey to collect data from 146 K-16 educators. Eighty percent of the participants said that students multitask, and 50% said they multitask a lot. Many of the instructors reported they are distracted by students' use of phones while more than half of the participants indicated they felt doing multiple tasks at the same time disrupts class time and may impact a students' ability to learn. These findings have important implications because classroom instructors may need to modify instructional strategies to keep students on task, have clear policies on the use of electronic devices, and ensure that students understand instructor expectations.

About one-fifth of the population in the United States in 2015 will be age 65 or older in 2050 and loneliness may be a contributing factor that inhibits their well-being and overall health. As the number of older adults continues to escalate, information and communication technologies such as smartphones and computers may create an increase in social connectedness leading to a decline in loneliness and social isolation. Results from this pilot study suggest that the older adult participants demonstrated some degree of loneliness. As the older adults used social media to connect with friends, family, and other information of interest, there was an increase in social connectedness for many of the participants, but the intensive iPad intervention was not significant in terms of reducing loneliness for either group over time.

Technology has become increasingly invasive and corporate networks are expanding into public and private spaces to collect unprecedented data and provide new services such as artificial intelligence and through unsettling human-like personas. The term "creepy technology" is appearing in the literature along with concerns for privacy, ethical boundaries, cybersecurity, and mistaken identity but is also in news articles to inform the public about technology advances that affect consumer privacy. Invasive technology provides the impetus for external adaptation for many organizations and current trends require rapid adaption to potential threats to security. Also, usability addresses how users respond and adapt to new technology. This chapter includes the presentation of an exploratory study of how the public responded to various

Preface

THE ROLE OF DISRUPTIVE AND EMERGING TECHNOLOGIES IN EDUCATION AND THE WORKFORCE

Technology continues to reshape the world we live in as academia and industry are being transformed with disruptive and emerging technologies such as artificial intelligence, machine learning, big data analytics, mixed realities, social media, and smart devices. These advancements in technology are associated with the Internet 4.0, part of the fourth industrial revolution, and have the potential to change the way we live, work, and conduct business. Such disruptive technologies hold great promise both for increasing access to information and as new developments in technologies take shape, they will create opportunities for innovative ways of teaching and learning which were never before thought possible.

LEVERAGING TECHNOLOGICAL SHIFTS TO TRANSFORM LEARNING

The 2018 New Media Consortium (NMC) Horizon Report noted that within the next five years, data analytics, makerspaces, adaptive learning technologies, artificial intelligence, mixed-realities, and robotics will be important trends shaping the future of higher education (Becker, Brown, Dahlstrom, Davis, DePaul, Pomerantz, & 2018). Such developments will require the rethinking and reshaping of traditional educational models. And, the recent International Society for Technology in Education (ISTE) Standards for Educators are challenging instructors to rethink their traditional approaches in preparing students to "drive their own learning" (ISTE, 2020, para. 1).

Many students are tech-savvy and expect their instructors to utilize technology in the classroom to enhance student learning. In a recent study, the majority of students want curriculum to move from traditional materials to digital tools (i.e., mobile technology) for learning (Dinsmore, 2019). As technology progresses, it will be vital for students to have opportunities to understand, create, and utilize emerging and disruptive technologies to develop critical thinking skills for solving real world problems. According to Backhouse (2013), emerging technologies in higher education are novel or innovative technologies purposed to change teaching and learning.

Institutions must begin to rethink how to meet the academic and social needs of their students, which will include finding ways to redesign learning spaces, advance cultures of innovation, personalize learning, and modularize degrees (Educause, 2019). These curricula updates will necessitate generating unique learning spaces in both formal and informal environments. For example, makerspaces, also called hackerspaces and FabLabs, are collaborative spaces where emerging technologies such as virtual reality, robotics, digital fabrication tools, 3D printers, and other tools may be found (makerspaces.com). Laura Flemming (2015) defined a school makerspace as "a place where young people have an opportunity to explore their own interests; learn to use tools and materials, both physical and virtual; and develop creative projects" (p. 5). Maker-based learning experiences build student "confidence, foster creativity, and spark interest in science, technology, engineering, math, the arts, and learning as a whole" (Byrne & Davidson, 2015, p. 1). Such spaces help to engage students and personalize learning.

Online learning has become more pervasive and sophisticated in the last five years. In fact, over six million students were enrolled in at least one distance education course in 2017 (Allen & Seaman, 2017). With the rapid advancement of online learning, sophisticated technologies such as the use of learning management systems (LMS) like Canvas (Canvas.com) integrate learning tools, gradebook, and collaborative discussion boards to engage students in both real-time and asynchronous learning experiences. Online learning will also incorporate mobile learning (m-learning) as over 67% of students will use a mobile device to complete their coursework (Magda & Aslanian, 2018). According to the McKinsey Global Institute (2013), mobile Internet devices are considered disruptive technologies that enable "users to go about their daily routines with new ways of knowing, perceiving, and even interacting with the physical world" (p. 6). Such mobile access will allow students greater flexibility and access to learning.

Also, Massive Open Online Courses (MOOCs) that enroll millions of users from across the globe in free courses are great for acquiring new knowledge open to nontraditional students (Delello & McWhorter, 2016). Additionally, flipped inclusion, a teaching method where students view lectures and materials outside of class so that time together is devoted to problem solving, has changed the instructor's role to one of facilitator and learning coach (Altemueller & Lindquist, 2017).

As educational contexts evolve, the use of big data, the thousands of data points generated in schools daily, could lead to a more personalized learning experience and in turn, lead to greater student success. However, there must be a human connection to the information, which will include training. A recent study by Dooilezel and McLeod (2019) concluded that the current usage of big data tools and technologies in the workplace was integral to the organization's success and called for more training to keep up with the demand for big data analytical skills.

NEW DIMENSIONS OF TECHNOLOGY ON BUSINESS AND INDUSTRY

As the job market consistently changes, higher education must prepare students for the next generation workforce dominated by Millennials and Zoomers (Generation Z) (Aldridge, Powell, & Harvatt, 2019). By 2020, researchers have predicted that 70% of the global workforce will consist of individuals from Millennials and "Nexters" (Bennett, 2019) and even Baby Boomers who are returning for an "encore career" (Bennett & McWhorter, in press). These generations will be of critical importance to business and industry as they will develop and shape the future of technology. Organizations will face challenges to identify innovative ways to do business.

In business theory, disruptive technologies or newly termed 'disruptive innovation' "are simpler, more convenient, and less expensive products that appeal to new or less demanding customers" (Christensen & Raynor, 2003, p. 2). To survive in today's global market, organizations must identify innovative strategies to achieve a competitive advantage. Oftentimes, it is the utilization of "multiple emerging technology that [become] the basis of modern competitive strategies that have a profound and lasting impact on businesses and create opportunities for new ones" (Coleman, 2015, para. 5). The challenge, then, is to utilize the emerging technology as a disruptive innovation (Christensen, 2020) and quickly integrate them into the organization's overall strategy to gain an innovative competitive advantage.

Data tracking has become both a boon and a curse for the modern consumer. Each time a user is online, they leave traces of their activity that can be tracked and collected as the user moves throughout the web tracking apps, browser, and even their emails that gather a myriad of personal preferences utilized to make suggestions for purchases and locate a loved one (Internet Health Report, 2018). However, the downside of data tracking is seen when it is too invasive and in fact, Google recently announced that its Chrome browser would eliminate the use of "cross-website cookies…following recent bans introduced by Apple, Microsoft and Mozilla, it would also aim to make 'obsolete' so-called third-party cookies that follow internet users from site to site, sometimes tracing their browsing for months at a time" (Barker & Murgia, 2020, para. 1-2).

THE CHALLENGES

With emerging and disruptive technologies comes costs, benefits, and risks. For example, the ubiquity of technology including the use of smart phones and tablets has resulted in an increase in multimedia data mining. Such mining is advantageous for storing customer preferences and giving insight to future product development; and, for offering management tips for employees. However, data mining can also be considered invasive and even creepy when customers and employees feel the data tracking has violated their personal privacy (Agarwal, Bersin, Lahiri, Schwartz, & Violini, 2018; Barba, 2016). Also, the risks of data mining include the theft of both sensitive personal data that could fall into the wrong hands or sold to future employers and healthcare companies (ConsumerNotice.org, 2020) and organizational entities.

Also, there is a lack of training and knowledge in the handling of sensitive data. Kohen (2019) noted that employees have accidently shared, misplaced or mishandled sensitive data and many senior executives attribute their most recent incident to these careless data handling behaviors. According to Cheng, Liu, and Yao (2017), threats to digital data occur inadvertently or maliciously and those handling data at all levels must be schooled in both prevention and detection of data privacy.

Distractibility is a major challenge for K-12 as well as higher education students. Educators should be knowledgeable of how and when mobile and other appealing technologies are advantageous in K-12 and higher education classroom so that the benefits outweigh the challenges. Students are not the only ones that have problems with distractibility. According to Hallowell (2015), employee activities such as social media addiction, multitasking, and unnecessary internet browsing are detrimental to productivity in the workplace.

SEARCHING FOR A SOLUTION

New technological developments are rapidly altering the business and social landscape, as we know it. We now control our homes from the touch of a button, ask chatbots for movie or restaurant recommendations, ride in autonomous cars, and have wearable devices, which allow us the ability to monitor our health remotely. Furthermore, mobile technology, social media, and cloud computing bring people together from all over the world, allowing equal access, and contribute to a global marketplace. In order to stay competitive, institutions of learning must provide an emphasis on the wide range of skills and experiences needed to contribute to a 21st century workforce. As new technologies emerge and even disrupt, there will be a demand for new forms of education and deeper learning.

In a short thirty-year span, we have seen the Internet develop and also mobile devices with applications and social platforms have connected individuals around the globe. Such innovations have disrupted life as we once knew it and it appears likely that the proliferation of such technological advancements will continue to change the status quo. It will be imperative that schools prepare students for the skills they will need in an ever-evolving world. However, Elon Musk, entrepreneur and businessman, remarked that "I think one thing that we can be quite certain of, is that any predictions we make today for what the future will be like in 50 years will be wrong" (Roylance, 2017, para. 12).

ORGANIZATION OF THE BOOK

This book focuses on new research findings in order to understand the role that the academy has in teaching the skills needed to keep up with the changing digital society and competitive global job market. This book is unique, in that it brings together a series of academic cases and empirical studies on the use of emerging and disruptive technologies rather than simply hypothesizing their utility by explanations and descriptions of the technologies. The book is organized into eleven chapters and the authors hail from three countries, each with their own distinctive cultures. A brief description of each of the chapters follows next.

Chapter 1 describes how a Targeted Online Learner Analytics (TOLA) application was utilized by educators to identify online learning behaviors at the individual-student level. Tools for data analytics are emerging to offer educators new ways to collect data and discover insights into students' preferences, skills, and abilities. The research attempts to map the potential for such a program to improve instructional practice and enhance teaching and learning. However, the authors point out that educators will need the time and training to know how to analyze and use such data.

Chapter 2 discusses the maker movement on school campuses as a means to improve student learning. From new fabrication tools like 3D printers and laser cutters to robotics and drones, the makerspace is advancing the development of 21st century skills by engaging students in hands-on learning and critical thinking skills. Drawing on qualitative data from schools and a community library, the author of this chapter describes the components and instructional practices taking place within educational makerspaces. This chapter concludes with a set of pedagogical strategies that can be adopted by other educators to provide students with new learning experiences.

Chapter 3 reviews a flipped inclusion model as an instructional strategy, in which, educators deliver content in non-traditional ways such as through online videos, the sharing of resources, and peer to peer networking through social media. The authors of this chapter provide evidence of how the flipped inclusion model was integrated into the education system in Italy. The results of the research seem to indicate flipped inclusion improves learning, student communication, and collaborative skills while enhancing prosocial behaviors.

Chapter 4 addresses how the use of robotics may promote computational thinking and provide practical hands-on approaches to learning in the classroom. Through the use of literature and concrete examples, the author aims to inform the reader of the limitless opportunities to promote 21st century skills in the educational environment and concludes with a framework for a future driven classroom.

Chapter 5 describes how e-service-learning (electronic service-learning) integrates meaningful community service within online instruction. Using a case-study methodology, the author outlines the benefits to students, non-profit organizations, and to the university. As the online student population increases, the author of this chapter suggests that such service-learning is not always available to students at a distance and higher education models should develop programming to meet this dilemma.

Chapter 6 addresses the use of digital learning management systems within six unique Jordanian universities. This study focuses on the distinctive perspectives of instructors including their perceived ease of use, the effectiveness, and overall quality of a LMS system. The authors conclude with the importance of having university strategies and policies in place to support faculty on the use of a LMS in order to improve higher education.

Chapter 7 provides a detailed look the development of the World Wide Web from the static, read-only Web 1.0 to the emergence of the symbionet Web 5.0. The author of this chapter describe how the Web has expanded from one of content to a Web of an Internet of Things (IoT). Furthermore, the author depicts the development of online learning environments including the growth of massive open online courses (MOOCs), which have led to various micro-credentials and greater numbers of online degrees and concludes with a set of best practices to serve new generations of learners.

Chapter 8 describes a world of pervasive technology use where there is a growing concern that students are always connected to their devices. However, is such use dividing students' attention in schools and disrupting learning? Do students multitask more than ever and are devices such as smartphones distracting to both students and teachers? What types of technology usage policies are in place? Should educators ban technology use all together? These and other queries are highlighted in this chapter as the authors describe the perceptions of educators around the world. Furthermore, this chapter discusses the implications of such use and strategies for better integration of devices into classrooms.

Chapter 9 describes the steady increase in adults, age 65 and older, living within the United States. And, although these older adults are living longer, loneliness may impact their overall health and well-being. The authors of this chapter describe an iPad intervention undertaken to decrease loneliness and improve social connectedness, which is a non-traditional and an understudied area of technology research, with two cohorts of older adults residing at senior living centers in the US.

Chapter 10 discusses the new generation of technologies which have become increasingly invasive. Allegations of creepiness have raised concerns across personal and professional boundaries as emerging and disruptive technologies such as artificial intelligence, cloud computing, and ambient social applications collect data across the Internet. The authors of this chapter present an exploratory study on the nature of such creepy technology and addresses how the public responds to such intrusiveness. The chapter concludes with a discussion of interventions critical to both public and private sectors.

Chapter 11 presents a research study detailing the invasive effects of technology on the personal lives of students in higher education. This chapter highlights the impact that over-connecting to technology has on work-life and the challenges individuals have in disconnecting from devices. The authors draw on narratives from participants to create a better awareness of such screen connectivity, and in turn, provide the reader with a better understanding of this phenomenon.

Julie Anne Delello
The University of Texas at Tyler, USA

Rochell R. McWhorter
The University of Texas at Tyler, USA

REFERENCES

Agarwal, D., Bersin, J., Lahiri, G., Schwartz, J., & Violini, E. (2018). *People data: How far is too far*? Retrieved from https://www2.deloitte.com/us/en/insights/focus/human-capital-trends/2018/people-data-analytics-risks-opportunities.html

Aldridge, S., Powell, M. M., & Harvatt, K. (2019). *Educating the next generation workforce preparing students to meet employers' needs and expectations.* Drexel University Online. Retrieved from https://online.drexel.edu/globalassets/next-generation-workforce-final-website.pdf

Allen, I. E., & Seaman, J. (2017). *Digital learning compass: Distance education enrollment report 2017.* Babson Survey Research Group. Retrieved from http://onlinelearningsurvey.com/reports/digtiallearningcompassenrollment2017.pdf

Altemueller, L., & Lindquist, C. (2017). Flipped classroom instruction for inclusive learning. *British Journal of Special Education, 44*(3), 341–358. doi:10.1111/1467-8578.12177

Backhouse, J. (2013). What makes lecturers in higher education use emerging technologies in their teaching? *Knowledge Management & ELearning, 5*(3), 345–358.

Barba, R. (2016). Data mining and the 'creepy' factor. *American Banker.* Retrieved from https://www.americanbanker.com/news/data-mining-and-the-creepy-factor

Barker, A., & Murgia, M. (2020). Google to phase out most invasive internet tracking. *Financial Times.* Retrieved from https://www.ft.com/content/1f56591e-36e1-11ea-a6d3-9a26f8c3cba4

Becker, S. A., Brown, M., Dahlstrom, E., Davis, A., DePaul, K., & Pomerantz, J. (2018). *NMC Horizon Report: 2018 Higher Education Edition.* Louisville, CO: Educause.

Bennett, E. E., & McWhorter, R. R. (in press). Digital technologies for teaching and learning. In Handbook of Adult and Continuing Education. Academic Press.

Bennett, R. (2019). *A breakdown of the five generations of employees in the workplace.* Retrieved from https://www.navexglobal.com/blog/article/formal-introduction-five-generations-employees-your-workforce/

Byrne, D., & Davidson, C. (2015). *Makeschools higher education alliance: State of making report.* Retrieved from http://make.xsead.cmu.edu/week_of_making/report

Cheng, L., Liu, F., & Yao, D. (2017). Enterprise data breach: Causes, challenges, prevention, and future directions. *WIREs Data Mining and Knowledge Discovery, 7*(5). Retrieved from https://onlinelibrary.wiley.com/doi/full/10.1002/widm.1211

Christensen, C. (2020). *Disruptive innovation.* Retrieved from http://claytonchristensen.com/key-concepts/

Christensen, C. M., & Raynor, M. E. (2003). *The innovator's solution: Creating and sustaining successful growth.* Boston: Harvard Business School Press.

Coleman, K. (2015). Tech innovation is the cornerstone of competitive strategy. *Computerworld*. Retrieved from https://www.computerworld.com/article/2991533/tech-innovation-is-the-cornerstone-of-competitive-strategy.html

ConsumerNotice.org. (2020). *Data mining*. Retrieved from https://www.consumernotice.org/data-protection/mining/

Delello, J. A., & McWhorter, R. R. (2016). New visual literacies and competencies for education and the workplace. In B. Guzzetti & M. Lesley (Eds.), *Handbook of Research on the Societal Impact of Digital Media* (pp. 127–162). Hershey, PA: IGI Global. doi:10.4018/978-1-4666-8310-5.ch006

Dinsmore, B. (2019). Contested affordances: Teachers and students negotiating the classroom integration of mobile technology. *Information Communication and Society*, *5*(5), 664–677. doi:10.1080/1369118X.2019.1568518

Educause. (2019). *Educause horizon report: 2019 higher education Edition*. Retrieved from https://library.educause.edu/-/media/files/library/2019/4/2019horizonreport.pdf?la=en&hash=C8E8D444AF372E705FA1BF9D4FF0DD4CC6F0FDD1

Flemming, L. (2015). *Worlds of making: Best practices for establishing a makerspace for your school*. Thousand Oaks, CA: Corwin Press.

Hallowell, E. M. (2015). *Driven to distraction at work: How to focus and be more productive*. Boston, MA: Harvard Business Review Press.

Internet Health Report. (2018). *The good, the bad and the ugly sides of data tracking*. Retrieved from https://internethealthreport.org/2018/the-good-the-bad-and-the-ugly-sides-of-data-tracking/

ISTE.org. (2020). *ISTE Standards for Educators*. Retrieved from https://www.iste.org/standards/for-educators

Kohen, I. (2019). 10 data security risks that could impact your company in 2020. *Forbes*. Retrieved from https://www.forbes.com/sites/theyec/2019/10/01/10-data-security-risks-that-could-impact-your-company-in-2020/#5c2d2eda0c06

Magda, A. J., & Aslanian, C. B. (2018). *Online college students 2018: Comprehensive data on demands and preferences*. Louisville, KY: The Learning House, Inc.

Makerspaces.com. (2020). *What is a makerspace*. Retrieved from https://www.makerspaces.com/what-is-a-makerspace/

McKinsey Global Institute. (2013). *Disruptive technologies: Advances that will transform life, business, and the global economy.* Retrieved from https://www.mckinsey.com/~/media/McKinsey/Business%20Functions/McKinsey%20Digital/Our%20Insights/Disruptive%20technologies/MGI_Disruptive_technologies_Full_report_May2013.ashx

Roylance, W. (2017). A conversation with Elon Musk. *Diplomatic Courier.* Retrieved from https://www.diplomaticourier.com/posts/conversation-elon-musk

Acknowledgment

The editors would like to acknowledge the help of all the people involved in this project and, more specifically, to the authors and reviewers that took part in the review process. Without their support, this book would not have become a reality.

First, the editors would like to thank each one of the authors for their contributions. Our sincere gratitude goes to the chapter's authors who contributed their time and expertise to this book.

Second, the editors wish to acknowledge the valuable contributions of the reviewers regarding the improvement of quality, coherence, and content presentation of chapters. Most of the authors also served as referees; we highly appreciate their double task.

Julie Anne Delello
The University of Texas at Tyler, USA

Rochell R. McWhorter
The University of Texas at Tyler, USA

Section 1
Disrupting and Emerging Technology for Teaching and Learning

The first section of this book specifically addresses the application of emerging and disruptive technologies in K-12 and higher education classrooms. This section contains empirical research, reviews of literature, and case studies that describe the benefits and the challenges of such technologies for educational use.

Chapter 1
Moving Targeted Online Learner Analytics Into the Hands of Teachers

Gregory Cottrell
North Hunterdon High School, USA

Isabel Christine Resende
Nazareth Area School District, USA

ABSTRACT

Every time that a student uses the internet on a school-issued computer, a trail of data about their online learning experiences is created. Until recently this data has largely gone unnoticed for educational purposes. This chapter will explore how research into the use of student online behavior data can be used to improve teaching and learning in digital environments. Techniques for the collection and analysis of targeted online learner analytics (TOLA) will be presented along with the findings for how TOLA impacted the teaching and learning process of real classroom teachers. The powerful and new form of data has the potential to disrupt the status quo of education, and this chapter will take a research-based look at that future.

INTRODUCTION

There has never been a time in the history of education when so much information about student online learning behavior has been collected. Every time a student browses the internet, accesses online content, Googles a question or types in a Google doc, a record of activity is created. This activity is essentially a descriptive pathway

DOI: 10.4018/978-1-7998-2914-0.ch001

of learning that is marked with the breadcrumbs of student internet browsing data. Unfortunately, much of this valuable data about student online learning experiences is not being utilized for instructional purposes.

The field of learning analytics has made some strides in helping educators make sense of this under-utilized data, but by in large learning analytics has had little impact on K-12 public education. Analytics has not had the transformative and disruptive impact on education as it has had in other fields of work.

This chapter will explore the groundbreaking potential of Targeted Online Learner Analytics (TOLA) in the design of instruction that optimizes student learning. Two research studies will be presented. Both studies looked to understand how information about student online activities could affect change within schools. Insights gained from the studies will be explored and a vision for how TOLA has the potential to disrupt the educational setting will be presented.

BACKGROUND

Defining Analytics and its Emerging Impact on Education

Using data to make decisions is an obvious best practice that improves organizational outcomes and productivity (Long & Siemens, 2011). The act of collecting data to measure, compare, and improve performance is at the core of analytics. Analytics has been successfully used in settings outside of education to support operational decision making. Ultimately, the goal of analytics is to provide the user with information and data to make informed decisions, with the goal of improving performance (Papamitsiou & Economides, 2014; Radu, 2017). According to the Society for Learning Analytics Research, "Learning analytics is the measurement, collection, analysis, interpretation, and reporting of data about learners and their contexts, for the purpose of understanding and optimizing learning and the environment in which it occurs…." (Wilson, Gochyyev, & Scalise, 2016, p. 2).

Unlike the sciences and industry, the educational community is late in its adoption of analytics to improve performance (Baker & Inventado, 2014). Only recently has it begun to explore analytics as a tool to gain insight into student-learning online (Papamitsiou & Economides, 2014). It has been referred to as a game-changer for teaching and learning (Drachsler & Greller, 2016). The use of analytics in education has gained more acceptance in recent years for four reasons. The first reason is that the quality of the data collected has improved in quality and quantity. Increased access to and use of technology in schools across the nation has led to an increase in the quantity of data available to educators. In most schools, each time a student uses a digital device to interact with content, that interaction is logged and made available for further analysis. The scope of the data that is collected during learning activities has increased exponentially as large textbook companies like Pearson and McGraw-Hill become interested in learning analytics (Baker & Inventado, 2014).

Second, the format by which the data is reported is more user friendly. Previously, educational data that was logged electronically was cumbersome to organize for the purposes of analysis and interpretation. Effective logging mechanisms of educational data have been standardized, which has led to more usable and manageable forms of analysis and interpretation for educators (Baker & Inventado, 2014).

The third reason the use of analytics has gained acceptance in education is the advances in computing. The increased processing and computational power of digital devices makes it much easier to accomplish tasks. Specifically, advances in computing have led to the ability to analyze large quantities of data from multiple sources in a shorter amount of time (Baker & Inventado, 2014).

The fourth reason is the sophistication of analytics tools, which now make collecting, storing, organizing, and filtering large quantities of data manageable, facilitating the analysis process. The number of tools that have been developed to assist with the analysis of data has also increased (Baker & Inventado, 2014).

Learning analytics in education has its roots in learning management systems (LMS) like Moodle and Blackboard, to name just two. Traditionally, these LMSs gave educators the ability to store course content, assess student learning online, receive and disseminate assignments, grade, and communicate with students through discussions forums, blogging, and messaging. However, learning analytics methods have evolved to offer much greater functionality. They can now personalize learning for students, measure engagement, and predict risk by collecting information on the interactions the student has with the LMS. For instance, learning analytics applications can collect information on the webpages students have accessed, forums posted, discussions with other users, time spent on a particular topic, answer to quizzes, and their grades (Radu, 2017). The availability of learning analytics is transformative and can provide a new model for decision making in teaching and learning. Students' electronic footprint can provide valuable insight for educators into what is happening during the learning process (Long & Siemens, 2011).

In short, the term learning analytics encompasses many different types of data and analytical procedures. This chapter is going to focus on a very specific and new type of data called Targeted Online Learner Analytics (TOLA). As the name suggests, this data is targeted or filtered to show activity for a specific assignment. It encompasses all online activity that was conducted for the targeted assignment. It is specific and unique to each individual learner and it is presented to teachers using analytics that describe the unique individual online learning experience of the student in a descriptive format, not prescriptive. TOLA provides a teacher with information needed to infer the unique online learning story for each of their students. This chapter will describe how groundbreaking research as well as continued advancements in the ability to collect and present TOLA will combine to disrupt the status quo of teaching and learning in a 1:1 classroom environment.

Finding a Place for Targeted Online Learner Analytics

Online browsing history contains extremely rich and personal data about a person. If in doubt, go to the history section of your web browser and take a look at your web activity for the last week. Now imagine that you were asked to share that data at your next job interview, to your current employer or on your next first-date. Most people would not be comfortable with the idea. Internet browsing history contains extremely personal information such as data about the times of day a person is online, how much time they spend online and where they are spending that time, the political affiliation of the news that people read and all of the oddities and curiosities that people explore online. While this information is very personal, it is also very powerful. Unfortunately, this data has had a very little impact on the teaching and learning process. The minimal impact of analytics on education pales in comparison to the impact of analytics on other industries. The authors of this chapter sought to investigate why this rich data has yet to disrupt the educational process and determine if it has the potential to do so.

This chapter will present the findings of two qualitative research studies conducted at East Stroudsburg University of Pennsylvania. These research studies were designed specifically to understanding the potential impact of TOLA on the instructional design process in K-12 public education. A related aim was to understand the value of TOLA as a formative practice in improving student performance. When developing a unit of study, teachers engage in an instructional-program-design process. They identify a set of learning targets, design a series of tasks, and identify resources and supports for students to engage in during the course of the learning (Lockyer, Heathcote, & Dawson, 2013; Reigeluth & Chellman, 1999). These components of the instructional design are interconnected and do not exist in isolation. TOLA has the potential to be a critical component in the instructional-evaluation process, thereby impacting the instructional-planning process (Reigeluth & Chellman, 1999). Stated plainly, these studies wanted to determine if student online browsing data had the potential power to impact education in the same way that data analytics has impacted other industries.

ETHICAL CONCERNS ABOUT LEARNING ANALYTICS

Although the availability of learning analytics for the educational community has been referred to as game-changing, it brings with it a set of challenges and concerns which, if not addressed, could limit the functional use of the metrics in teaching and learning. Historically, many of the concerns raised about the collection of learning analytics have risen out of a deeper, more pervasive concern about privacy and commercial exploitation of data collected. In the educational community, the acceptable use of learning analytics for learning design and instructional design will depend on the transparency of the model for collecting the data as well as the reliability of the metrics. Data-privacy issues are at the core of establishing learning analytics in education. Despite concerns and laws governing personally identifiable information, there is little guidance in policy and regulations that addresses privacy issues or establishment of best practice for the use of learning analytics in education on a large scale and with third parties (Drachsler & Greller, 2016).

That being said, Drachsler and Greller (2016) make the case that learning analytics and privacy safeguards can exist in the same space and have developed a DELICATE checklist to assist in the use of learning analytics. Similarly, Khalil and Ebner (2016) have developed a conceptual approach to allow privacy safeguards and learning analytics to exist in the same space for educational purposes. Their methodology combines anonymization strategies with learning-analytic techniques to establish privacy safeguards. Khalil and Ebner (2016) use a de-identification process of students to keep personally identifiable information confidential. The de-identification ensures anonymity and compliance with the Family Educational Rights Privacy Act (FERPA). FERPA clearly allows for the release of educational records, without consent, if personally identifiable information, such as, but not limited to, name, Social Security number, and birthdate have been removed. Together, the work of Drachsler and Greller (2016) and Khalil and Ebner (2016) is helping to provide frameworks and guidance allowing learning analytics to be used in educational settings while still addressing student privacy.

TOLA RESEARCH IN ACTION

Cottrell Study Design and Findings

The first attempt to understand how the availability of analytics about student online learning behaviors affect the teaching and learning process was conducted in 2017 by Greg Cottrell in a study titled *Thinking Outside the Box: A Descriptive Study Focusing on the Application of Learning Analytics on Instructional Design*. At the inception of this research project, Cottrell was an assistant principal working in a school that had recently adopted a 1:1 Chromebook initiative. The school had also purchased monitoring software to filter and restrict student activity online. As an assistant principal, Cottrell would receive a notification from the filtering software every time that a student went somewhere online that was inappropriate, such as a website that promotes adult content, drugs or violence. Cottrell's job at the time was to reprimand the students for accessing inappropriate websites, but as he would look through the record of online activity he noticed all of the data about student online learning that he had available to him. He realized that the data that he and his teaching staff could access had to potential to tell a powerful story about the unique online learning experience, style and preferences of the students. Cottrell hypothesized that somewhere buried in the recorded data of hundreds of webpages views for each student every day was a story about an individual learning experience that, if told to teachers, could improve their teaching.

The problem was figuring out how to uncover the data necessary to tell a story about specific learning experiences of individual students. Cottrell discovered that it was not uncommon for a student to have over 500 different webpage views recorded in one 24-hour period. Students were online at all hours of the day and in multiple locations. The data that was available was in spreadsheet format for each student. There is an example below in Figure 1. The spreadsheet simply lists the URL of the website and the timestamp for when the webpage was opened and closed. This record of activity for 24 hours presented way too much information.

It is difficult for any person to make meaning out of something that looks like Figure 1. The spreadsheets had several intertwined learning experiences for the various assignments that students worked on throughout their day. These experiences where also intermixed with social and entertainment activities that the student engaged in online. The sheer volume of data collected presented a serious roadblock to any single teacher who wanted to figure out exactly what a student was doing online throughout the day for a specific subject area. In order to address this roadblock while testing the hypothesis that this data can improve teaching, Cottrell designed a study where he manually sifted through the online behavior data of students in participating teachers' classes. Through a slow and painstaking process, Cottrell was able to provide each teacher in his study with analytics about the online learning behaviors of each student in the class for a targeted assignment. This is the type of data that we are now referring to as Targeted Online Learner Analytics (TOLA).

Cottrell designed and conducted a research study that was intended to determine if a teacher's analysis of student online activity influenced their instructional decisions. His study was based on the following research questions.

Figure 1. Sample student internet history.
Illustrates data collected of students' online activity

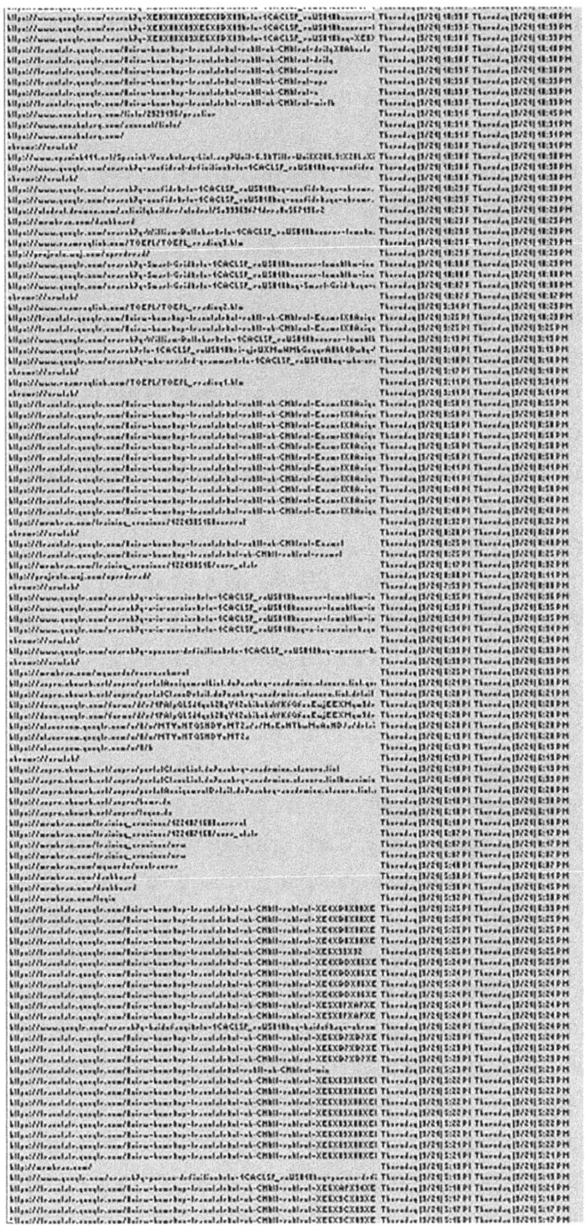

1. Is there a consensus pedagogy amongst the teachers within the study on how to use Chromebooks within their instructional design?
2. Are teachers' instructional design decisions changed as a result of receiving a comprehensive learning analytics log? If so, in what way?
    a. What student internet usage data do teachers find most influential for making instructional design decisions?
    b. What relationship exists between teachers' perceived value of educational technology before and after receiving detailed feedback about how students used technology to complete a class assignment?

To accomplish this research endeavor, Cottrell designed a qualitative study that utilized a participatory action research method. Cottrell conducted his research in the school where he was working as an administrator. This is significant for several reasons. First, the fields of data mining and learning analytics have been working to make sense out learning data, yet so far they have had little impact on the classroom practices of 1:1 teachers. This is mainly a result of the fact that analysts do not have complete access to student information and are rarely focused on the context in which learning takes place (Reyes, 2015). The proximity of data origination to the data consumer is important. The epistemological approach taken by this study was that the teacher is the one with a rich understanding of the student, knowledge of the subject content and complete picture of the learning context and intended outcomes. It was the goal of this study to put TOLA in the hands of teachers and see what insights they produce.

A second significant point about the participatory action research method of this study has to do with student privacy. Concerns for student privacy have contributed to a gap in K-12 learning analytics and data mining research (U.S. Department of Education, 2012; Buckingham, 2012). The Family Educational Rights Privacy Act (FERPA) established federal guidelines that protect student records, including server logs. Student records may not be disclosed to outside organizations without written consent from a parent. However, student records may be disclosed to a school official as long as it is for a legitimate educational purpose (Family Educational Rights Privacy Act, 2015). It is for this reason that many educational researchers have difficulty conducting research that involves student data, but due to Cottrell's role as both an educator and researcher he was able to use an action research methodology and work with the teachers within his school to access and view student data without jeopardizing the student privacy concerns that have hindered the advancement of learning analytics research in the past.

The study was organized using a purposive sampling method to identify five participating teachers who all either taught English or Social Studies. These subject areas where selected because there are many similarities between the standards for these subject areas such as citing evidence in written documents, reading comprehension and writing statements to support a claim (Common Core State Standards Initiative, 2016). The five teachers formed a Professional Learning Community (PLC). At the first PLC meeting Cottrell asked each of the teachers to identify an assignment what would begin during class and take 1-3 days for the student to complete, with some of the work expected to take place independently outside of regularly scheduled class time. Once each teacher had identified an assignment, data was collected using the cycle of feedback illustrated below in Figure 2.

Figure 2. Cycle of feedback
Illustrates the process used to provide feedback and collect data.

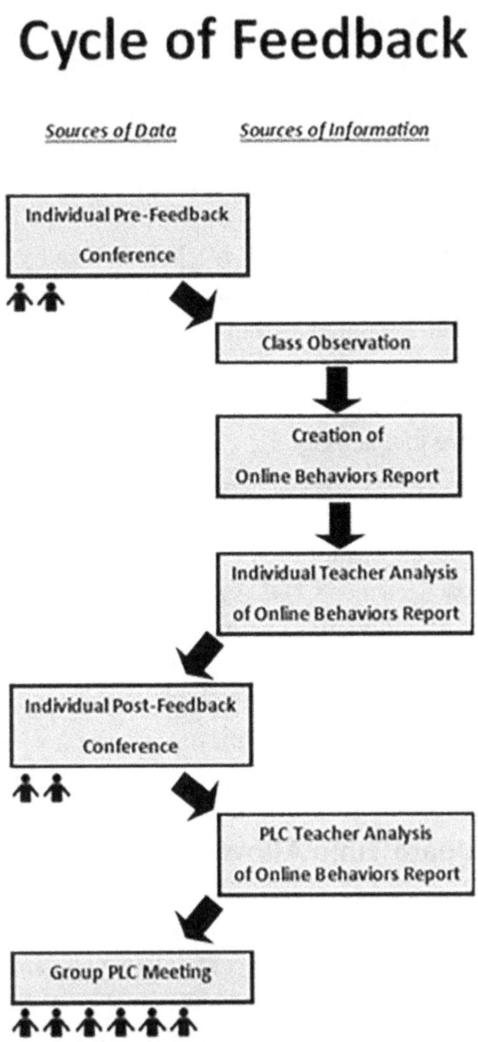

This cycle involved the researcher meeting individually with each teacher prior to the assignment to gather information about their expectations for the assignment. The researcher then observed the class to gain a better understanding of the context of the assignment. After completion of the 1-3 day assignment, Cottrell used his access to online browsing history of school-issued Chromebooks to download a record of all activity that occurred during the days of the assignment. Cottrell then manually sifted through the record of each student to identify the page views that matched the topic of the assignment. After identifying the page views for when a student was on task for the targeted assignment, Cottrell was able to provide some basic analytics and a short summary about each student's behavior back to the teacher. Instead having a spreadsheet with 500 different page views for each student, the teacher was presented with a much smaller list of the website activity that was conducted for the targeted assignment. Through the use of timestamps, Cottrell was also able to provide the teachers with a report about how much time each student spent in class working on the assignment as well as how much time they spend outside to class and what time of day the work was done. Also, Cottrell provided the teachers with the google search terms that were used by each student as they completed the assignment.

The TOLA feedback report that each teacher received was tailored to match the teacher-reported questions that they had about their student online behavior. For example, one teacher was interested in how often students accessed the list of resources that he provided to them, so the TOLA report showed that information. Another teacher wanted to know how long the student spend working after school, so the report highlighted that activity.

After receiving the TOLA report, teachers were given 24 hours to review the content individually, they then met with Cottrell to report and discuss their insights. A final step in the cycle of feedback was for the PLC group of all 5 teachers to get together to review and discuss each teacher's TOLA report. This process was repeated for all 5 teachers in the study. All individual conferences and PLC meetings were recorded and transcribed. The transcriptions were coded using the data analysis spiral. In this process, the researcher collects, reflects, categorizes and organizes data at various times throughout the study (Creswell, 2007). As a result of this analysis several themes about the impact of TOLA on the teaching and learning process emerged.

Theme 1: Re-Evaluate Time Allowances of Assignments

The most widely discussed topic in this study was the use of time. Every teacher wanted to know how much time students spent both in class and out of class on their assignment. It was clear during each teacher's pre-interview that they had very clear expectations about how much time that a student should spend on an assignment.

Access to the TOLA showed that teachers had many misconceptions about when students are working online, how long they are spending online and the frequency and duration of the time intervals for which many students work.

Theme 2: Model Use of Resources

There was a strong interest amongst the teachers in this study in learning how students are interacting with online resources. Every teacher in this study incorporated some degree of independent student research. In each of these assignments students needed to find information online in order to complete the task assigned to them by the teacher. The teacher did not expect students to do this totally on their own. In every instance, teachers gave the students a list of resources to use to complete their assignment. Also in every instance, students were permitted to find their own resources as long as they cited the source. As a result of looking at the TOLA, teachers reported that they were surprised by how little students used the provided resources. Teachers stated that for future assignments they would stress to the students why they should use the websites that the teacher selected for them and how to evaluate quality websites.

Theme 3: Increase Control of In-Class Behaviors

Classroom management is a critical part of every teacher's ability to provide effective instruction. Unfortunately, student online off-task behaviors are not always easy for a teacher in a 1:1 classroom environment to spot. This was certainly the case for some of the teachers in this study. An analysis of online student behavior during class helped these teachers to gain an understanding of how often students are straying off-task. Several teachers in this study reported that they were surprised to see off task behaviors occurring online during class time. As a result they planned to increase the use of Chromebook monitoring software to restrict access to certain websites during class time.

Theme 4: Promote the Flow of Learning

The sequential listing of the resources, documents and other websites accessed by the students provided the teachers with an interesting look at the flow of the online pathways students are taking to complete the assignments. This also revealed the breaks and gaps in those pathways. These breaks in flow were particularly noticeable in classes where teachers let students listen to music. Students were frequently clicking back and forth between on-task activity and online music websites in order to select songs. Teachers viewed these frequent song selection breaks as a distraction that prohibits students from engaging in a deep metacognitive learning experiences. Teachers reported that they would have preferred to see longer stretches of time where the student is engaged solely in the assigned learning experience. As a result, several teachers reported that they planned to talk to their students about reflecting on the metacognitive approach that they take to completing assignments.

In summary, this study titled *Thinking Outside the Box: A Descriptive Study Focusing on the Application of Learning Analytics on Instructional Design* was the first academic research focused on finding applications for analytics about student online behaviors into real classroom settings. This action research showed that TOLA does have the potential to influence instructional design decisions but it also revealed the access and reporting barriers that prohibited TOLA from having a broad disruptive impact on the educational system.

TOLA Software Development

This study conducted by Cottrell indicated that TOLA had a positive impact on the teachers within his study, but unfortunately there was no feasible way to continue to provide meaningful TOLA data once the study ended. The manual sifting and quantifying of individual student online browsing data in order to create the TOLA report was very time consuming. It took approximately 15 minutes per students or 8 hours per class. This time consuming process was acceptable for the limited scope and focus of the study but it was not scalable for more frequent use. As a result, when Cottrell's research ended, so did the use of TOLA. It simply was not practical for teachers to spend that amount of time to create a TOLA report.

In order to promote the advancement of TOLA as both a research tool as well as a tool for the improvement of teaching and learning, an organization called Learnics was formed in 2018. Learnics developed a piece of software called *ThinkingApp*. The *ThinkingApp* allows students to record and submit back to their teacher a record of their online activity for a specific assignment. The students control when to turn on and off a recording of their online activity and they have the ability to delete data that that they accidentally recorded. The record of online activity is submitted to the teacher, by the student, using a unique assignment code. The student is able to start, stop and continue recording of online activity so that a student can start an assignment in school and finish it at home.

The process by which students control the recording of their activity serves several purposes. The first is that it serves as a way to get "targeted" data. In this process, the student in the person who is essentially filtering their online activity so that the teacher is provided with only the data that they need about student learning activity for a specific assignment. This is the step that was taking Cottrell 15 minutes per student.

The second major benefit of having students control the collection and submission of their data is that it provides them with ownership of their data. The student is able to see their own TOLA be aware of the experience that they had online. Future research studies are currently being designed to explore how a student's awareness of their online learning behaviors affects their future learning behaviors. Through this process of student self-submission of data, the students also become aware of how their online activity generates data that is being used by others.

Through the use of Learnics' *ThinkingApp*, teachers are able to access a dashboard that displays a both class wide and individual TOLA. Teachers can see the average amount of time that was spent on the assignment and a chart that sorts the students based on the amount of time that was submitted. Teachers can also see the most popular websites utilized for the assignment and the google search terms that were used. Teachers cannot see data that the students delete from their record prior to submission. The *ThinkingApp* has allowed for the use of TOLA to be scaled for use outside of research projects. Teachers can collect TOLA data on a regular basis by creating a new assignment and sharing the assignment code with their students.

The creation of the *ThinkingApp* has allowed for more frequent use of TOLA but more research is needed to help educators fully incorporate TOLA into their practice. Additionally more feedback from teachers is needed in order to determine how the *ThinkingApp* can best collect, quantify and present TOLA to teachers and students. The next section of this chapter with discuss how another research out of East Stroudsburg University of Pennsylvania advanced the use of TOLA and the development of Learnics' *ThinkingApp*.

Resende Study Design and Findings

Building on the research study conducted by Greg Cottrell in 2017, Isabel Resende conducted her research through East Stroudsburg University in 2018. At the time, Resende was the Assistant Superintendent of a mid-size suburban school district. As Assistant Superintendent, Resende provided leadership and support for the development and monitoring of a district-wide digital-learning-environment plan. Under this plan, all 7-12 students were provided with an individual Chromebook to use in school and at home. Given the district's financial investment as well as time dedicated to professional development, Resende became interested in gaining an understanding of how technology impacts teaching and learning. Adding to this interest were the results of a research-based perceptual survey that was administered by the district to all teachers, students, and parents. The results of the survey indicated that the teachers shared some common interests in understanding how online learning behavior can inform instructional practices. As a result, her study, *An Action Research Study of Teachers' Use of Learning Analytics as a Formative*

Practice, sought to gain a deeper understanding of how the availability of TOLA influenced teachers' instructional practice at the individual-student level. Cottrell's findings supported the notion that TOLA served as formative feedback for teachers between student use of technology and the pedagogical design process for teachers. In addition, time on task in and outside class was an important piece of formative data for teachers when they thought about the design or redesign of an activity.

In short, Cottrell's (2017) study showed that "the incorporation of learning analytics into a teacher's instructional design toolkit will allow a teacher to make more informed pedagogical decisions" (p. 140). However, the data collected on student TOLA was not used to proactively intervene in the learning process at the individual-student level. The TOLA data were collected and analyzed to gain an understanding of its influence on the instructional design. Therefore, this action-research project's aim was to build on Cottrell's work and understand whether TOLA could be used in the instructional evaluation process, through formative assessment, to modify teachers' instructional planning in a proactive manner throughout the student learning process (Reigeluth & Chellman, 1999).

This study was based on the premise that teachers evaluate their instructional practices and student learning experiences based on the formal and informal feedback that they receive from students while they are actively engaged in the learning process. Feedback through formative assessment or through direct observation of student behavior is used by teachers to adjust teaching and learning to improve student achievement of the intended learning goals (The University of Michigan, 1996; Popham, 2008). The challenge with online assignments is teachers lose the ability to receive feedback on the learning process within the instructional design. Without insight into the student learning process, teachers are unable to fully evaluate where students are in the learning continuum and intervene with responsive changes to instruction that positively influence the intended learning outcomes. The availability of TOLA presents a unique opportunity for teachers to evaluate how technology as a resource interacts with other aspects of the instructional design to impact student learning (U.S. Department of Education, 2012). The data collected in real-time can provide immediate feedback and potentially guide teacher instructional-planning (Chan, Konrad, Gonzalez, Peters, and Ressa, 2014; Lockyer et al., 2013). Most studies involving learning analytics have been conducted within virtual learning environments and or learning management systems (Papamitsiou & Economides, 2014). This study used participatory action research as a methodology, in a middle school setting, to examine the impact of TOLA on teacher instructional practices at the student level using an application called *ThinkingApp*.

The research questions that guided this participatory action research study were the following:

1. What influence does the availability of learning analytics have on teacher instructional practice?
2. Do instructional practices change on an individual student level based on the results of the learning analytics?
    a. Will teachers use learning analytics as part of their formative practices to adjust individual student learning?
3. What factors contribute to the use of learning analytics and what factors hinder the use of the learning analytics?

The study required participants to use the learning analytics tool *ThinkingApp* in at least one assignment that had students use their Chromebooks inside or outside the classroom.

The district used a web filter, *Securly*, which collects limited analytics on student computer usage. However, teachers did not have access to the information collected through this application. Moreover, the information collected through *Securly* was limited, unlike *ThinkingApp*. *Securly* collected information at the individual-student level on keyword searches in Google, websites accessed and blocked, time accessed, and the date on which it was accessed. *Securly* did not provide information on the amount of time the student spent on a website (Cloud-based Web Filtering for Schools, 2017). Time on task is an important piece of feedback for teachers in their instructional planning (Cottrell, 2017). On the other hand, *ThinkingApp* collected information on the websites that students accessed, total time on each website, top websites accessed, and time on task. *ThinkingApp* also was capable of sorting and filtering websites accessed by most time spent and order in which they were visited in an easy to interpret format. For instance, Figure 3 shows how the *ThinkingApp* presented TOLA in a user friendly manner, an improvement from the first study conducted by Cottrell (2017).

The ability to use a tool that made the collecting, organizing, sorting, and filtering the data manageable was critical to the success of the study. It allowed participants in the study to focus on the analysis of the data rather than spending their time organizing the data into a readable format. The participants reviewed, analyzed, and reflected on the aggregated and disaggregated TOLA reports provided by *ThinkingApp* collectively as a team as well as individually by participating in a pre-learning analytics semi-structured interview, two inquiry-group meetings, and a post-learning-analytics semi-structured interview. The collective analysis and reflection lead to the following results in this study:

Figure 3. TOLA Chart. Sample individual student data summary generated by ThinkingApp.

Research Question 1

The first research question sought to determine whether the availability of TOLA had an impact on teacher instructional design, a process of analyzing the learning needs of students through the use of formative practices and develop a plan to address these needs and meet instructional goals (Reigeluth & Chellman, 1999; University of Michigan, 1996). The two guided-inquiry group sessions and the post-interview sessions provided the greatest insight into how instructional design was impacted by the availability of TOLA. The study found that the availability of TOLA clearly made teachers pause and reflect on instructional design. Specifically, teachers reflected on the design of assignments and how students tackled the completion of assignments online. Four themes emerged through the inquiry group sessions and post interviews that were of common interest to the teachers in the study: 1) time spent on an assignment, 2) amount of time deleted from logs, 3) student approaches to learning, and 4) sequence of websites visited.

In short, teachers concluded that TOLA could be a targeted source of data that provides insight into the instructional design. The analysis of the TOLA led teachers to reflect not only on the quality of their assignments, but also on the quality of the questions they asked students during the learning process. Teachers self-reported in the post interviews that the leaning analytics made them "take a closer look" (Resende, 2018) at their assignments and how students conduct research. The teachers also reported that the analytics also helped them to understand how students approach an assignment differently in class and out of class and the time of day they complete assignments.

Research Question 2

The second research question sought to understand whether teachers would intervene in the learning process with students as a result of the data collected. Would teachers use TOLA as part of their formative practices to adjust student learning? Teachers overwhelmingly responded that TOLA was useful in order to help students with assignments. Analytics gave them an awareness of how students conduct research, which was helpful in intervening when students were struggling with the research process. Teachers also stated that TOLA provided them an understanding of how students were accessing the available resources to support their learning and become more independent in problem solving. Finally, they reported that TOLA provided some insight into student effort and focus when completing assignments online.

Research Question 3

The final research question was concerned with understanding the factors contributing to the use of TOLA versus the factors that hinder its use. The teachers overwhelmingly found value in using TOLA to inform their instructional design and practice. They reported that the most valuable metrics were the sequence of websites visited, average time on task, and understanding how students approach an assignment toggling between multiple tabs. On the other hand, the *ThinkingApp*'s functionality did hinder their use of the application. The submission errors led to incomplete data sets for analysis, hindering the data analysis for patterns and trends. The teachers also felt that placing the responsibility solely on the student to submit logs led to fewer submissions. The teachers would like to have more control over the beginning and ending the data collection. Finally, the teachers found value in the metric and would have liked to use the *ThinkingApp* with other assignments that were not tied to Google Docs. Therefore, they would prefer to not have the *ThinkingApp* tied to Google.

In summary, this study set out to understand the impact of the availability of TOLA on instructional practice at the individual-student level. Clearly, the results indicate that the availability of TOLA impacted teacher instructional practice and design. However, it became much more than just TOLA and instructional practice. It was an experimental journey for all involved by breaking ground in an area of teaching and learning that had not been explored. This was the first time that teachers had access and insight into TOLA, that teachers and an administrator were jointly studying a professional problem of practice focusing on instruction, and most important, that teachers' input and needs were considered in the development of a tool designed specifically for their use. Throughout this study, the worlds of technology and education collided in unique ways. Teachers had to think deeply and reflect on what information was most useful for the deconstruction of instructional practice. Then they had to articulate that need to the developers in clear language to improve the functionality of the application. As they uncovered patterns in student online learning behavior and asked more questions, the questions led to further discussions with the developers. This partnership lead to real and impactful change in how educators viewed teaching and learning and the use of technology that had never been explored previously.

TOLA RESEARCH INSIGHTS, SOLUTIONS AND RECOMMENDATIONS

The combined findings of both research studies indicate that TOLA has the potential to disrupt the teaching and learning process by providing teachers with a rich set of data that was previously either nonexistent, unavailable or diluted within big data sets. The studies described in this chapter provided insight into three areas of influence: 1) the need for student feedback 2) the value of placing TOLA data in the hands of teachers, and 3) the need for teacher training in the analysis of TOLA.

Feedback is a critical component of the instructional design process. In a traditional classroom setting, embedded within the learning tasks are assessments to measure how students are interacting with the resources and content. This information is then used to monitor and adjust instruction at the individual-student level to assist students in meeting the learning targets. When teachers employ the use of online-learning activities the design process presents a greater challenge. Teachers lose the ability to gather information on how students are engaging with the tasks in the course of the learning progression. Without this insight, teachers cannot intervene to support students in meeting the intended learning outcomes. TOLA has the potential to bridge this gap by providing immediate and specific data on students' engagement with online-learning activities. The feedback available within the TOLA data has the possibility to disrupt every step of the instructional design process starting with instructional evaluation. TOLA can be presented at the time the learning occurs. As a result, teachers receive daily feedback from students to improve instructional practice. The ability to make student learning processes visible through the collection of data in real time on what students know and how they are engaging with online-learning assignments opens up opportunities for teachers to gain insight into student performance. This can help teachers adapt their teaching and tailor assignments to meet students' needs.

Similarly, these studies both reaffirmed the epistemology that teachers are best positioned derive meaningful insights from the analysis of TOLA data. Much of the work in the field of learning analytics has been centered on predictive analytics which use data points to make assumptions about what students have learned in order to make predictions about what learning experiences would be best to prescribe next. Prescriptive learning analytics may have potential benefits in a controlled learning environments such as a specific sequential digital library of curriculum resources, but in reality, 21st century schools are much more dynamic and harder to predict. Local control of school boards enable each school to develop their own curriculum and many teachers exercise ample academic freedom in the way that they deliver the curriculum. In many of these classrooms, teachers are designing authentic learning experiences that have students search, scavenge, explore and exploit the internet for information and ideas that will meet their learning objectives.

The amount of variables that come into play when determining if a learning objective is met for an authentic online learning experience is beyond the practical scope of predictive analytics modelling. An in depth knowledge of the individual student, intended learning outcomes and context of the learning environment is a critical factor to include when finding meaning in TOLA. Multiple times throughout this research, the classroom teacher made insights that the researchers and others overlooked. They did this because of their rich understanding of the learner and the learning environment. Many of these insights resulted in instructional changes. Therefore, putting TOLA in the hands of teachers in a manageable and interpretable fashion has the potential to transform how we look at teaching and learning. Teachers could more readily make informed decisions about how students are learning and what supports they need to be successful. TOLA provides a student learning profile that can be used to truly personalize learning.

Finally, the combined experiences of conducting both research studies described in this chapter, pointed to the need for providing educators with professional learning opportunities in the areas of: 1) defining TOLA data, 2) analyzing TOLA data, 3) redefining instructional practices based on TOLA. Some educators claim to have perfected the science of using data to guide instruction and develop curriculum. However, data analysis has been confined to mastery type metrics. For example, in schools today, educators use the results of state standardized test to measure the effectiveness of their programs and instruction. Similarly, educators use curriculum based measures to determine if students have mastered a concept. These curriculum based measures are also used to generate student grades. Theoretically, grades communicate students' acquisition of knowledge and skill level, but do they? At what point do educators evaluate the process of learning and students' metacognition as opposed to an end product. In a global society whose access to information is limitless, measuring the mastery of knowledge that students can easily access and research online should not always be the goal. On the other hand, measuring students' ability to reflect on their learning, and understanding how students engage in authentic problem solving is a much more significant endeavor.

Imagine a school system in which educators, have the ability and the tools to collect data on the process students participate in while engaging with online content. Educators would have the ability to truly personalize student learning. They could create a learner profile that focuses on an individual's continuum of learning as opposed to a grade that reflects an end product. TOLA, has the potential to disrupt how we think about the use of data to inform instruction and develop curriculum. However, it requires a mind-shift. Educators need to move away from the concept of measuring static benchmark data to more valuable real-time ongoing data that reflects an individual student's process of learning. In order to achieve this mind-shift, educators need to be afforded the opportunity to collectively learn and understand the value of TOLA. They also need to be afforded the time to think about how TOLA can be used inform their practice at the individual student level. To achieve this goal, educational systems need to rethink how they train teachers in the use of data to inform instruction and develop curriculum.

FUTURE RESEARCH DIRECTION

It has been reported that data analysts at Facebook can predict with a high degree of accuracy when a romantic relationship of one of their users is about to end. They do this by looking at how user behaviors change. People that are going through a rough patch in their romantic relationship will likely show signs of disengagement from their partner. For example, their posts transition from using words like "we" and "us" to "me" and "I". Pictures with both partners are not posted as often. The amount of times a user visits other user pages or adds new non-mutual friends may change and the amount of time a person spends on social media may be impacted. These are all examples of how data shows signs of disengagement. Students who drop-out of school also go through a process of disengagement. This disengagement from learning is a behavior change. When learning occurs in a digital environment, evidence of this behavior change can be found within the student's TOLA. The student may be putting in less time engaged in learning outside of school. They may show an increase in social media, games and streaming video use in school. The time of day that they are doing their work may change or the academic quality of the websites that they are using may change.

All of the behaviors listed above are indicators that an educator adept in the use of TOLA may use to identify and suggest interventions for a student who is becoming disengaged. Conversely, TOLA can also be used to identify when a student is engaged in their learning, displays excellent digital learning behaviors but fails to submit a high quality assignment that accurately reflects their level of knowledge. For example, consider a struggling student who decides to put forth the effort needed to complete a homework assignment by accessing relevant academic sites and following the suggested activities but when it came time to articulate their thoughts, the struggling student gives up. A teacher with a knowledge of this students' engagement in the learning process would be able to praise the struggling student for their valid learning process and provide interventions to help with tackling the written portion of the activity.

The incorporation of TOLA into the field of education is just starting to emerge. When fully embraced, educators will be able to assign and value the learning process, not just the summative end product. The syntheses this of the research in these studies revealed that when teachers create an assignment, they usually have an expectation in their mind for how long the assignment will take, the websites used, how long spent researching versus writing and the approach that the student will take to complete the assignment. In other words when a teacher creates an assignment, they are essentially trying to create a learning experience. TOLA research had revealed that often the experience that a teacher expected to occur was not the same as was actually occurred. In a visionary look at an educational system that has been disrupted by TOLA, a teacher gains the ability to not only assign an output product to be submitted, they also now have the ability to assign clear expectations for the process by which that product should be created. For example, a teacher could assign that students simply spend 30 minutes of time researching a topic. They could set specific parameters for the types of websites visited but essentially to goal is simple for students to spend time learning online, without having to submit any type of written product to prove it.

Aside from simply helping teachers to produce higher quality learning activities, TOLA also has significant potential to help teachers understand individual students. The unique approach that every student takes to an assignment is essentially their online learning style. This online learning style, or preferred approach to learning may help educators diagnose and intervene with struggling students. Further research into the online learning styles of students based on gender, learning disability or socio-economic status will provide new insights and strategies that can help students become successful digital learners.

CONCLUSION

Robert Marzano (2010) in his book *The art and science of teaching: a comprehensive framework for effective instruction,* makes the case that instructional practices should be grounded in sound research, but understanding which practice to use, when to use them, and with what student, is the art of teaching. Educators that have access to research-based practices are best poised to make instructional decisions that support student learning. Based on the research studies presented in this chapter, TOLA data has proven to be of value to educators in their instructional practice. However, the use of TOLA in education is still an emerging field resulting in its underutilization. As applications like the *ThinkingApp* begin to emerge, simplifying the collection and organization of "big data", educators will become more inclined to employ the data in their practice. A concerted effort needs to be made to move TOLA into the hands of teachers to help in their instructional decision making. In summary, the use of TOLA in the instructional design process to support student learning is an emerging and innovative technique in the field of education that has been underutilized. It has the potential to open up new avenues in designing instruction that is authentic and personalized to students' needs. TOLA metrics as an instructional evaluation tool has the potential to disrupt and revolutionize instructional practices.

REFERENCES

Baker, R. S., & Inventado, P. S. (2014). Educational data mining and learning analytics. In *Learning analytics* (pp. 61–75). Springer New York. doi:10.1007/978-1-4614-3305-7_4

Buckinham Shum, S. (2012). *Learning Analytics Policy Brief.* UNESCO. Retrieved from http://iite.unesco.org/files/policy_briefs/pdf/en/learning_analytics.pdf

Chan, P. E., Konrad, M., Gonzalez, V., Peters, M. T., & Ressa, V. A. (2014). The Critical Role of Feedback in Formative Instructional Practices. *Intervention in School and Clinic*, *50*(2), 96–104. doi:10.1177/1053451214536044

Cloud based Web Filtering for Schools. (n.d.). Retrieved November 18, 2017, from https://www.securly.com/

Common Core State Standards Initiative. (n.d.). English Language Arts Standards. Retrieved September 26, 2016, from http://www.corestandards.org/ELA-Literacy/RH/11-12/

Cottrell, G. J. (2017). *Thinking outside the box: A descriptive study focusing on the application of learning analytics on Instructional design* (Doctoral Dissertation). East Stroudsburg University.

Creswell, J. (2007). *Qualitative Inquiry & Research Design: Choosing Among Five Approaches* (2nd ed.). Thousand Oaks, CA: Sage.

Drachsler, H., & Greller, W. (2016, April). Privacy and learning analytics-It's a DELICATE issue. In *Proceedings of the Sixth International Conference on Learning Analytics & Knowledge* (pp.89-98). ACM. 10.1145/2883851.2883893

Family Educational Rights and Privacy Act (FERPA). (2015, June 26). [Guides]. Retrieved January 25, 2016, from http://www2.ed.gov/policy/gen/guid/fpco/ferpa/index.html

Glazatov, T. R. (2012). Applying Instructional Design System Theory to Mobile Learning Environments. *Journal of Applied Learning Technology*, *2*(2), 29–35.

Khalil, M., & Ebner, M. (2016). De-identification in learning analytics. *Journal of Learning Analytics*, *3*(1), 129–138. doi:10.18608/jla.2016.31.8

Lockyer, L., Heathcote, E., & Dawson, S. (2013). Informing Pedagogical Action: Aligning Learning Analytics with Learning Design. *The American Behavioral Scientist*, *57*(10), 1439–1459. doi:10.1177/0002764213479367

Long, P., & Siemens, G. (2011, September/October). Penetrating the fog: Analytics in learning and education. *EDUCAUSE Review*, 31–40. Retrieved from http://er.educause.edu/

Marzano, R. J. (2010). *The art and science of teaching: a comprehensive framework for effective instruction*. Alexandria, VA: Association for Supervision and Curriculum Development.

Papamitsiou, Z., & Economides, A. A. (2014). Learning analytics and educational data mining in practice: A systematic literature review of empirical evidence. *Journal of Educational Technology & Society*, *17*(4), 49.

Popham, W. J. (2008). *Transformative assessment.* Alexandria, VA: Association for Supervision and Curriculum Development.

Radu, I. (2017). Emergent technologies for learning analytics. *ELearning & Software for Education*, 2509-524.

Reigeluth, C. M., & Carr-Chellman, A. (1999). *Instructional-design theories and models. A new paradigm of instructional theory*. Academic Press.

Resende, I. C. (2018). *An action research study of teachers' use of learning analytics as a formative practice* (Doctoral Dissertation). East Stroudsburg University.

Reyes, J. A. (2015). The Skinny on Big Data in Education: Learning Analytics Simplified. *TechTrends, 59*(2), 75–80. doi:10.100711528-015-0842-1

Stake, R. E. (1995). *The art of case study research.* Thousand Oaks, CA: Sage Publications.

University of Michigan. (1996). *Definitions of Instructional Design.* Retrieved from http://www.umich.edu/~ed626/define.html

U.S. Department of Education. (2012). *Enhancing Teaching and Learning Through Educational Data Mining and Learning Analytics: An Issue Brief.* Retrieved from http://tech.ed.gov/wp-content/uploads/2014/03/edm-la-brief.pdf

Wilson, M., Scalise, K., & Gochyyev, P. (2016). Assessment of learning in digital interactive social networks: A learning analytics approach. *Online Learning, 20*(2).

KEY TERMS AND DEFINITIONS

1:1 Initiative: A program in which every student is issued a mobile learning device such as a laptop, iPad or Chromebook.

Instructional Design: A process of analyzing the learning needs of students through the use of formative practices as part of the Instructional-Evaluation Design Theory to develop a plan to address the needs of the students and meet instructional goals (Reigeluth & Chellman, 1999; University of Michigan, 1996).

Instructional Evaluation: A component of the Instructional-Evaluation Design Theory that relates to "the process for evaluating instruction, (summative and formative)" (Reigeluth & Chellman, 1999, p. 8).

Instructional Planning: A component of the Instructional-Evaluation Design Theory that relates to "the process of creating the instructional plans" (Reigeluth & Chellman, 1999, p. 8).

ThinkingApp: A Chrome extension that enhances the functionality of the Chrome browser by allowing for the collection, review, and submission of student-online-learning behavior associated with a specific online learning experience.

Chapter 2
Teaching in a Makerspace:
The Pedagogical Practices of Makerspace Instructors

Catherine Otieno
Aims Community College, USA

ABSTRACT

This chapter provides an in-depth study of the teaching practices of instructors who primarily guide and facilitate learning in a makespace. With a close look at the pedagogical practices that govern teaching and learning in the maker classroom, this study presents instructors who modeled these frameworks. In addition to their own knowledge base and expertise, they were able to efficiently and effectively integrate multi-resources in a unique learning environment while helping learners succeed and adopt the maker mindset. Makerspaces are changing how we perceive learning and teaching. Instructors highlighted in this chapter put forth activities and learning goals that were learner centered and interesting to various learning needs. They designed and created a learning environment that safeguarded learners and allowed them to experiment with ideas and materials, creating different iterations of learning and redefining what success and failure means.

INTRODUCTION

The call for all educators to create learning environments that implement teaching strategies, which not only helps student master content, but also learn essential 21st century skills has been echoed by many (Partnership for 21st Century Learning (P21), 2019; Wagner, 2008; Washor & Mojkowski, 2013). Makerspaces are learning

DOI: 10.4018/978-1-7998-2914-0.ch002

environments that proponents argue will help educators tackle such issues and hopefully reform education (Dougherty, 2013; Honey & Kanter 2013; Martinez & Stager 2013; Petrich, Wilkinson, & Bevan, 2013). Defined as spaces where individuals can freely make, tinker, innovate, and create, makerspaces have become a phenomenon (Dougherty, 2005). Makerspaces call for educators to foster the maker mindset among learners; the mindset that students can turn ideas into creative realities, and use the knowledge gained to be innovators (Honey & Kanter, 2013). A making and tinkering mindset, within these spaces, strives to help learners learn through making. As a rising trend, makerspaces are positively pushed in order to provide spaces that bring out and sustain the curiosity of young minds instead of stifling them by extrinsic goals and expectations as called for in many schools settings (Honey & Kanter, 2013).Makerspaces are designed to be supportive, challenging, and engage learners in meaningful activities, which in turn, rekindle their natural motivation to learn (Honey & Kanter, 2013).

Pure Instructionism is a pedagogy, which equips students with skill sets needed; however, with the evolution of learning spaces and technology integration, skills required of learners has changed. With Instructionism pedagogy, there exists two set of students: 1. Ones who succeed academically in areas they were told and who have been limited to explore and reach their full potential and 2. Those that struggle in specific academic areas. Those students may have given up due to a low self-esteem or due to others' opinions, which may hinder their ability to reach their full potential (Martinez & Stager, 2013). Making in the makerspaces strives to remedy this and provide opportunity for growth in one's full potential for all students. With the evolving technology, not only have opportunities for integration become imminent, making and tinkering has also become much easier.

Many studies have focused on the connection between learning and making in the makerspaces, but none have really delved into the teaching strategies within makerspaces that makes them so successful. The purpose of this qualitative case study was to examine the pedagogical practices of makerspace instructors in both formal and informal makerspaces.

A REVIEW OF LITERATURE

The Maker Movement

The maker movement is characterized with the idea that people are makers and creators rather than consumers and it has recently become a social, technological, and economic development (Brahms & Werner, 2013). The maker movement has given rise to makerspaces, tinkering studios, technical shops, and FabLabs which are generally spaces where making takes place. The way educators, students, and instructors view and see learning is being reinvented and transformed by the maker movement (Brahms & Werner, 2013). The evolution of new technology such as 3D printing, microcontrollers, laser engravings, robotics, physical computing, and fabrications has recently seen the movement rise in the digital setting allowing users and learners to innovate, create, and tinker (Petrich et al., 2013). According to Martinez and Stager (2013), the maker movement is "an active process of building, designing, and innovating with tools and materials to produce shareable artifacts…as a naturally rich and authentic learning trajectory" (p. 32). Additionally, makerspaces, inspired by the do it yourself (DIY) maker movement phenomena, have loosely been defined as "physical location where people gather to share resources and knowledge, work on projects, network, and build with technology" (Hackerspaces, 2015, para. 1). Makers generally take things apart, put things together in a new way, and find it intrinsically rewarding to do so (Kalil, 2013, p. 15). It is, therefore, a very hands-on, learn by doing movement.

Honey and Kanter (2013) observed that the maker movement has the potential to transform education and foster a "make mindset" among students (p. 5). The movement has the potential to promote creativity, problem solving, collaboration, and self-expression; skills that are vital to succeed in the 21st century environment (Kalil, 2013, p. 16). Makerspaces provide tools and space in a community environment--a library, community center, private organization, or campus (Honey & Kanter, 2013). Most makerspaces are cross-disciplinary and provide materials and room for physical learning. Students from a variety of fields can use them and find technical assistance from mentors and experts as well as collaborate with peers to build authentic artifacts (Petrich et al., 2013). Due to its inquiry and problem-based pedagogy, makerspaces provide opportunities for students to not only get hands-on use of emerging technologies, they also take control of their own projects that they have not just designed but defined as well (Simon & Brown, 2013).

Learning Theoretical Frameworks and Their Contributions to Pedagogy

Pedagogy involves selecting strategies and methods to use in a learning environment that best meets the learners learning needs and objectives set (Driscoll, 2000; Gagné, 1977). Over the years there are many learning theories that have influenced pedagogy. For example, According to behaviorists, we acquire new knowledge by responding to reinforcements, both positive and negative, from the environment and this shapes our behavior (Driscoll, 2005; Keller, 2010). Instructors for years have employed behaviorist principles for instruction and reinforced certain desirable behaviors from the students, identifying and clearly stating instructional objectives, and determining achievements from observable behavior after instructions (Driscoll, 2000; Gagné, 1977). The instructor's role in teaching is that of authority of knowledge and students understanding is checked through change into the desired behavioral outcome which is often teacher directed or determined.

Cognitive information processing learning theory (CIP) proponents believed that information is processed as it passes from sensory, short-term, and long-term memory like a computer processor (Atkinson & Shiffrin 1968; Ausebel, Novak, & Hanesian, 1978; Driscoll, 2005; Gagné, 1977). The information we sense is stored in memory, retrieved when needed, and outputted in the form of learning outcomes desired. In understanding the learners CIP, instructors can design instruction and employ strategies that reduce cognitive overload and instead help learner process new information into the long-term memory for retrieval and connection making. These strategies include organizing and scaffolding learning if learners are to make sense of the new information, signaling what information is important to increase selective attention, using multimodal resources, creating learning environment that allows for "perfect practice" and context where these can be applied, and help learners be more metacognitive of their learning process and self-regulate (Driscoll, 2005; Gagné, 1977). Instructors must also activate learners' prior knowledge, so they can relate to the new knowledge being learned providing better chances for long term memory and reducing learner cognitive overload if learning is to be meaningful.

In situated cognition learning theory, learning happens in the sociocultural setting and we acquire knowledge from our interactions with other people in our society (Bruner 1997; Driscoll 2005; Vygotsky, 1978). Its influence in pedagogy includes the need for instructors to provide apprenticeship opportunities as well as "projects in which the instructors can model desired skills and coaches learners as they attempt to follow suit" (Driscoll, 2000, p. 175). In addition, instructors can provide real-world problems that students can solve and instead of being the knowledge dispenser, they work collaboratively with the students to meet the learning goals set. In this case, the classroom becomes a learning community where students can bring their interests and experiences and instructors provide support and resources. For this framework, summative testing is not enough to determine students' success in the social practice. Instead the teacher employs formative assessment throughout the course period by identifying progress during activity, collaboration with peers, and learning in the classroom community (Bruner 1997; Driscoll 2005).

Gagné's (1977) theory of instruction proposed conditions of learning that instructors and educators must allocate if learning is to be successful. Gagné identified multiple learning outcomes desired, which include verbal, intellectual, cognitive, attitudes, and motor skills. Once an instructor has identified the skills, they must provide learning conditions that helps the learner to be fluent in those learning outcomes (Gagné, 1977). He designed nine events of instruction that instructors and instructional designers have referenced and used for years. These include gaining attention, informing the learner of the objective, presenting stimulus, providing learning guidance eliciting performance, providing feedback, and assessing performance (Gagné, 1977; Gagné & Medsker, 1995).

In a constructivist framework, learners "construct knowledge as they attempt to make sense of their experiences" (Driscoll, 2000, p. 376). In contrast to behaviorists, learners are "not empty vessels waiting to be filled but rather are active organisms seeking meaning" (Driscoll, 2000, p. 376). For learning to occur, certain conditions must be provided by the instructors in the learning environment. These conditions, according to Driscoll (2005), could range from;

Embedding learning in complex realistic and relevant environments, providing social integration as an integral part of learning, supporting multiple perspectives and the use of multiple modes of presentations, encouraging ownership in learning and finally nurturing self-awareness of the knowledge construction process. (p. 393).

Instructors need to provide rich student-centered learning environments where authentic experiences are emphasized, creating a safe learning environment where learners can collaborate and work together freely, design effective goal- based and problem-based learning allowing learners to think critically, and problem solve and work through tasks efficiently.

Finally, building upon the theory of constructivism, constructionism theory emphasizes learning through making. Constructionism goes beyond building knowledge to incorporate learners consciously building and constructing a "public entity or artifacts" (Papert, 1993a, p. 142). It helps educators understand how different media can be used to express and transform multiple ideas in different context (Ackermann, 2001). Learning as such, happens when the learner's thinking is worked out through the making of external artifacts that "can be shown, discussed, examined, probed, and admired" a distinction from constructivism where leaners build knowledge (Papert, 1993a, p. 142). To Papert (1993b), we learn by doing and instructors need to redefine success in the classroom as learning is a process and iteration is key. Collaboration while making, as well as seeing technology as a tool to create and build plays a big role in successful learning. Instructors must provide innovative learning environments where students can create, make, and identify ways to assess this learning process. Martinez, and Stager (2013) stated that instructors can use constructionism principles to implement constructivist learning.

Teaching and Learning in the Makerspaces

The desire and need to make and create comes from this inherent nature of humans as each person is born curious, eager to understand the world around them (Papert, 1993a). Defining learning and knowledge is a complex matter that psychologists and philosophers have considered for centuries. Generally, the acquisition of knowledge has been defined as the, "persisting change in human performance or performance potential" (Driscoll, 2005 p. 9; Papert, 1993a). Driscoll (2012) further states that "to be considered learning, a change in performance or performance potential must come about because of the learner's experience and interaction with the world" (p. 9). According to Reiser and Dempsey (2012) learning can be many things but must, however, be distinguished from human "maturation and human development," "changes in ability that are temporary" and "learning as an instruction process and learning as a cultural process" (p. 35).

The ideas behind making and the maker movement are at least a decade old as people have always made, but with the new advanced technology such as digital fabrication tools, making in the formal and informal learning settings has become more popular and common. Making unites, inspires, informs, and entertains a growing community of resourceful people of all ages who undertake amazing projects (Honey & Kanter, 2013). The maker movement celebrates the right of individuals to tweak, tinker, hack, and bend any technology to their will (Honey & Kanter, 2013; Petrich et al., 2013). Making "develops an alternative and powerful way of knowing and of thinking things through that contrasts with mere abstract analysis" (Washor & Mojkowski, 2013, p. 208). With the younger people, making provides opportunities to use their hands and minds, stimulating them to develop their imaginative, creative, entrepreneurial, and scientific abilities (Honey & Kanter, 2013; Washor & Mojkowski, 2013). Characterized by the iterative process of developing a personally meaningful idea, becoming stuck in some aspects of physically realizing the idea, persisting through the process, and experiencing breakthroughs, individuals involved in the tinkering/making process must enhance their creativity and problem solving skills to find solutions to problems (Petrich et al., 2013). Petrich et al. (2013) further stated that the iterative process in tinkering and becoming "unstuck" is a fundamental process in learning as students develop "authentic authorship, purpose and deep understanding of the materials and phenomenon" (p. 55). Making, or tinkering, as such "centers on the open-ended design and construction of objects or installations, generally using both high- and low-tech tools" (Blikstein, 2013; Dougherty, 2013, p. 9; Petrich et al., 2013).

In the late 2000s, most researchers and educators seriously started considering the influence makerspaces have in education. For example, in 2008, Stanford University launched the FabLab@School project and started building FabLabs in K-12 schools around the world. In 2009, the MC2STEM High School in Ohio (USA) opened its first digital fabrication lab. In 2011, the Maker Media launched the makerspace project. In recent years, countless museums, schools, universities, community centers, and libraries have established and started many digital fabrication and "making" facilities. Making, as a learning trajectory, has been a discussion for centuries by previous learning psychologists and philosophers and as such supported by learning theories like constructionism, constructivist, sociocultural, and situated cognition (Driscoll 2005; Honey & Kanter, 2013; Papert, 1993a).

The benefit of learning and working with our hands cannot be overemphasized. To engage the hand is to engage the mind and education must provide students with a hands-on, minds-on approach. Educators should engage the learners and deepen their understanding of what a quality product is. In addition, there should be an appreciation for artifacts, allowing for a model that will produce thinkers and creators who can apply what they learned (Wilson 1998; Washor & Mojkowski, 2013). The development of emerging technologies that make different learning possible has risen over the years and calls for the need to reevaluate the benefits of learning by making (Dougherty, 2013; Honey & Kanter, 2013; Petrich et al., 2013). In a learning environment where making is the sole practice, learners engage with the materials, peers, instructors, practice, and ideas present in the makerspace, and since the challenges or problems are not assigned they discover and engage in an exploratory manner (Petrich et al., 2013, p. 59). This is in line with Papert's (1993b) constructionism learning view that builds upon Piaget's constructivism and states that learners construct knowledge when they build, make, and share their artifacts with not only their peers but the public as well. In the makerspace, learners can make variety of items depending on the technology or tools available in the space. These range from but are not limited to phone cases, home amenities, toys, mugs, wheels, parts of equipment needed all using 3D printers; use laser tag engraving tools to engrave their metals or plastics personalizing them in the process; robotics where the learner puts together robots based off problems to find solution or perform a task using the put together robot; programming where the learners program avatars to perform tasks and create games; and many more.

Within the makerspaces, instructors are mostly mentors whose role is to guide the learner during their making sessions (Brahms, 2013; Honey & Kanter, 2013). They facilitate learning rather than dispense it something that constructivists propose in a learning environment. New technology in the makerspaces provide opportunities that put advanced and innovative tools in the hands of the learners who in turn use these to explore, create, build, design, and program (Blikstein, 2013). In fact, previous studies have shown that learners who have interacted with advanced technology in a makerspace have led to a powerful and generative creation of "ideas and artifacts" (Berland, 2008; Blikstein, 2013, p. 6; Petrich et al., 2013).

These studies further expand that the difficulties experienced by the learners on tasks within the makerspace "were due to deficient design rather than learner's cognitive deficiencies" (Berland, 2008; Berland & Wilensky, 2006; Blikstein, 2013, p. 6; Brahms, 2014; Petrich et al., 2013).

With makerspaces, disciplinary boundaries are reconfigured as the making process taps into STEAM (science, technology, engineering, arts, and mathematics), leading to a more diverse learning opportunity for the students (Blikstein, 2013). Resources in makerspaces often vary depend on the facility, space, and funding available, but quite common in such areas are equipment such as microcontrollers, 3D printers, fabrication tools, video and media, and robotics. Such spaces also allow learners to explore questions, fail and retry, bounce ideas off one another, and build something together. Since the makerspace learning environment is not like a traditional classroom, instructors in the makerspace must redefine their teaching strategies and what learning looks like. Maker movement teaching strategies have been influenced by bits and pieces of pedagogical frameworks discussed in this chapter but heavily borrow from Piaget's constructivism and Papert's constructionism. The research presented in this chapter specifically addresses the following question: How does pedagogy manifest itself in the makerspace?

MAIN FOCUS OF THE CHAPTER

Methodology

A multi-case study methodology was used in this study as stipulated by Creswell (2007) and Merriam (1998). Each makerspace location was treated as an individual case. The goal for the study was to extensively understand the makerspace itself and how the makerspace instructors teach within the space. Using a case study allowed for a better understanding of the process, context, and discovery taking place in the makerspace.

Context

This study took place at three sites; a library makerspace and two middle schools makerspaces. These sites were purposefully sampled because they had access to diverse technology resources and offered lessons/instructions to students as well facilitated the making sessions.

Library Makerspace

The library makerspace also referred to as a studio. The studio was open to various programs or sessions ranging from 3D design and printing, sewing, gaming and game design, programming, and stop motion animation. Tools available in the space included a MakerBot 3D printer, sewing machines, a digital photography lab with multiple digital SLR cameras, three Mac desktop computers, about twenty laptop computers, software programs like Sploder for videogames and 3D model designs, toys for younger age groups such as KIBO robotics kits, and other supplies for do it yourself crafts and textile projects. Except for the laptop computers, the rest of the tools were strategically located around the room. The majority of the resources were located on a central wall while in the middle of the room movable tables and chairs allowed for collaboration. Laptop computers were also set on the tables to be used as needed. The studio was available to walk in and open to any age groups.

Middle School Makerspace One

The middle school makerspace was comprised of four sections. The: first section was the classroom study area with round tables and chairs facing the smartboard and the teacher's desk. Section two, was a student display and work in progress station. The third section was a small library with reference resources and a quiet area with 3 desktop computers and workout space and finally section four was a computer lab with thirty-three individual desktop computers facing a teacher station. Apart from the desktop computers, the space also had MakerBot 3D printers, a laser cutter and engraving tools, iPads in a cart, and software for designing, gaming, and coding. This makerspace was used in the school's "Advanced Robotics", "Science/ Technology", "Computer Science", "Introduction to Engineering", "Digital Media", and "Electronic Publications" courses. Students had access to all these resources. For safety reason, the teacher monitored access and gave approval for high end tools.

Middle School Makerspace Two

This makerspace at the second middle school was located in an arts classroom. The room had tables and chairs centrally located where students would work and collaborate facing the teacher's projector. Makerspace resources including iPads in a cart for activities done digitally, art tools, painting resources, a class station with pottery materials, pottery throwing tools, sewing stations with fabrics, and other materials for sewing were strategically located close to the walls around the room.

Participants

Participants for this study were adult makerspace instructors from K-12 and library makerspaces. There were four makerspace instructors; two from same library and two from middle school makerspaces that were observed and interviewed, and one library guide who was only interviewed and not observed. Table one shows demographic information of all the participants.

Table 1. Participant demographic information

Instructor	Demographic Information
Ms. Josephine*	Library Guide Age: 30 Years as a guide: 7 Users catered for: All age groups Program observed: 3D design and printing (3 classes) Guides: All age groups Age group observed: k-8th graders
Mr. Jackson*	Library Guide Age:30 Years as a guide:6 Users catered for: All age groups Program observed: Gaming-video game creation Guides: All age groups Age group observed: K-8th graders
Ms. Anne*	Art Teacher Age: 31 Years teaching:7yrs Years teaching in a makerspace: 2yrs Grade level: 6, 7th & 8th (middle school) Classes observed: 3 Art making sessions
Mr. Mike*	STEAM science Teacher Age: 31 Years teaching: 5 yrs. Years teaching in a makerspace: 2 years Grade level: All grades (Elementary through high school) Classes observed: 4 middle school classes (Two with 7th graders, 2 with 6th graders)
Ms. Ashley*	Library guide Age: 29 Years as guide: 5 Programs: Art and designing Age group: All ages No observation done

Note. All instructors were interviewed but only four were observed as they taught in their makerspaces.
* Pseudonyms used at all times.

Data Collection

In order to understand the instructor's practices in the makerspaces, multiple data were collected. These ranged from observations, interviews, and field notes.

Observations

Observations were an integral part of the data collection to see making and tinkering in action and how these instructors interacted with the space, students, and technology. According to Merriam and Tisdell (1997), observation together with other data collection documents like interviews allows a researcher "a holistic interpretation of the phenomenon being investigated" as they are in the setting and can witness and record the events and phenomena as they happen (p. 111). Observing the makerspace instructors in their natural environment while they guided and taught allowed the researcher a firsthand experience and view of the phenomenon in the space. A total of three session were observed at the library; two 3D modeling and printing and one video game creation. Four class sessions were observed at the first middle school makerspace which involved two 3D modeling session with 7th graders, and two engineering class sessions with 6th graders. Lastly, three classes with 6th, 7th and 8th graders was conducted in the second middle school in the art makerspace. All observations were recorded as field notes using observation protocol as proposed by Creswell (2007), then typed and transcribed.

Interviews

Interviews were conducted, ranging from 28-45 minutes, with all instructors in the makerspace after the final observations. For the library makerspace, a focus group session with four library guides was chosen. During the interview, open-ended questions were posed to the participants to allow for a discussion and probing to occur. Creswell (2007) and Merriam (1998) state that interviews are necessary to help the researcher find out about behaviors, feelings, and past practices that cannot be observed. Trustworthiness of this study was ensured through careful negotiation of entry to and exit from site, conducting long repeated observations, collecting multiple data, using epoch technique to suspend researcher's bias, and seeking a qualitative expert review.

Data Analysis

Data collected was analyzed throughout the study; a strategy that helped make sense of the data and ensure coding was a continuous process. Creswell (2007) stated that data analysis in qualitative research allows for "preparing and organizing the data, reducing the data into themes through the process of coding and condensing the codes, and finally representing the data in figures, tables, or discussions" (Creswell, 2007, p. 148). Using this guideline, data was analyzed on a regular basis in order to answer the research question: how does pedagogy manifest itself in the makerspace?

Constant Comparative Method

Immediate analysis was conducted after each observation. Field notes, journal entries and artifacts were coded as data was collected. Interview transcripts followed as they were collected at the end of the observations. Continuously looking and coding the data into categories was vital in describing the process and context of the discovery that happened in the makerspace. Codes were written out as data was read and reread and compared to the previous codes each time to identify relationships and gaps. It also provided for insight to probe and look further to answer the research question.

Within Case Analysis

Within-case analysis was used at the end of all data collection as proposed by Creswell in analyzing data from more than one case. This method allowed for a keen look into each makerspace as a case to better understand the phenomenon process and uniqueness of each case. In this stage, Creswell's (2007) lean and focused coding method was used to recode the data and identify relationships with codes from the initial constant comparative coding method. When this initial coding was completed, common themes were identified that came out and were believed to have answered the research questions posed. This process was repeated for each case. After within-case analysis process was complete, the relationships in the similarities and differences across cases was examined.

RESULTS

The following themes were identified after a thorough analysis of the data: (a) Some scaffolding and structure; (b) multiple pathways/entryways to activities; (c) Do It Yourself: A self-taught mindset; (d) More about the student, less about the teacher- Paradigm shift; (e) Guiding Iterative process; (f) Encouraging artifacts creation through iteration; and (g) Feedback. These themes represent the pedagogical strategies that these specific makerspace instructors used in their individual spaces to guide learning.

Scaffolding and Structure

Across all instructors, some scaffolding and structure was observed. Instructional scaffolding occurs when an instructor provides supportive tools, resources, and guidance as they construct knowledge and gain new skills to meet the learning objective (Vygotsky, 1978). Instructors provided some sort of prompts and setup of the projects at hand for the users and students. This gave students some basis for entry to the projects and gave them ideas allowing room for enhancements. This basis helped the users to relate easily and find some starting point. Some instructors built on those starting points and some deviated and explored their own pursuits and ideas, which was just as welcomed in the space. The guidance provided at the beginning of these sessions played a major role in helping learners find their bearing in the sessions as they could be seen exploring freely on their own afterwards. In each session, there were introductions done, mentor texts provided, as well as chunked steps. In the library makerspace, the instructors provided multiple avenues to the projects to cater to the varied abilities in the space so students that were advanced could dig in and start where they want while first timers could follow through the guides to get where they want to be. In one instance Ms. Josephine commented;

It is a good way to get people through those mental blocks to be really successful at one thing and ones they can attach something to their Gmail then we can talk about all other things on Gmail, but they need to know that they are capable of doing it so whether that's finishing the first section of stitching a project or I guess that's a good example or if we are doing embroidery project they got their first color of red embroidery done, they got their stitch done and when they are beginning to feel like they are kinda hitting their groove and so it's like making it feel like formal instruction for that first bit but we are doing it instead of seating in front of the class on a okay now do this, now do this, so we are kinda pushing them a little further into the pool before you shove them into the deep end.

In the art class makerspace, the instructor provided a painting image projected on the whiteboard after the students settled in class. Students reflected on the image, identifying every element that stood out for them, what they noticed, thought about the image, and why the image was important. Students then had the opportunity to share their observations with class and this led to discussion on hues, dimensions, story narrative of the art. After this, the students broke into their individual projects and they had to ensure that their work incorporated some of the elements they discussed from the painting if they did not already have them. If their work did have these elements already they had to reflect on what that meant for their overall project and it was upon them to polish on those skills.

When asked why she did so, this was her response:

It is important for the students to see work of art out there and look keenly at the pieces that bring the art work out and how this might help inspire them in the paying attention to little details and elements in their own creation as makers.

Mr. Mike started his classes by posing questions to the students about their previous class sessions after which he projected that day's session activity guidelines on the board. Students were then given go ahead to start on their projects. Using these guidelines, students were given the cue to create their 3D models and CubeSats to meet the objectives. For the 6th graders' CubeSat sessions, the learners were guided through the problem which was to create a CubeSat to help the National Aeronautics and Space Administration's (NASA's) mission of making the world a better place. Students were given five themes to choose from after which they selected their own goals and designed CubeSats that met those goals. How they designed the CubeSats were up to them. If the CubeSat served that objective and survived the tests of the fall, the students were then allowed freedom of design.

In the library, there is a starting and end time as well, however, users can come and go as they wish. At the beginning of every session, the guides provided about five minutes of introduction to the sessions, including what the sessions were about, the stations available, and how printing, 3D modeling, and gaming design worked. Afterwards, users created what they wanted, and the instructors answered questions asked. The explanations varied between the makerspaces, obviously, due to the formal and informal nature of both spaces. Through scaffolding, instructors teaching in the makerspaces can model guide students into taking control of their project designs. The instructors allowed for students to rise from novice to apprentice to experts in their creation becoming more autonomous.

Multiple Pathways to Activities

The instructors set up the makerspaces in ways that allowed learners to create differently using multiple resources and techniques in the space. They set up the space to be flexible to multiple activities but be rigorous at the same time. The library instructors for instance employed this technique often as they get people from varied age groups and abilities. This flexibility was highlighted in the following excerpt:

We have a really form of informal learning in the studios and we pick things that are interesting to different age groups. So, something like 3D printing if I was just teaching a class I am going to teach it differently if its teens or adults even if they come in, am going to teach it a lot differently. But a lot of things is having an understanding of what is the easy entry points to this project. So, if it's a tote bag project to sewing and crafts day, the very easy entry point is coloring on it with markers or fabric crayons, the next step is maybe adding working on stitching, maybe sewing in their first buttons. And then you have different kids that are doing a very elaborate design rather than marker and stitchery. And then you have adults that are comfortable with sewing and really just having a good time and are at a completely different self-guided level and adults can fall anywhere in that same spectrum. But I think having projects that are easily adaptable so a lot of time we have skills we want to teach but the projects are the starting points for easily adaptable.

In another instance, one library instructor described a scenario where they are creating video games but there was a user that wanted to sew instead. In creating multiple avenues, that user was still able to sew while interacting and collaborating with the users creating video games in the same space. This account was reported by Mr. Jackson who stated:

And you know that reminds me, I had a videogame program it was really fun, we had a lot of people creating video games and someone came in and said I wanna work on a dress for my daughter. And I don't have much experience with sewing but we did have someone in hand to help actually Beth was there to help out and it was this crazy experience because we had a parent working on a dress while next to people working on this completely different thing and talking to each other and engaging in a different weird way, I don't think they would have otherwise been doing that, it's a weird melting pot sometimes.*

The same observation as above was made by Ms. Ashley who said:

A lot of, like we have a theme for every open studios, but being able to adjust and be flexible on who shows up, and what skills levels, sometimes I would have a theme but someone shows up but they brought their own project that kinda relate to that but it's not what I was kinda thinking about so then being able to just jump in and go with that and that inspires somebody else um I think that's something to strive for kinda that virtual experience, so people are self-teaching and creating on their own without that really guided help.

In this instance, the flexibility in the space allowed users to find different pathways to projects. Users could create and invent multiple projects and even though some were not related, they could discuss, talk to one another, and share ideas. In addition to multiple pathways, this flexibility provided a safe environment for the students to create comfortably and take charge in this endeavor. This strategy is widely applicable across many formal and non-formal makerspaces.

Do it Yourself: A Self-Taught Mindset

Makerspaces were founded on the concept that everyone can learn, make, and tinker and given the right environment, guidance, and resources, we can all be innovators. Across all cases noted above, none of the adult instructors could say they had any experience prior to starting their makerspace teaching. That meant they had to constantly be reading, researching, reevaluating, and troubleshooting technical difficulties in and outside the classroom sessions. According to Mr. Mike, a teacher in the middle school makerspace, it was a constant learning curve as illustrated in the following excerpt:

Last year I was basically told here are the names of classes you are teaching, good luck. Huh. My background is biology so things like computer science and engineering like I got some knowledge about, like that's not what my schooling is about so I have had a lot of learning. . . . But there is a lot of great things online as far as what other instructors are doing but I have had pretty limited connections with other instructors.

During the first observation in a 3D modeling and printing session, Ms. Josephine's attention was called by one of the students who had already designed their 3D model on Tinkercad, the software used by all instructors to teach 3D modeling. However, the model did not print. The instructor immediately started troubleshooting the computer and the 3D printer looking online for solutions. This problem resulted in a loss of 20 minutes of classroom time. This is a common occurrence in many makerspaces where staff are constantly troubleshooting most things themselves. This instructor further explained that one of the things they learned is that anything can go wrong in the space, but experience teaches one to be ready to figure it out, find resources, stay calm, and hopefully use that as a teachable moment. These instructors were constantly learning on their own, digging through resources to troubleshoot, and learning new ideas to make the space more interactive, engaging, and learner centered. They also involved students in this process so they too learned to take charge of their own learning and help each other figure out things through research and inquisition.

More About the Student: Less About the Teacher

Within each of these spaces, teacher-student roles shifted. The instructors worked towards granting more control, responsibility, and urgency to the learner, a strategy that allowed these users and students to push their understanding and ideas through making and tinkering. The spaces became safe environments where the users and students were guided through their ideas and instructors become facilitators rather than managers. Within the library makerspace, making and tinkering is very informal and users come and go as they wish and sometimes come in with different project ideas than what was being offered. In another instance, Ms. Josephine stated that:

I think that's what's nice about the studio; we all have things we are interested in and different expertise we have but knowing that we don't have to be the experts and I think a lot of the patrons here identify that we are going to help them find the resources to become the experts themselves but they are not leaning on us to help them in every step but we can point them to different directions. You know okay Sewing machine is showing this error, okay let's look up online what is that error mean, kinda showing people means to become successful.

The instructors' strategies further involved helping users and students learn by being experts themselves as noted by Ms. Ashley:

And I think that's a challenge we run into coz a lot of people did grow up with that structured setting and they come in here and expect that you are going to go step by step and project and it is something we want people to have that experience.

These instructors shared control of learning in the space and gave the students and users more ownership of their learning as they assumed more responsibilities. In this regard, Ms. Anne said:

I have also learned a lot that having choice in the curriculum gives students more engagements, that's in itself is a structure is a lot better. I have had the students write objectives. For about the last year and half I went to a conference and the teacher there had the students write objectives and I was like 'why am I doing all the work? Why am I trying to write overarching objective that had nothing to do with their projects, so then that's when I was introduced to the studio habit of mine. Why do we not utilize these for their objectives?

Having students write and pursue their own learning objectives for their projects was something different from what one might see in a regular classroom. Students had to think critically about what they wanted to do and utilize the available resources and tools in the space. The students worked harder, smarter, and were more driven to be successful in seeing their goal completed. They asked questions and were more engaged in the making process resulting in authentic learning.

Guiding the Iterative Process

The goals of the instructors were to help their learners and users understand that creation and innovation take time. The instructors created a safe place for students to try different designs while making mistakes and retracing their steps. The instructors also guided the users throughout the tinkering process and allowed them to reflect upon their progress. It was important for the users to understand that they may not be perfect the first time. Trial and error is part of the process. In the library makerspace, users came and made whatever they wanted using the material provided; this meant two people could be making something entirely different using similar materials or alike. It also meant that varied abilities and age groups may need different types of guidance. Within the space, success is defined and not looked at as completion of tasks but rather success is personal. The users redefine what success is depending

Table 2. Excerpts from instructors on guiding the iterative process

Excerpts from Instructors on Guiding the Iterative Process
• I think just also teaching them ways to define what they are seeing rather than saying ooh I made this thing is a failure, you could say, what did you make, a 15 second thing you told a story with it, or you made a playable game with a clear ending, and once they have those words to define it instead of when they explain it to somebody's else is another way to describe it. • To a degree and it's something we have explicitly said and agreed on as a class. You know we talked about the first time we made the CubeSats, do we expect them to fail, choral respond yes, is that okay yes, what are the things we learn if they fail, my favorite response is, 'we learned what not to do' so we are talking about it, but am not sure for every kid it translates to prototyping. But we practice that message over and over, yeah, its prototype yeah, it's the first time let's go with it and see what happens. • I think just also teaching them ways to define what they are seeing rather than saying ooh I made this thing is a failure, you could say, what did you make, a 15 second thing you told a story with it, or you made a playable game with a clear ending, and once they have those words to define it instead of when they explain it to somebody's else is another way to describe it. • To a degree and it's something we have explicitly said and agreed on as a class. You know we talked about the first time we made the CubeSats, do we expect them to fail, choral respond yes, is that okay yes, what are the things we learn if they fail, my favorite response is, 'we learned what not to do' so we are talking about it, but am not sure for every kid it translates to prototyping. But we practice that message over and over, yeah, its prototype yeah, it's the first time let's go with it and see what happens. • And it is okay to fail. I mean I want a quality product at the end, but it's okay that the color mixing over here in 8th grade it didn't work but in a sense of failure it was a fail, but then what did we learn from that mistake? Okay we learn that I need to ask help from Ms. Anne, and Ms. Anne was able to give me a suggestion and now I can try this suggestion.

on the project they are working on. The instructors further echoed these sentiments in the following excerpts noted in Table 2.

In addition to redefining success and learning, in that making mistakes is an accepted practice, the instructors further fostered the understanding that learning is a process of exploration. Ms. Anne for instance noted that:

And that's another thing that's a hard thing for the new kids coming in from the elementary, because elementary its teacher directed, and am not saying that's wrong, am just saying it's a huge transition from that to this environment and they wouldn't want to take those risks, they wouldn't feel comfortable with that. They would just wanna rush rush rush, and am to get it done, and am like wait wait art is a process it takes time to go through this. I know you have a great start, and it's a wonderful sense of exploration, but now let's go back and refine and put our best into it. And so that can be a challenge, because they are middle schoolers and they are always just like we wanna be done, right we just gotta slow down a little bit and focus, the key word there is FOCUS.

Encouraging Artifact Creation Through Iteration

Every makerspace session observed involved students creating and making something either tangible or digital that they could share and reflect on. Students come to the space with an idea of what they want to make and left with knowing there was an instructor to guide and encourage them through the creation process. Specifically in the library makerspace, users left with something tangible and digital that they created. These ranged from printed 3D models or video games they made that they would share with friends and family. This ownership and sense of fulfillment from their creation made them come back often to further polish their works, or create a fresh again. The same happened in the middle school makerspaces, even though the making was not one session but over multiple sessions, students still had access to their designs and creations. Completed artifacts were shared through different avenues such as art galas and on different online platforms. Instructors encouraged students to participate in sharing their creations and obtaining feedback and ideas on their designs as noted in the following statement:

I think that's a working progress right now, but we have like an annual art gala, third quarter we would do an art show in conjunction with the high school and it's like a band choir and art show together, so when their pieces are judged and they get the reviews is a big profound moment and that on a grand scale.

Similarly, another of the instructors stated:

So, another element is we got a new display system in the hallway, yeah walker display system, very expensive but very pleased with it, worth it. Before I was confined to this little bulletin board and displaying system in the cafeteria, and I was like, it's too hard to choose what piece of art goes up, and what if I repeat a student's art work then another student doesn't get a chance to celebrate their art work, and it was just like this is not gonna work and we are no longer doing the same thing of a guided drawing and am gonna choose the best one. I want every voice to be out there, and if it's out there and you are like ooh my God am slightly embarrassed. Then maybe that's telling you that there is something you need to go back and address. Right, but if you are like this is mine this is me, and the instructors comes to you and say hey you know what you did a great work of art on this one there. It's more of a dialog, right you are going to be accountable.

Providing Constant Feedback

Providing constant feedback occurred during the iterative process. Making and tinkering is a continuous process and the instructors created an environment where peer to peer feedback as well as instructor feedback was always happening. Instructors are often asking questions and offering suggestions here and there that helps the learners through their making process. Within the school setting, these makerspace instructors used both summative and formative assessment to guide student's learning. Feedback varied from person to person, some students asked questions, and others needed some probing and guidance, some worked through walking around and finding inspiration through peer to peer conversations, while still others seemed afraid to ask for assistance. In all these scenarios, the instructors still provided a safe environment allowing imagination and creation to take place through feedback. Mr. Jackson while discussing some users needing guidance referenced how this guidance and feedback occurs in the library makerspace:

They are afraid to ask for help sometimes, I use an almost delusional amount of enthusiasm I found that usually helps, I think I always respond with ooo what is that, and just if you hit somebody with that you can kinda go with the flow, I lead into asking things I think.

Students and users also gave each other feedback facilitated by the instructors. They shared their creation whatever level they were in and discussed ways to make changes and ideas the other could try.

And then the Self-assessment has a section where they do a class critique; And basically, that means they have their artwork and assessment with critique side by side and we all come around and write a sentence, I appreciate your use of color and value, Anne, then on the critique side, have you considered adding texture to your composition? And then that way the students have something to think about, it's not giving them like you have to do this, but it's a choice.

In addition, they created a learning space that gave students opportunities to self-reflect on the learning process and their creations. By using Seesaw, a mobile application for example, Ms. Anne gave students a way to track their progress and reflect on it. It also provided a way for her to give constant feedback. She noted regarding this:

Students documenting their artwork, beginning, middle and end of their art work. Seesaw app I use. So, they are all in the same class. They scan the code. They take a photo of their artwork and they assign it to their name. They can also add a caption, here is what I'm working on and go here is what I am really succeeding in and here is what we are struggling in. And so, it helps me to say I can give you feedback and help guide you with your struggling but also help celebrate your strength. So, I try to do that and I require them to do a posting on it 3 time in their project, beginning middle and end.

In another instance, feedback is provided during the iteration process where the instructor holds students finished artwork in class, without interrupting progress, and shares that with everyone to help inspire each other as noted by Ms. Anne:

But I also think like when I was holding the, drawing of a car up for the students to look at I think it kinda gave them a chance to see at it in a different new eyes and say hey am headed at the right direction, this is good and hearing the wow from the other students is encouraging, and just knowing that some students like our robot student who creating a robot with paint the fact that he is willing to seek criticism and constructive feedback is huge for him. So that's the moment of hey Ms. Betty could you let me know that you appreciate what I have done? Absolutely, I say rock on good work, now how can you go back and refine it, well this is the most paint I have used, okay well I get that can you let it dry and clean up some of your edges. So, things like that so applauding what they do but also suggesting and we do that with the critique, so like I really appreciate this is cool.

Self-Instructor Assessment

Every instructor in this study commented on the importance of self-assessment. Because everyone had been teaching in the makerspace for less than three years, the instructors constantly assessed and reevaluated their strategies for the space and made improvements as well as learned from each encounter. Instructors had to constantly identify their progress and how that helped making and tinkering for the learner, what worked and what did not, and how could they make the next years' sessions better. All instructors agreed that while there was a learning curve; they were constantly improving and adjusting. For example, one instructor remarked:

So you know last year was a lot of coming up with stuff for the first time and seeing what worked and what didn't and so this year I have a little bit better idea and I can guide them more but I am kinda seeing where the spread is as far as student abilities, what kind of scaffolding they do need, so like in developing more cohesive unit lesson plans this year, what you saw is pretty much teaching everybody and it's not broken much by individual student needs, but am really developing a lot of curriculum this year, and once I have all that my goal for next year will be to have like computer science for example, would be to have a whole unit buildup of materials then I could just set kids on their next project coz they will already have all the instructions and resources available and huh, so that way they can kinda pace themselves right now am not quite there yet so.

Another reflected:

I have gotten better at asking questions, I have gotten better at helping them scaffolding the thinking from here is a concrete notion of football, how can we take it more abstract. What is it that do you like about football? Hang out with friends. Okay I can respect that's awesome. Why are friends important? How is that friendship impacting your field playing on football? Okay how do we show this in your piece? And am not gonna lie, there are moments I go, ooh my God I had an awesome idea ever, but the moment I voice my idea I take away their ownership and ideas. So, I have to focus at editing myself, am like don't say it, don't say it.

CONCLUSION AND IMPLICATIONS

Schools must evolve with the ever-changing needs of the learner. Instructors in the makerspaces put forth activities and learning goals that are user-centered and interesting to various learning needs. They design and create a learning environment that safeguards learners to experiment with ideas, their own goals, and materials and iterate learning in the process to redefine what success and failure means, becoming inventors and pioneers in creations that are big and small. The pedagogical strategies discussed in this chapter can be adopted by other makerspace instructors as well as general instructors teaching in non-makerspace classrooms.

In a traditional classroom, students have less chance to refine and tune the struggles in each project as feedback is mostly summative either as percentage or words like “need more evidence,” and instructors move onto the next project or segment with the expectation that the learner will improve on the next project. Little is done to ensure a meaningful learning process through these using an iterative process. Additionally students most often check out of the projects once it is submitted and graded and rarely have time to reflect and rework those their products.

Instructors in makerspaces provide activities to learners that helped broaden their minds and put into practice what they imagined and learned in the classroom by fostering creativity, problem solving through iteration, and transfer those skills to a very different form of learning. Learners take ownership of their creations and have more autonomy in return. Instructors facilitate and share in the responsibility. Through emphasizing iteration, learners can grow and problem solve as they tinker, create, and make. Instructors further identify the variability in skills in the space and provide for multiple pathways to projects that allow students to further iterate, share ideas, and experiment.

There are several implications of this study. Educators need to create learning environments that allow learners to hone their 21st century skills as well as revisit their teaching strategies if we are going to prepare our students for the global world. Most instructors may think the fancier and bigger the tool the better outcome and teaching I can do but this is not the case as instructors in this study used readily available materials such as cardboard, tape, and straws, in addition other sophisticated tools to achieve meaningful learning. Instructors must, therefore, select strategies that would effectively integrate available resources to help meet learning goals and engage their students. Makerspaces encouraged students to experiment and create while still learning and acquiring those important skills using not only technology but also other non-expensive tools available every day. Instructors can borrow from this and be ready to design lesson activities using resources available to them to help learners meet not only content skills but also 21st century skills and reimagine their classrooms. Instructors must provide supportive learning environment through guidance to allow students to take ownership of their learning which in turn helps them critically think and problem solve throughout their design process. Administrators must also support instructors who are willing to venture out of the old teaching zone; instructors that are ready to reinvent the learning and provide them with resources, guidance, support, and education opportunities to carry out a successful makerspace within their classrooms. Revise and revisit is a consistent skill needed if teaching strategies and experimentations with are to be fruitful.

REFERENCES

Ackermann, E. (2001). Piaget's constructivism, Papert's constructionism: What's the difference. *Future of Learning Group Publication*, *5*(3), 438–448.

Atkinson, R. C., & Shiffrin, R. M. (1968). Human memory. A proposed system and its control process. In J. Spense & K. Spence (Eds.), *Psychology of learning and motivation* (Vol. 2, pp. 89–189). New York, NY: Academic Press.

Ausebel, D. P., Novak, J. D., & Hanesian, H. (1978). *Educational psychology: A cognitive view* (2nd ed.). New York, NY: Holt, Rinehart and Winston.

Blikstein, P. (2013). Digital fabrication and "making" in education: The democratization of invention. In J. Walter-Herrmann & C. Büching (Eds.), FabLabs: Of machines, makers and inventors (pp. 1-21). Bielefeld, Germany: Transcript-Verlag Publishers.

Brahms, L. J. (2014). *Making as a learning process: Identifying and supporting family learning in informal settings* (Order No. 3582510). Retrieved from ProQuest Dissertations & Theses Global. (1609201673)

Brahms, L. J., & Werner, J. (2014). Designing Makerspaces for Family Learning in Museums and Science Centers. In M. Honey & D. E. Kanter (Eds.), Design, make, play: Growing the next generation of STEM innovators (pp. 12-16). New York, NY: Routledge.

Bruner, J. S. (1997). Celebrating divergence: Piaget and Vygotsky. *Human Development*, *40*(2), 63–67. doi:10.1159/000278705

Creswell, J. W. (2007). Qualitative inquiry and research design: Choosing among five approaches. Thousand Oaks, CA: Sage publication Inc.

Dougherty, D. (2005). *Maker faire founder Dale Dougherty on the past, present, and online future of the maker movement*. Retrieved from https://goo.gl/wRkKxF

Dougherty, D. (2013). The maker mindset. In M. Honey & D. E. Kanter (Eds.), *Design, make, play: Growing the next generation of STEM innovators*. New York, NY: Routledge.

Driscoll, M. P. (2000). *Psychology of learning for instructions* (2nd ed.). Needham Heights, MA: Allyn & Bacon.

Driscoll, M. P. (2005). *Psychology of learning for instructions* (3rd ed.). Boston, MA: Pearson Education Inc.

Gagné, R. (1977). *The conditions of learning*. Boston, MA: Thomson Learning.

Gagné, R. M., & Medsker, K. L. (1995). *Conditions of learning: Training applications*. Belmont, CA: Wadsworth Publishing.

Hackerspace. (2015). *Hackerspace*. Retrieved from http://hackerspaces.org/

Honey, M., & Kanter, D. E. (Eds.). (2013). *Design, make, play: Growing the next generation of STEM innovators*. New York, NY: Routledge. doi:10.4324/9780203108352

Kalil, T. (2013). Have fun-learn something, do something, make something. In M. Honey & D. E. Kanter (Eds.), Design, make, play: Growing the next generation of STEM innovators (pp. 12-16). New York, NY: Routledge.

Keller, J. M. (2010). *Motivational design for learning and performance: The ARCS model approach*. New York, NY: Springer. doi:10.1007/978-1-4419-1250-3

Martinez, S. L., & Stager, G. (2013). *Invent to Learn: Making, tinkering, and engineering in the classroom*. Torrance, CA: Constructing Modern Knowledge Press.

Merriam, S. B. (1998). *Qualitative research: A guide to design and implementation*. Hoboken, NJ: John Wiley & Sons.

Merriam, S. B., & Tisdell, E. J. (1997). *Qualitative research: A guide to design and implementation*. San Francisco, CA: Jossey-Bass.

Papert, S. (1993a, February 1). Obsolete skill set: The 3Rs--Literacy and literacy in the media ages. *Wired Magazine*. Retrieved from http://www.wired.com/1993/02/1-2-papert/

Papert, S. (1993b). *The children's machines: Rethinking schools in the age of computers*. New York, NY: Basic Books.

Partnership for 21st Century Learning (P21). (2019). *Framework for 21st century learner*. Retrieved from http://www.p21.org/

Petrich, M., Wilkinson, K., & Bevan, B. (2013). It looks like fun, but are they learning? In M. Honey & D. Kanter (Eds.), *Design, make, play: Growing the next generation of STEM innovators* (pp. 50–70). New York, NY: Routledge.

Reiser, R. A., & Dempsey, J. Y. (2012). *Trends and issues in instructional design and technology* (3rd ed.). Saddlebrook, NJ: Pearson.

Simon, M., & Brown, G. (2013). Raft: A maker palace for educators. In M. Honey & D. E. Kanter (Eds.), Design, make, play: Growing the next generation of STEM innovators (pp. 138-162). New York, NY: Routledge.

Vygotsky, L. (1978). Interaction between learning and development. In M. A. Gauvin (Ed.), *Reading on development of children* (pp. 34–40). New York, NY: Scientific American Books.

Vygotsky, L. S., & Cole, M. (1978). *Mind in society: The development of higher psychological processes*. Cambridge, MA: Harvard University Press.

Wagner, T. (2008). *The global achievement gap: Why even our best schools don't teach the new survival skills our children need--and what we can do about it*. New York, NY: Basic Books.

Washor, E., & Mojkowski, C. (2013). *Leaving to learn: How out-of-school learning increases student engagement and reduces dropout rates*. Portsmouth, NH: Heinemann.

Wilson, F. R. (1999). *The hand: How its use shapes the brain, language and human culture*. New York, NY: Vintage.

Chapter 3
Flipped Inclusion Between Educational Emergencies and Transformative Socio-Semiotic Didactics

Tonia De Giuseppe1
University of Salerno, Italy

Felice Corona2
University of Salerno, Italy

ABSTRACT

This chapter analyzes aspects of the information and communication society, which affirm the power of knowledge in the continuous recognition and reinterpretation of signs. Deviant aspects of technology use related to cognitive capitalism and consumer needs that are expressed in knowledge are addressed and new forms of educational emergencies are outlined. The flipped inclusion model is based on epistemological constructs, which characterize the cross-media society and on the value of the circular economy, understood as self-regenerative, through investments in circular education as a permanent re-education process. The research, carried out in Italy, proposes an existential design model, focused on the perspectives of a permanent and inclusive system of learning. The aim of the research was to highlight how a trans-educational pedagogical-didactic model formed personalities with a plural-prosocial identity and favored the processes of co-management and the co-construction of inclusive contexts.

DOI: 10.4018/978-1-7998-2914-0.ch003

INTRODUCTION

In the age of information, the power of knowledge lies in the recognition and continuous reinterpretation of signs. However, the use of technology for communication may produce forms of cognitive capitalism (Paulré, 2000). Globally, society is consuming technology at unprecedented rates. Much of this will generate new forms of knowledge and transform contexts of culture, receptors and emitters of signs, languages, and meanings. Pervasive communicative exchanges derive from increased accessibility and will generate a transformation of our everyday lives. Such communication enhances forms of global democratization and a breakdown of social hierarchies, but it does not always guarantee an equitable distribution, recognition or respect for roles in an inclusive, ecological-systemic cooperative society. An unconscious management of technology, with an interweaving of perceived meanings and projections, increases the production of perverse semiotic configurations, mixed with situated symbolic phenomenology (Merleau-Ponty, 1945). Society is witnessing new and continuous transformations of subjectivity and intersubjectivity, in cross-media interrelations. Iconic and symbolic signs are transformed into conventions (Jung, 1999), through increased times and spaces, which exceed the contextual dimensions. In such a socio-cultural context, additional educational interventions in the use of technologies are needed for lifelong learning (Simondon, 2008). This includes the use of a Flipped Inclusion model, which has been shown to improve students' skills and competencies.

BACKGROUND

Signs, sensations, and performative representations become the fields on which the formative avant-gardes are articulated. The new educational challenges are placed in the management of communicative acts, as semiotic decoding of signs and phenomena of signification (de Saussure, 1970). The signification is intended as an expression of the relationship between the significant (form) and meaning, with an issuer that is not present but is a projection of the recipient. The educational investment in the interpretation of the relations of significance allows to grasp the correlations between material resources and intangible resources, such as human capital. This avant-garde educational formative approach allows us to recognize the phenomenological repercussions, expression of the concrete processes of communication at a psychological, anthropological and cultural level, which are reflected in the choices and the social and political directions.

Through the promotion of competent self-reflexivity, it is possible to control the risks of antisocial modeling, through educated investments in self-awareness

(Mead) including perception and recognition in comparison (Merleau-Ponty, 1945). In the educational process, a pedagogical value is assigned to the semiotic act and its centrality of inclusive, socio-prospective direction. It is about promoting educational actions, aimed at strategically encouraging changes and stimulus situations. Using incremental change processes, through the use of cyberspaces, it is possible to promote transformations that intervene on the perception of the self and its own effectiveness.

Through permanent evolutionary, socio-technological processes (Parag, 2016), we see structural and super structural transformations, deriving from the redundancy and diffusion of interconnected infrastructures, data platforms, which combined with creative capacity, produce dynamic socio-economies. Furthermore, the process of the mechanization of life (Bauman, 2005) which sees technological tools becoming receptors and emitter of signs and language, increasingly makes culture a process that can be associated with the mechanical-technical place of virtual learning contexts. In the postmodern society, there is no single, homogeneous and linear sense of episteme, which traces explicit rules in the construction of the sense of civilization. The new technologies favor communications full of symbolic meanings, with continuous flows of meanings and intrinsically disruptive contradictions. The people are in constant presence of a virtual collective co-construction of knowledge, with transformative sub-cultures, which are self-generated through told cultures and virtual projections of experiences. This highlights the social nature of the cross-media language that transcends communication which is full of signs, "senses and projection of its meanings" (Ortiz, 2016, page 32) and involves the knowledge and understanding of multiple conformations of reality, sense and perspective of human society.

Data, Pervasiveness, and Internet Addiction

The unconditional use of technologies is infiltrating every field (WHO, 2015) and generating relevant data related to Internet dependency. Jerald and Block (2008) noted that this dependency is a "compulsive-impulsive spectrum disorder that involves online and/or offline computer usage and consists of at least three subtypes: excessive gaming, sexual preoccupations, and e-mail/text messaging" (p. 306). All of the variants of this dependency share the following four components: 1) excessive use, often associated with a loss of sense of time or a neglect of basic drives, 2) withdrawal, including feelings of anger, tension, and/or depression when the computer is inaccessible, 3) tolerance, including the need for better computer equipment, more software, or more hours of use, and 4) negative repercussions, including arguments, lying, poor achievement, social isolation, and fatigue" (Jerald & Block, 2008, pp. 306).

Issues, Controversies, Problems

Globally, Internet addiction and associated problems is on the rise. For example, in Asia, specifically in Korea, this dependence has been identified as the greatest health problem experienced among children. In China, the highest rate of Internet addiction and other related problems was in individuals ranging in age from 18-23 (Goldberg, 1995). In Europe, specifically Switzerland, there was problematic use of the Internet (defined as Compulsive Internet Use), which reached 6.5% between the ages of 15-19, but only 0.9% among the general population (WHO, 2015). The use of cross-media technologies is growing even among older populations and therefore population data are becoming increasingly relevant (ISTAT, 2018). The ISTAT survey (2018) shows that in Europe, Internet users constitute an average of 70% of the population. In particular, in Italy, Internet usage drops below the European average to 48.9% of the population, where only 26.4% connect daily.

The daily use of the web (ISTAT, 2014) is more frequent among the 15-24 years (over 70%), but the strong increase in the daily use of the Internet is evident between the 25-34 years (+7 percentage points) and among young women between the ages of 15 and 17 (+9.2 percentage points).The daily use of the web (ISTAT, 2014) is more frequent among 15-24 year olds (over 70%) but the strong increase in the daily use of the Internet among 25-34 year olds (+7 percentage points) and between young women aged 15-17 (+9.2 percentage points). Compared to 2013, the share of people who use personal computers for work remains stable, at around 54% (33.5% make it a daily use); instead, the percentage relative to the use of the Internet is on the increase and passes from 54.8% to 57.3% (36.9% use it daily). Compared to 2013, the share of personal computer users remains stable at around 54% (33.5% make it a daily use) while that relating to the use of the Internet increases from 54.8% to 57.3% (36.9% use it daily). The greatest growth in network use is recorded among people aged 60-64 (41.6% compared to 36.4% in 2013). The proportion of people who use the Internet on a daily basis also increases significantly, from 33.5% to 36.9%. To these data is added the extent of the video game phenomenon and the impact on the habits of young people aged between 3 and 38, in Italy and in the world, which sees 18.7% of children aged between 3 and 5 years choose video games as a favorite pastime; between the ages of 6 and 10, video games clearly emerge as favorite gambling tools, with 70.4% of children and 39.9% of girls, with regular and preferential play.

In 2013, Internet Gambling Disorder was included in the classification of mental and behavioral disorders of the American Psychiatric Association (DSM-5, 2013) as a condition to be explored. Since 2014, the World Health Organization (WHO, 2015) has conducted activities related to the implications of excessive use of the Internet, computers, smartphones and electronic health devices. For instance, gaming disorders, in online-offline variants, represent new dependencies, as noted in the 11th edition of the International Classification of Diseases (ICD-11, 2018) on mental pathologies. It is intended as a clinically recognizable and significant disorder that manifests itself in uncontrollable gaming behavior. It is evident for at least twelve months and generates impairment in socio-psycho-relational, personal, educational or occupational functioning, with neurological complications and epilepsy in the most serious stages. However, "the absence of universal definition of purpose for this relatively recent phenomenon, including the lack of universal terminology and diagnostic criteria, hinders international communication and research in this field" (DSM-5, 2013, p.8).

Unfortunately, today the estimates vary according to age, between geographical areas and with diagnostic screening and thresholds, which do not allow international research to have totally reliable data. In fact, the prevalence estimates vary up to 0-90%, depending on the country, age group and epidemiological method in question (Jiménez-Murcia et al., 2014). For example, the prevalence of problematic Internet use within Europe ranges from 1% in Norway to 18% in the UK. In contrast, the range in the United States of America is reported as 0-26% and 7-23% in Hong Kong (Kaye & Farrell, 2014). It states (WHO, 2015) the need to conceptualize and define areas, phenomenology and types of disorders associated with the excessive use of the Internet, computers, smartphones and devices, to outline signs and symptoms, disorders and behavioral addictions (Yen, et al., 2007). The definition of a clinical taxonomy, based on the intensification of adequate studies, defining the boundaries, the diagnostic limits and the possible elements of comorbidity associated with an excessive use of the Internet and cross-media devices is emergent. At the same time, social and training policies must be launched, aimed at prevention and strategic interventions. Moreover, socio-educational-systemic investigation policies must be prepared. This includes periodic data collection through standardized tests using validated tools; internal registry systems for health care services, follow-up data to document and monitor the scale of the problems, age and gender distribution of cases, health and social consequences of medical conditions and expenses; surveys, and phenomenological investigations on the courses and interventions. This type of data will reveal the pervasive cross-media incidence that can determine forms of degenerative dependence upon technologies and the Internet.

In the presence of this pervasive phenomenology, the pedagogical dimension becomes strategic and represents the anchor to which the boundaries of an aware, cooperative, accessible and prosocial competitiveness are bound. The centrality of communication in the knowledge and information society indicates the epistemological core of emergency tendencies, to be experimented with a view to democratic re-educational circularity (Dewey, 1984). The use of technology requires a responsible and conscious application. In fact, individuals give moral value to actions and consequently to the use of instruments but the technology is neutral from the moral point of view, until the person applies it (Gibson, 1994). The emphasis placed on the aspects of permanence and the process of formation is reaffirmed in the expression knowledge society, which considers every social dimension of life that characterizes the media world. The promotion of a healthy society is based on the re-humanization of relationships, with a centrality placed in the values of the person and in systemic ecological development, to stem the capitalist risks of a technocratic dictatorial society.

The indissoluble interweaving between the cross-media transformation of society and personalities, as constitutive and necessary aspects of the functionality of a global system, requires individual skills capable of moving towards innovation and the qualitative well-being of life projects. For the learning society, as a metaphor for a social system based on knowledge, a driving force for socio-economic development and learning as a conditional condition for the functioning and health of the system, the structural dimension intersects with improvement of intangible assets focused on human and social capital. Technology does not dominate, but it needs a co-responsible government, centered on the individual and collective needs of post human society. the current society are witnessing a change in the focus on the centrality of the person in lifelong learning and on educational perspectives of lifelong learning, located in experimental social and uniqueness structures in the continuous search for intercultural biographical mediations. In this sense, the subjective co-construction of meaning by context is reaffirmed, with a process of signification as a presupposition of development and formation, centered on the dynamic, situated and manifest human relationship: the process of producing meaning influences action, the individual beliefs and emotions.

In this perspective, learning takes on differentiated social identities for categorical systems and formal, non-formal and informal contexts, which represent unavoidable pieces of the integrated training system. The expansion of the learning space and the macro-categories of learning outline the new theoretical-praxic space in which democratic form-actions of substantial educational citizenship are carried out. The educational challenge is not so much about the instrumental use of technologies, but about inducing a critical reflection about the inconveniences and potentials inherent in cross-media society. The aim is to train the necessary management and re-understanding skills in a constantly hyper-connected cross-plural-global world. The new declinations of the augmented classroom, as contexts in which to favor new educational centers, represent the places of the manifestations of development potentials, characterized by the formative dialectics on subjective and intersubjective experiences. It is a matter of initiating processes for a mobilization of subjective internal resources, so as to make them interact dialectically with the context.

In this perspective, the conceptual category of attribution-meaning constitutes the paradigmatic dimension that is culturally located and connoted, which assumes the global socio-educational perspective of lifelong learning. This reiterates the unrepeatability and authenticity of the training process, centered on the ability to discover, attribute and modify meanings in the comparison between the self and the other. Therefore, there highlight the need for defined operational contexts and proactive and prosocial cognitive-motivational and affective-relational repertoires, within which to experience citizenship. Learning is placed in a holistic dimension, as a circular continuum and complex of subjective potentials which, through paths of strategic quality, invest in shared attributions of meaning. The use of the web increases opportunities and raises the need to face incessant re-generative processes. In this sense, the correlation between democratic processes and the function of prosocial competences in the knowledge society is highlighted. People are subjected to continuous situational stimulation and situational problems, which permanently solicit the management resolving abilities. This shows the educational need to acquire prosocial skills and the formative pursuit of socially significant goals.

In this society, the typical role of knowledge is the constitutive strategic element of systemic inclusiveness, a structural necessity that guarantees the democratic socio-political order and can be experienced in the cognitive dimension through the exercise of the rights of participatory citizenship. The integrated system in the perspective of lifelong learning presupposes the recognition of the emancipatory value of knowledge, which value only in relation to the significance for the individual in a specific context. An educational emergency is highlighted that integrates the global perspective and the holistic dimension, intended as systemic synergies characterized by subjective and institutional conditions. This dialectical-strategic role also includes the temporal-quantitative dimensions, the pervasiveness of training opportunities, which is also connected to the distribution of power and wealth, both globally and locally. The new opportunities that emerging technologies afford and the dimensional multiformity of learning, highlight the need for a rethinking of training policies. The synergic-design dimension of educational systems must be considered as to the preparation of conditions and tools to train people capable of becoming lifelong learners at every stage of their life. With the advent of the 4.0 industrial revolution, technology plays a central and critical role in its presence in organizational contexts. Technological, scientific and epistemic mediations converge on the construction of worlds de-hierarchized by democratic accessibility to information. The resulting transformations, which affect the socio-political-contextual variables, bring out the need for mediated reflective regulation about the multimodal modalities of communicative languages and intersubjective exchange. This social need involves education, leading us to reflect on the relationship between virtual and real context, communication, society and users (Ortiz, 2016, p.74).

HYBRID SOCIETY AND DIALECTICAL FLIPPED INCLUSION, BETWEEN NEO-GENERATIVE SYSTEMS AND TRANSFORMATIVE EDUCATIONAL RESEARCH

The Flipped Model of Inclusion

The incessant and emerging needs of the knowledge society, as new permanent research challenge, represent the propulsive-investigative elements of the Flipped Inclusion literacy model, researched at The University of Salerno (Italy). The concept of learning is understood as a process of life, which goes beyond the limited dimension of the education and training paths, with a permanent transformation through potential educational experiences situated for formal, non-formal, and informal areas. Investments must be made to integrate the political-educational systems in order to provide adequate responses to needs of technology use. In education, knowledge and skills constitute the structural capital of the economy and social development. Self-realization and autonomy, such as access to information, to knowledge because information becomes knowledge, constitute the founding elements of the new sense of freedom.

The Flipped Inclusion model was born from the socio-educational need to experiment with systemic inclusion models. The interpretative schemes of the post-humanistic vision put questions about the incidence of information technology and biotechnology advances on the changes induced by natural human traits, in their integration with non-human ones, up to the creation of hybrid individuals, from new physical and cognitive abilities. All this requires a transduction of process perspectives (Ortiz, 2016) with signs of imprinting which, in representing, 1) produce enunciations in subjective actions; 2) they allow to re-recognize, de-construct and convert signals; 3) denote and connote with transductor circular movements; 4) guarantee transformational, experiential communication processes. Flipped Inclusion, therefore, has the programmatic objective of negotiated and conscious participation in media processes, in a prosocial-cooperative ecological perspective. The model focuses on the socio-educational value of formal, non-formal and informal learning environments. In fact, within them, knowledge and skills are negotiated through different interaction plans that are not always controllable (Jonassen, 1995). Skills of conscious are needed for decoding and co-mediation as well as to encourage the construction of inclusive contexts and personalities. Education must promote continuous critical reflection on the concept of the person, on the modalities of co-constructing and thinking, from a systemic ecological point of view, using technology training. For the prosocial management of media interconnectivity, through the promotion of inclusive management of empowerment, the intent should be to influence the being and the perception of oneself, as an individual and as a community, to favor the formation of plural identities and the construction of contexts, guarantors of ecological-systemic-inclusive well-being (Bronfenbrenner, 1979).

The Flipped Inclusion model starts from an analysis of new learning contexts, which require new skills aimed at decoding practices and relationships, in the light of the changes determined by the new dimensions produced by the globalized cross-media (Bauman, 2001). "The media are environments [...] in which we discover, model and express our humanity in particular ways" (Postman, 1968, p.154). In fact, the technological information environment, which interfaces systems, contexts and subjectivity, has transformed relationships. We are witnessing new forms of correlation, characterized by a space-time beyond, unlimited and intangible. Therefore, one of the new training challenges in the global context is represented by the semiotic decoding of cognitive products, to adequately manage the new forms of communication. These in fact affect through intersubjective modalities (UNESCO, 2009), both on the permanent and evolutionary trans-formation of individual personalities and on the co-construction of new socio-systemic perspectives.

In the society of informational communication, the recognition and interpretation of signs as a category is based on the knowledge of the objects of the world, for the bond that has been generated between the projection of meaning and meaning. Society is in the presence of a semiotic closure (Verón, 1993) with new forms of subjectivity trapped in a network of signs, such as dynamic and media units, necessary to allow a connection with society (Ortiz, 2016). The range of semiotic domains, which characterize cross-media, make it a place of learning experiences, in which meaning is the image of action (Gee, 2005). All this requires media-literacy processes (Gee, 2005), with a continuous re-understanding of the symbolic alphabets on which the dynamics on the web unfold: this is what Flipped Inclusion involves. Virtual places, as places of experimentation, with ordered simulations, represent the didactic-strategic avant-gardes used in the activity of Flipped Inclusion. For this reason, the logic of gamification or game-based learning (Barab, et al. 2009) has been applied in Flipped Inclusion, with complex blended learning modalities. The aim is to facilitate simulation processes with digital technology (Falloon, 2011) and educational-transformative paths. Two types of formal and structural interventions are combined which, in being places of practice and interpretative theory, allow for a phenomenological understanding of reality. All this makes it possible to constantly verify the changes made in the experiment's *per-course* (intended as a traveling process) through a comparison between the simulated empirical data observed in the virtual computerized reality, and those observable at the end of the process in the concrete reality. In fact, in the presence of a digital simulation, it is possible to understand the effects, as measurable and comparable results with empirical data. The predictability of events and the modification of variables, processes and factors postulated are thus strengthened in the event of failure to achieve the expected results. The digital simulation games-based learning of Flipped Inclusion allows to experiment the effects of decisions, through an interactive-simultaneous operation simulated that it proved to be a valid didactic support to prosocial inclusiveness.

In the knowledge society, we are witnessing a non-linear process of change, which that has an impact on the communication methods, on creative and re-creative forms, and on the configurations of subjects and subjectivity, on thinking and culture. This highlights the decisive weight of self-generativity, which is determined in the virtual communication environments of social and cultural interaction, through semiotic acts (Ortiz, 2016). We are witnessing processes of 1) coding, 2) decoding, 3) deconstruction and 4) conversion of the signal, which is transformed in relation to the significance of the context. New ways are introduced to symbolize and express phenomenology, through multimodal signals, images, visual and auditory resources, which determine clear separations from linguistic forms, socially normalized. We are witnessing a tendency towards homogeneous standardization of knowledge, which is perceived as an economic phenomenology, with consumerist configurations of commodity-product. We are in the presence of a collective hypermedia co-construction, in a distinction of sign and symbol-meaning and symbolic sign by convention (Hierro, 1980), which characterize post-modern culture. With the advent of new technological realities new aphasic forms are manifested; they involve languages, meanings, redesign contexts, broaden the communicative complexity and affect the socio-psycho-relational domains.

The new communication processes are characterized by simultaneity and multiplicity which, through semiotic redefinitions, generate transductions of perspectives and relations with multimodal cyber-society trans-educational processes (Ortiz, 2016). Society is in the presence of synchronous, fast and simultaneous spaces (Bauman, 2010) which, by eliminating the geophysical limit, become increasingly difficult to sustain in the real world. The multiple open cyber-cultural fields, in fostering representations and communicative formations, promote forms of social anonymity and imaginative-projective processes: they involve identities, in a representative union of virtual reality and intersubjectivity. New epistemologies of the knowledge society may be characterized by a socio-semiotic-performative paradigm with a reflexive-transformative action, centered on dynamism and the change of languages and inter-corporeality.

Technological logics and the dialectic of life, through new and original intertwining and contamination, produce inexorable transformations and incessant questions on the symbiotic combination of biological and mechanical aspects; all this is leading to the hybrid evolution of the human race. For postmodern technocratic men (Fromm, 1994) there are risks of alienation from the world. We are witnessing the establishment of vicious circles, characterized by the unconditional and uncontrolled formation of fictitious and alienated personalities which embody the discomfort of distancing oneself from the context-community, with a zeroing of one's roots. In the world of cyber-society, there are new cyber-cultural, relationships subjective and intersubjective configurations, as a result of the process of cross-media globalization. In fact, the symbolic-cybernetic exchanges in the cyber-society world constitute incessant flows of socio-semiotic practices. They are characterized by continuous mediation of information, communication and knowledge, which permeate and modify social, cultural and educational practices.

Cyberspace is a border world; it generates projected forms of virtual presence, socio-semiotic-expressive mediums, and multi-present incorporeal subjectivities in symbolic online spaces. The cyber-society constitutes the expression of social semiotics, characterized by epistemic configurations, temporal variables and technological tools, as a result of a symbolic process of cyber-cultural globalization, with continuous modifications of socio-cultural-educational domains. We are in the presence of projected virtual presences, which intervene with cybernetic epistemic trans-formations on objects, subjects and subjectivity. Languages, as a form of knowledge, trace the structure of the episteme, which refers to the historical sense (Foucault, 1967) of knowledge co-constructed in socio-economic-cultural and political contexts-space. The emphasis on knowledge, technology and communication in the hypermodern society (Lipovetsky, 2006), is rooted in the recognition of the empowerment that these factors have at a socio-economic-political level. Many times, new epistemologies of knowledge are produced as a result of cyber action; they model and transforms subjectivity in cross-media intersubjectivity (Ortiz, 2016). Therefore, the dizzying acceleration of communicative technocracy highlights the tendency to a disintegrating human / technical hybridization.

There is an atrophic disproportion of the biological body in the pursuit of models of efficiency and effectiveness, aimed at a technologization of man. This is accompanied by a propensity to pursue the humanization of technological prototypes, which intend to retrace the forms and systems of the human. The identity of man, in joining with technology on a biological level, changes and diverges from the essentialist traits of the uniqueness of being a person, for a technological plurality capable of adapting to the body. The advent of post humanist thought (Marchesini, 2009) introduces a new perspective of corporeality, not as an object to be interpreted, but as a subject-representation of a permanently transformative locus complex. The Corporeality is intended as an interface of technology (de Kerckhove, 1997). It is capable of being a scenic representation of life (Goffman, 1974) and of assuming identity (Le Breton, 2016) through a conscious grounding in the values of being-Person. This vision considers the investment in a transformative evolutionary process, induced by technology; it is no longer understood only as an external tool for improving living conditions and general well-being, but becomes part of man himself. Technology becomes part of the human body and changes the way we relate to the outside; sensory activity, supported by technological mediation, generates a new interface on the world. Changes are produced that affect the perceptive and cognitive level, through the introduction of nanotechnologies and hybrid computing in the body. The union of the human body / technology, created within the organism, lays the foundation for the birth of individuals, endowed with new attributes and abilities. The man and the modification of his psycho-physical characteristics become the object of direct manipulation and make necessary a re-evaluation of the humanistic, philosophical and biological concept of man. The biological and technological evolution implies the scientific principle of the construction law (Bejan, 1997), which provides for the natural predisposition of the systems to become complex and at the same time to promote facilitation processes in the flow between components. In this it is possible to see the role of technology both in shaping and reshaping the prevailing order, and in accelerating the changes between orders, which leads to a rethinking of levels of intellectual supremacy, in a perspective of geopolitics and geo-economics. We are in the presence of a hybrid geo-technological age, which has generated a new paradigm of empowerment, from power centered on the breadth and incidence of centric info-states / cities.

The skills and processes, generated by the union of man / technology and made explicit in the German term Technik, combine the determinism of the techno-mechanistic dimension, with the socio-constructivist interest, connected to the effects on people and society. The term geo-technology refers to the techno-socio-cultural construct of power, with Technik the concept of accessibility-adaptability is reaffirmed, in a permanent balance between technology gaining and quality of life. The neo-social generative systems (Chomsky, 2002) of the hybrid society are characterized by the inexhaustible capacity to promote connections, facilitating the creation of values and products. Such generativism, whose maximum expression can be found in language and the Internet, represents the driving force of paradigmatic changes in generative systems of peer to peer learning. All this inexorably involves every sphere of information, information, economics and governance of power in general. The pedagogical-didactic importance of the interpretation of the sign lies in being an indicator of actions that, in permeating the behavior, allow representational and transductive knowledge; it allows interpreting meanings and values, understanding the symbols of the state and recognizing interconnection signals (Cárdenas, 2015).

The Flipped Inclusion model, taking up the collaborative view that characterizes the theory of structuring (Giddens, 2000), reaffirms the instrumental-pedagogical and socio-material value of technology as a medium. According to this theory, human action contributes to the creation of an understanding of information systems. In fact, in a continuous and recursive re-adaptation between instruments, actors and structures, the institutional properties of social organizations are established, in a situated interchange of knowledge (Orlikowski, 1992). The actors control the flexibility or the rigidity of a system in order to build further meanings in the organizational context. The priority role played by technology for its influence on the social and institutional structuring of organizations is reiterated (Leonardi, 2012).

The theoretical roots of the model of Flipped Inclusion is based onthe existential design model (Dewey, 1961b), with an inclusive socio-semiotic approach.

The model is based on epistemological constructs, which characterize the cross-media society and on the value of the circular economy, understood as self-regenerative, through investments in circular education (Merleau - Ponty, 1942), as a permanent re-education process (Dewey, 1961).

Flipped Inclusion derives from the permanent need to manage the socio-psycho-anthropic-cultural processes, permeated with computational logic, which characterizes postmodern society. Perspectives of the learning are proposed, which aims at promoting conflict management in a cooperative manner. The system is strengthened through the prevention of individual and contextual risks; integrated methodologies are applied transversally in the Flipped Inclusion model. The goal of Flipped Inclusion is to provide a cooperative, reticular and procedural organization of learning, with a permanent capacity to solve the management of emerging problems, in a formative perspective of lifelong learning (Alberici, 2002). Such a model may benefit society as a whole as it may empower students. It may also allow for the the formation of plural identities and the construction of life contexts, which promote a type of ecological-systemic-inclusive well-being (Bronfenbrenner, 1979).

Flipped Inclusion characterizes its interventions on the transformative power of shared co-production and on the concept of accessibility by levels of practices. The premise of Flipped Inclusion is the promotion of a democratic education (Dewey, 1984) through a systemic inclusion (Caprara, & Cervone, 2003) with training, leading to improved awareness and empowerment. The complex idiomatic expression Flipped Inclusion (De Giuseppe, Corona, 2016), combines two visions. The inclusive perspective of International Classification of Functioning, Disability and Health (ICF) (OMS, 2001) understood as a social construct and the Logic of inversion of spaces and roles of flipped learning. The education in the conscious management of the liquid semiotic stimulations of the cross-media network are activated through experiential paths; they are structured on self-planning based on micro, eso-meso and macro-systems (Bronfenbrenner, 1979).

The Flipped Inclusion design model invests in the recognition of the primary role of technologies, information processes and communication codes in human affairs (Strate, 1999). The aim is to promote a structured educational systems perspective, interdisciplinary action (Lum, 2006) and metadisciplinary (Nystrom, 1973), with educational programs of media ecology (Postman, 1968). The goal of Flipped Inclusion is to solidify "the management of the dynamics that generate themselves in the technological habitat and to foster a healthy and conscious use of the techno-communicative-cultural medium and on the incidence "[...] on perception, knowledge, the emotions and human values "(Postman, 1970, p.161). A theoretical-methodological orientation of the New Literacy Studies of the Knowledge Society has been undertaken (Castell, 1997), as a neo-form of literacy as an understanding of the communicative modalities of a society (UNESCO, 2005, p.1). The multi-perspective ecological system of the Flipped Inclusion model aims to promote media-literacy (UNESCO-EFA, 2005), understood as rational and moral accessibility to information and knowledge (Street, Lefstein, 2007), through four levels of scope (Gee, 1990):

1. Personal - with the development: a) of trust, b) of potential, c) of creativity, d) of critical thinking;
2. Politics - with the promotion of the capacities of a) participation, b) promotion, c) reconciliation, d) sharing
3. Cultural- to a) preserve identity, b) recognize the differences, c) promote comparison, d) reflect on attitudes.

Figure 1. Flipped Inclusion and the cross-structural design architecture

THE CROSS-STRUCTURAL DESIGN ARCHITECTURE of flipped inclusion is organized in EIPE.

MODULAR AND RECURSIVE STRUCTURE IN FRAME ANALYSIS			
1) Esplore = E	KEY	It is a SINGLE WORD that describes our interest / **problem to investigate**	1) **WORD=** Example: **Uniformity**
2) Idea = I	FRAME	To the previous WORD (interest-problem) we add the DEFINITION OF AN AMBIT, **a concept discovered.**	2) WORD + **SPECIFIC CONCEPT** Example: **Massifying** Uniformation
3) Project = P	FRAMING	At the WORD and the CONCEPT Discovered previously, we need to add a SPECIFIC, circumscribed PROBLEMATIC, **to structure intervention routes**	3) WORD + SPECIFIC CONCEPT **+CIRCUMSCRIBE PROBLEMATIC ELEMENT** Example: **The discrimination** in Massifyng Uniformation
4) Experiment= E	FRAME WORK	A DETAILED METHODOLOGICAL RESOLUTION, added to the investigated problem, discovered and circumscribed, **must be experiment in the field.**	4) WORD + SPECIFIC CONCEPT + CIRCUMSCRIBE PROBLEMATIC ELEMENT + **METHOD OF RESOLUTION OF THE PROBLEMATIC ELEMENT** Example: **The flipped inclusion to manage** the discrimination in massifying uniformity

4. Socio-economic with phases of: a) prevention; b) support, c) guarantee d) sustainability to: education, equality, opportunity, *well-being*

The Flipped Inclusion model is structured for levels of increasing complexity, according to the simple logic (Berthoz, 2011). Four phases are used, involving different domains from the individual (micro-systemic) to the collective (exo, meso and systemic-design macro). It uses multimodal re-literacy (Gee, 2008), which is based on the decoding and re-understanding (Gee, 2005) of meanings for semiotic domain (Street, et al., 2012), with significant signs and symbols / communicative by context of practice (Gee, 2008). Languages are considered, from the iconic to the visual, in a perspective of flipped inclusive of lifelong learning of the system (Alberici, 2002). The didactic actions trace the theory of operant conditioning (Skinner, 1953), in a permanent perspective of re-literate learning (Gee, 2008). The didactic actions by phases in Flipped Inclusion model retrace the operant conditioning (Skinner, 1953), in a permanent perspective of re-literate learning (Gee, 2008). These activities, however, take into account elements of the social learning (Bandura & Walters, 1964), organized by levels of growing cooperative problematization (Sibilio, 2014) as follows:

1) Explore is the phase of prompting-stimulation and fading-support;
2) design is the phase of modeling-imitative modeling;
3) Designing is the phase of shaping-modeling meta-behavior;
4) Experiment is the phase of chaining-linking.

The phases of Flipped Inclusion follow the *principles of learning* (Gee, 2005); they are organized by retracing the computational logic, modular and recursive of the EIPE module:

1. *Explore* (E) is the phase of self-knowledge, based on the principle and semiotic domain, aimed at identifying a stimulus-word (key), which supports the intrinsic motivation for the activation of educational investigative pathways (Kuhn et al. 2000);
2. *Idea-* (I) is based on the Principle of Amplification, Input and Communicative Competence, (frame) based on the method of discovery-Inquiry learning (Bruner, 1960);
3. *Project designing* (P) - with forms of meta-level Thinking, is the phase characterized by the Principle of design and Active Control, (Framing) which allows the principle of Identity to be declined: the methodological structure of Mastery Learning is retraced (Bloom, 1964).;

4. *Experiment* (E) - is characterized by the application of committed principles and continuous learning (Framework), focused on experiential learning (Kolb, 2014).

The model shows the systematic organization of thought, as a situated understanding of the learning context: the strengthening of behaviors and an attribution of significant meaning is applied (Gee, 2005). Flipped Inclusion promotes modeling processes (Bandura, 1977), through incremental stimulations for levels of increasing complexity. The areas of solicitation involved are organized in: 1) basic knowledge and basic skills; 2) communication skills; 3) cognitive and metacognitive skills; 4) social and prosocial skills. The model makes use of facilitating pathways which, considering the areas, are aimed at promoting the ability to select stimuli. The choice of the stimulus represents the propulsive element on which to center the educational processes of imitative learning (Bandura, 1986), adopted initially at a latent level. Thus, starting from the choice, which is rooted in intrinsic motivations, an emotional training path is promoted, through an education in the selective search for the resolutions located. In the Flipped Inclusion pathways, the process of generalizability (Baer, Wolf and Risley, 1968), focused on transformative-proactive teaching, presupposes a programmatic intentionality (Kadzin, 1975). It is structured according to *simplicity levels* (Berthoz, 2011) of cooperation in *increasing complexity*, with a systemic group and systematic organization of micro eso, meso and inclusive macro-planning and inclusive evaluation.

In Flipped Inclusion, the intentionality of generalizability interventions guarantees: controllable results (Stokes & Baer, 1977); accessibility and extensive transferability of the model to other conditions; applicability of reinforcement contingencies, which are not easily identifiable. We are in the presence of an analytical procedure, which represents a facilitating and activation strategy for prosocial behavior, even in the presence of specific situational difficulties. The expansive control paths of the original stimuli are subjected to phases of: 1) motivated and motivating identification, 2) selective and analytical research; 3) programmatic re-elaboration with problematization starting from the stimulus; 4) experiential application of resolutions, for situated contexts. The Flipped Inclusion invests in the co-construction of multi-stimulus virtual environments, to promote prosocial generalization processes (Baer, & Deguchi, 1985), also through processes of metacognition paths on the mass video-media cultural phenomenon. At the base of the Flipped Inclusion are pursued the four focuses of gamification such as: 1) Commitment (Engagement), 2) Autonomy (Autonomy), 3) Mastery (Mastery) and 4) Progression (Progression), organized according to the EISE matrix, with objectives for levels of increasing complexity (Kim et al, 2009). The model traces the principles of learning (Gee, 2005), focused on motivation and perseverance in action by levels and context (Gee, 2008). It is characterized as a media-peer education of design experience, which makes use of cooperative peer & video education, organized with groups of increasing complexity.

Figure 2. Flipped Inclusion and the design for systemic levels in modularity-frame (De Giuseppe, 2018)

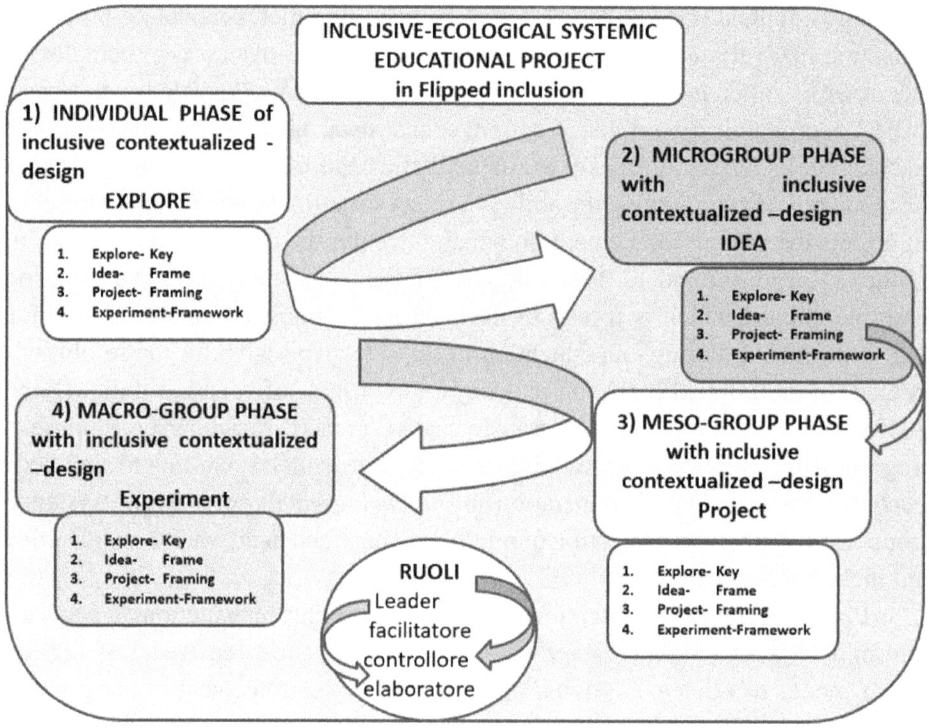

The learning by gaming (Neulight et al., 2007), proposed in the Flipped Inclusion, provides: 1) Gamified learning paths, micro learning also of virtual reality and video-learning modeling; 2) Gamified assessments with rankings and performance statistics; 3) Social learning; 4) Learning gateway in cooperative intergroup learning actions. The gamification applied in Flipped Inclusion uses two levels of verification, one within the simulation process the other at the empirical level. In both cases the production of the expected results is evaluated. The activities of Flipped Inclusion follow the mechanics of gamification, with courses organized in phases of: 1) Challenges, as learning objectives; 2) Levels of the learning path; 3) Instant feedback, with Rankings for a level analysis for a group perspective; 4) Intergroup and level cooperation. The Flipped Inclusion model, which is characterized by the formation of an ethnographic decentralization approach, invests metacognitive-reflective abilities to reinterpret the phenomenologies of meaning and meanings in continuous motion. For this reason, the model incorporates the use of video-modeling educational activities with evidence-based practices, skills, abilities and knowledge, for a phenomenological decoding, which involves the communicative, social and behavioral domain. With the audio-video productions it is possible to educate to a personalized translation of subjective and intersubjective phenomenologies and to make re-interpretable for experiences of context. The participant observing variant and an immersive approach are therefore privileged, in order to return data on relational dynamics. The purpose of the model is to guarantee the validity of the results on the knowledge of social phenomena, considering the risks of unpredictability due to the flexibility of the processes.

In the evaluation process, the participant observation is used, and frameworks are avoided in descriptive categorizations, which could limit the understanding of the intrinsic complexities to relational dynamics.

The use of visual sociology in Flipped Inclusion derives from the importance assigned in learning, to the meta-reflexive processes of self-representation. They allow a re-education to cross-media and the adaptive-conscious re-production of contents, actions and reflections. The Flipped Inclusion proposes teaching strategy, such as video modeling, focused on social learning by imitation, whose intervening components are: Attention, Retention and Reproduction. The videos must offer facilitating and modeling target behaviors (Bellini & Akullian, 2007), which aim to manage antisocial behaviors and favor the acquisition of languages, life skills and prosocial skills (Buggey, 2005). Video recordings activate visual learning, self-assessment and social coaching, which facilitate focusing on visual patterns of behavior, skills or competences. In this way, moments of inattention and anxiety are reduced and the intrinsic motivation and self-awareness processes, individual and group, are strengthened (Apple, et al., 2005).

The EIPE modularity, which characterizes the Flipped Inclusion design logic, is also applied to the different types of video-modeling. In fact, the videos are organized according to the analysis of the frame by Goffman (1974)) and structured in phases of: 1) video-base-structural (key), to delineate the stimulus elements related to a context; 2) video auto-modeling and self-modeling (frame), with self-recording to subsequently explore behaviors and to highlight ways of processing sensitive information; 3) video-modeling (framing), with which the target skill is recorded from the point of view of other students; 4) product-video (framework), designed and implemented by levels (individual, micro, meso and macro group), which involves the systematic deepening of target behaviors. Therefore, the Flipped Inclusion video-modeling activities are structured taking into account the EDDE modularity (De Giuseppe, 2018) with: Explore (E), focusing on sociological observation with video-stimulus-base and elements of video framing; 2. Design (D), which involves the elaboration of texts and the collective production of multimedia hypertexts; 3. Designing (D), characterized by the subjective and cooperative production of images and videos; 4. Experiment (E), which presupposes the collective vision of the video-product, to guarantee metacognitive processes.

In the Flipped Inclusion model, experiments of inclusive communities are proposed through a didactic organization, which provides for systemic and interdependent interventions by levels. Learning and practice communities are activated, to acquire knowledge, skills, competences and to find solutions to social problems detected by the context. The aim is to integrate the two community levels, with a didactic application for formal, non-formal and informal contexts and with virtual off-line and on-line learning environments, designed to promote inclusive prosocial meta-competence. For the design of activities in Flipped Inclusion mode, virtual platforms such as Blendspace and Spiral are used, but above all the flippedinclusion.it website. It allows a choice of communication tools, exchange of materials and audio-video productions. The flippedinclusion.it site uses four types of formal learning environments (Koster, 2004): 1. Scheduling, to publish and define calls and forward targets; 2. Conferencing, on which topics of discussion are defined and folders of materials and messages are managed; 3. Document sharing and Link sharing, to share online documents and individual and group links. 4. Personal messaging: to send and receive individual, micro, meso and macro-group messages. On the site flippedinclusion.it there is also a reserved area, in which the calendar of courses, the structures of the courses and the activities by phases are explained. Still in the reserved area it is possible to load the products, made according to the EISP structure of the Flipped Inclusion model. Links are provided that refer to the self-assessment and evaluation monitoring, both initial and final online, by stages and levels (Wagner, & Srivastava, 1989).

Research Example

This study on the Flipped Inclusion model (De Giuseppe, 2018) carried out over a three-year period investigated the potential connections between cross-media applied in the classroom and beyond the classroom using a didactic flipped learning method applied through an inclusive logic of system learning. A multi-method, interpretative-descriptive-transformative research was carried out (Cox, Geisen, Green, 2011) to examine data based upon empirical evidence. Using a framework from the assumption that scientific understanding is a process of ones' quality of life (Dewey, 1984), the research focused on a data collection, which helped answer questions related to understanding society (Bailey, 1982). The research questions were: Can Flipped Inclusion teaching promote the formation of individual prosocial, inclusive-plural profiles through simple didactic practices? and 2: Does Flipped Inclusion teaching affect the promotion of inclusive contexts?

The research was organized by levels: macrosystemic (inter-institutional), systemic meso (university course / school); at the microsystemic level (academic context, class context). Furthermore, the research was undertaken in the formal spaces of university courses, school, in non-formal contexts (on Blendspace and Spiral) and in informal environments (campus of the University of Salerno- Department of human and philosophical sciences and training, Department of Medicine and Surgery and with WhatsApp).

Figure 3. Research contexts, research subjects, research sample

RESEARCH CONTEXTS:
FORMAL SPACES (academic courses - Department of human and philosophical sciences and training, Department of Medicine and Surgery, Comprehensive Institute of Montoro-Avellino);
INFORMAL SPACES (Facebook, campus of Salerno's University)
NON-FORMAL SPACES(Blendspace, Spiral)

RESEARCH SUBJECTS- RESEARCH SAMPLES- RESEARCH YEARS 2014/2017			Age of users involved in the experimentation
Anno di ricerca	TOTAL STUDENTS involved in the experimentation	SAMPLE NUMBER (simple random sampling)	
2014/15	318	159	
2015/16	802	401	between 12 and 55 years old
2016/17	652	326	
University	1772	886	
Secondary school	50	25	
TOTAL	**1822**	**911**	

The research proposal verified the hypothesis formulated on a sample consisting of academic students, trainee teachers and compulsory school students, with a didactic-laboratory methodological framework. The students involved ranged in age from 12 to 55 years and were also part of Active Training Internship courses. The subjects involved in the research and the research samples were:

The choice of the students of young people from the first level secondary school examined is connected to the criticality of the adolescent phase, a period considered critical in the process of personality development and in the formation of personal identity, but also for the frequent use of this age group. Recruitment was based upon the location of the school, the disciplines, the sensitivity towards action research, the teachers, the percentage and type of user, and the registry criterion. Therefore, the choice of compulsory education of Montoro took into account its institutional contiguities, as compatibility of leadership on vision and mission, but also spatial proximity to the University of Salerno: fundamental elements for establishing experimentation activities and obtaining necessary information for research and to obtain the effective information. In the second year of research, the data were collected and tabulated through data collected using the flippeddinclusion.it website. A multi-method research study was carried out, deriving from interdisciplinary fields. The aim was to avoid interpretative problems, with low scientific rigor. Cross-checks and varied data collection procedures were carried out to examine the same issues. In fact, the complexity of the object of investigation and of the quality of the pedagogical-educational processes brought out the need to integrate quantitative and qualitative research models (Creswell, 2003). The aim was to understand and manage the dynamic complexities of the teaching-learning-process, with a view to lifelong learning. The tools identified allowed to contain the risks of an intuitive and subjective evaluation of the process. For this reason, qualitative and quantitative approaches have been integrated to deepen the aspects of process dynamism of the observed event and to provide at least a first answer to the research questions. Qualitative investigation techniques (Zammuner, 1998) used included focus groups and participant observations with descriptive assessments. Quantitative tools were used, such as structured questionnaires and structured evaluation forms, to guarantee a sufficient quantity of empirical data and to enable a critical reflection and attribution of meaning.

The data were collected through the administration of incoming and outgoing questionnaires, structured assessment sheets for monitoring levels (individual phase, in micro-group, mesogroup, macro-group) and processing by role; a re-test 3 years after applying the methodology to the sample of 80, including pupils and interns, Toronto Empathy test administered to the 20 students of the compulsory school. The research was followed by an alternative authentic evaluation methodology (Wiggins, 1993) with the aim of using verification tools not only of knowledge, but above all of skills (Benzoni, 2004). This type of evaluation required an active role and a conscious involvement in concrete experiences, with applicative tasks of knowledge and evaluative tasks. Therefore, a self-assessment of learning self-efficacy was used (Bandura, 1991) and a co-evaluation based on pro-sociality. The following were proposed: a socialized evaluation, aimed at groups, with a comparison of performances;

a mixed external evaluation, centered on self-esteem, which considered individual and comparative progress; personalized assessment, to parameterize the individual goals achieved, with respect to the starting levels. Evaluation rubrics were used, relative to the object of research / development, with details about the dimensions, indicators and general objectives; descriptors-specific objectives and expected levels of performance, with a scale of value and taxonomic criteria. All this has had the purpose of curbing the risks of an evaluative subjectivity of the competences.

The *evaluation of social skills* has been structured through (Bandura, 1975): 1) discussion, to examine and define the skills to be learned and the recognition of new behaviors; 2) observational objectivity, relative to prosocial behaviors, to induce imitation; 3) reinforcement, understood as any event that increases the probability of issuing behavior; 4) reflection on the experience. From the observation of the repertoires of skills and difficulties, the behavioral assessment (Garb, Lilienfeld, & Fowler, 2008) made it possible to apply a behavioral early intervention program, with strategies to strengthen positive behaviors to reduce problem behaviors. Programs of early behavioral intervention and assessment of skills have been proposed with three methods of analysis: 1) qualitative (Sorzio, 2005) of skills and difficulties; 2) quantitative behavioral problems; 3) Functional (De Giuseppe, & Corona, 2017). The qualitative analysis was carried out using check-lists with a global focus, focused on the observational-descriptive assessment of communication, cognitive / metacognitive and social / prosocial skills and competences, to ascertain their presence and acquisition levels. Thematic focus groups were used, through modules in frames and levels of increasing complexity, aimed at the production of texts and videos.

The quantitative assessment of behavioral problems of interaction and isolation from the group (Cottini, 2016) was based on the systematic observation of the frequency, duration and intensity of goal behaviors, positive meta-behaviors to be increased, and on target behavior, target-behaviors, inadequate to extinguish or reduce (Moore, Cheng, & Dainty, 2002). The moments of meta-evaluative synthesis were characterized by self-perception and self-awareness of the competences achieved, with cross-actions between self-evaluation and evaluation. Therefore, the research used proposed synoptic design sheets and evaluation sheets to monitor and evaluate products and processes activated by inclusive system learning. The *functional analysis* (Larson, Maag, 1999), aimed at understanding the motivations behind problem-behaviors, completes the behavioral assessment, as it aimed to highlight the relationships between observed behavior, the environment and the interrelations produced. The investment of research on social cognition, which is connected to the development of empathic-emotional skills and prosocial behaviors, has seen the useful application of the Toronto Empathy Questionnaire (TEQ) (Spreng, McKinnon, Mar, Levine, 2009), a measure psychometric of empathy. The TEQ allows to validate a factorial-analytical solution to several empathic measures, to verify the evidence of the research hypothesis according to which: The Flipped Inclusion model promotes prosocial behavior (Roche, 1995), mainly reflects an emotional process or an accurate affective intuition in the emotional state of others.

The questionnaire was administered to a sample of 20 students; it produced a calculation of the average relative to the tested behaviors. A t-test was used to verify the difference. It was necessary to validate the adherence to the theoretical research framework (Trinchero, 2004), to arrive at an evaluation and meta-evaluation of the research object. Everything was, therefore, realized with a triangulation of data, to avoid internal logical contradictions and always controlling adherence to possible implicit meanings (Perla, 2010). Finally, in order to verify the research process, its potential and its critical points from different points of view, it was decided to adopt a specific analysis technique, S.W.O.T. (Strengths, Weaknesses, Opportunities, Threats) (Hill, & Westbrook, 1997).

A triangulation of the three-year data shows that almost all students (79%) attended the course regularly. From the questionnaire addressed to students and interns, to the question concerning the transformative incidence of Flipped Inclusion on learning styles (42%) expressed itself favorably, while only six percent of sample considered study method unchanged and learning. The data emerged in the post-test questionnaire allowed to detect the transformative incidence of the Flipped Inclusion's model on learning styles. Forty-two percent of research sample of the analyzed students expresses a positive evaluation of the model, while six percent of the sample students does' not detect significant personal changes, related to the study and learning method, after the experiment. Fifty-eight percent of the students believe that the Flipped Inclusion's model has transformed its behavior by improving the cooperative competence-style. Furthermore, fifty nine percent of the sample students believes that the model has also influenced the relational style, favoring an empathic opening to the needs of others. Furthermore, from the questionnaire it emerges that the experimentation of the model, organized by groups and by increasing levels of complexity, has allowed the acquisition of an inclusive modus operandi. In fact, students reported a change in their attribution styles also within the micro-group, the meso-group and the macro-group, attributing to each item the highest score 4 on 4 (detected in Flipped Inclusion through focus on learning, communication, relational and cooperative styles). In fact, as many as 44% of those interviewed said that, despite the experimentation of the Flipped Inclusion method had ended, they continued to work in the micro-groups. The results were collected in relation to the four areas observed, such as: 1) knowledge and skills; 2) communication skills, 3) cognitive and metacognitive skills, 4) social and prosocial skills

From the data collected on a sample identified with simple random sampling (in figure 4), the forty-six percent of the subjects involved in the experiment perceived the inverted inclusion model as determinant for its influence on the learning style; only

Figure 4. Post test data analysis on styles

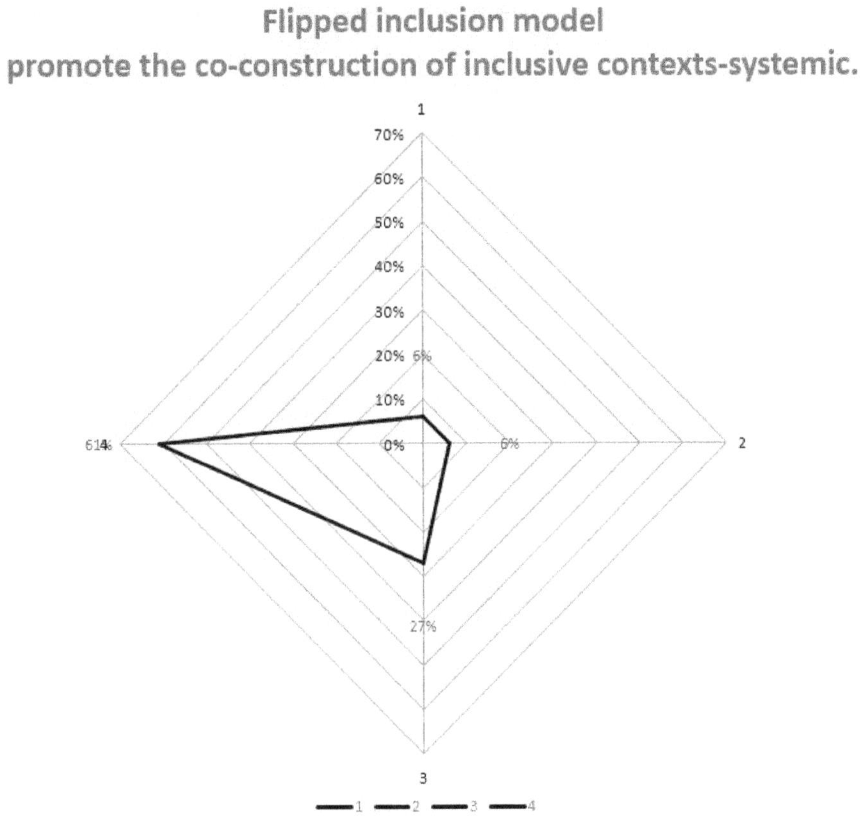

eight percent% noted they the site had influenced their learning style. Forty-five percent of the research sample reported a significant influence in changing communication style, while only five percent% of the students reported a low incidence of Flipped Inclusion on personal communication style. Fifty-nine percent of the sample say they have transformed their relational style, while eight percent% say they have not undergone major changes in the relational style. Finally, fifty eight percent of the sample showed improvements in the cooperative style; while for seven percent of the participant at the experimentation has noted the transformative incidence of the tested model was minimal on cooperative style.

For forty five percent of the research sample the Flipped Inclusion model contributed greatly to broadening basic knowledge and skills; only eight percent of the sample perceived that an impact. In relation to cognitive processes, five percent of the research sample reported an impact from the application of the Flipped Inclusion model; while forty two percent highlighted that there was a

Figure 5. Posttest-data analysis on knowledges and skills

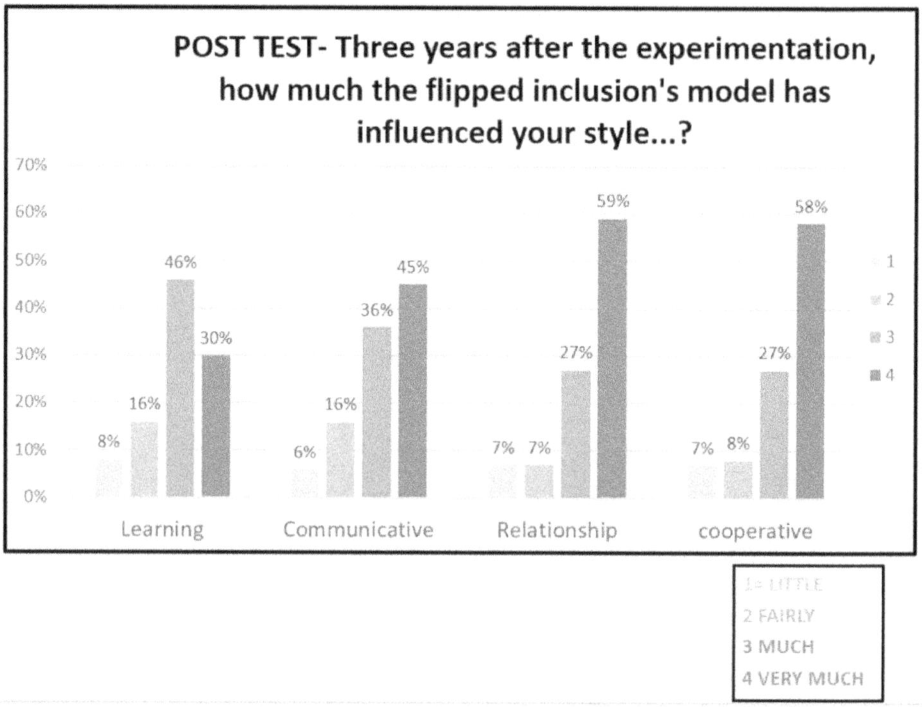

significant progress achieved from the experimentation. From the percentage data it is possible to find, in relation to the item synthesis and data processing, very positive percentages, expressed by forty four percent of the sample and minimum positive results, explained only by five percent of the research sample. Finally, an improvement in the resolution of problems, following the experimentation of the model, is recognized as very incident by fifty six percent of the sample and not very low by only eight percent of the research sample.

Forty percent of the students' participant at the experimentation believe to have improved control over grammatical structures through Flipped Inclusion; while only thirteen percent the research sample do not record improvements on this item. For forty two percent of the sample the mastery of the linguistic repertoire was also greatly improved and only ten percent of the research sample did not show many improvements. Regarding the organization of the coherent discourse from the questionnaire, good developments emerged for forty three percent of the sample, while for six percent of the research sample, there are only a few improvements. The interaction is improved with great progress for fifty two percent of the sample overall.

Figure 6. Posttest-data analysis on communication skills

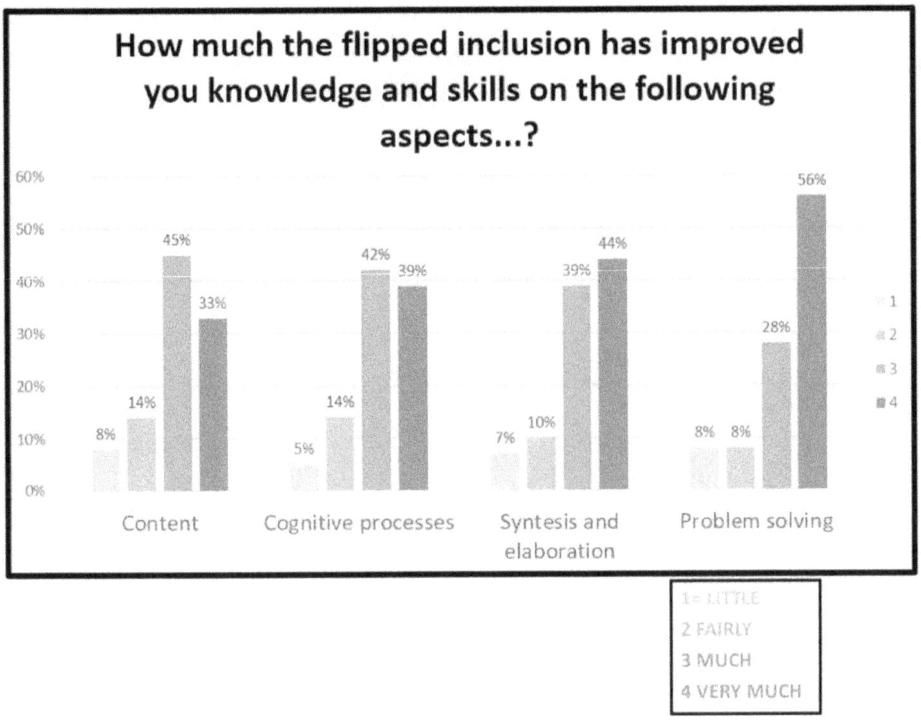

Forty-eight percent of the participant at the experimentation has a high incidence of the Flipped Inclusion model in relation to the ability to describe their mental processes. Forty-six percent of the students highlights the large incidence of the model in relation to the awareness of its choices. Forty-seven percent of the students recorded a positive transformation in the ability to attribute success / failure to the performance through the application of the model. Finally, forty three of the sample students note improvements in the capacity for critical analysis and reformulation of processes, thanks to the experimentation of the model. For the percentages of the students that oscillate between 7 and 12% the experimentation of the Flipped Inclusion model did not affect much on the cognitive and metacognitive area.

The most significant results are those recorded in relation to the promotion of social and prosocial skills following the application of the Flipped Inclusion model. From the graphs it is possible to observe that very positive percentage data are recorded in the tested students. In particular, the graphs shows that are confirmed: from fifty-four percent of the students, incidents positive transformations, related to the ability to listen actively and accept the rules; from fifty five percent of the students the transformative incidence of the Flipped Inclusion model on empathic

Figure 7. Posttest-data analysis on cognitive and metacognitive skill

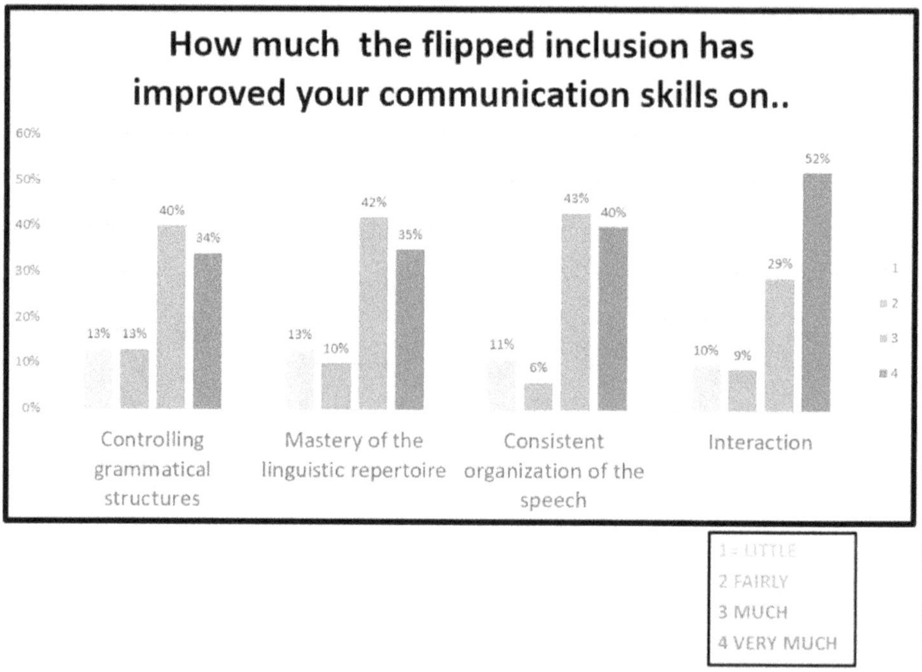

competence; from fifty one percent of the students an improvement in the ability to address conflicts. ; Four percent of the sample consider the action of Flipped Inclusion to be improper in relation to the improvement of active listening skills, and nine percent of the students, considers the Flipped Inclusion model as being of little use on the ability to accept the rules

Forty-five percent of the students noted that the Flipped Inclusion model promotes the construction of inclusive personal profiles. The students that fluctuates between 6% and 7% confirms this assertion, which however shows a lesser influence of the model on the promotion of inclusive processes.

The Flipped Inclusion model has greatly promoted the co-construction of inclusive, ecological and systemic contexts for 61% of the students; even for 27% of the students the model tested has greatly favored the promotion of inclusive contexts. Only for 6% of the specimen tested, the Flipped Inclusion model has favored little co-construction of inclusive contexts.

Figure 8. Posttest-data analysis on social and prosocial skills

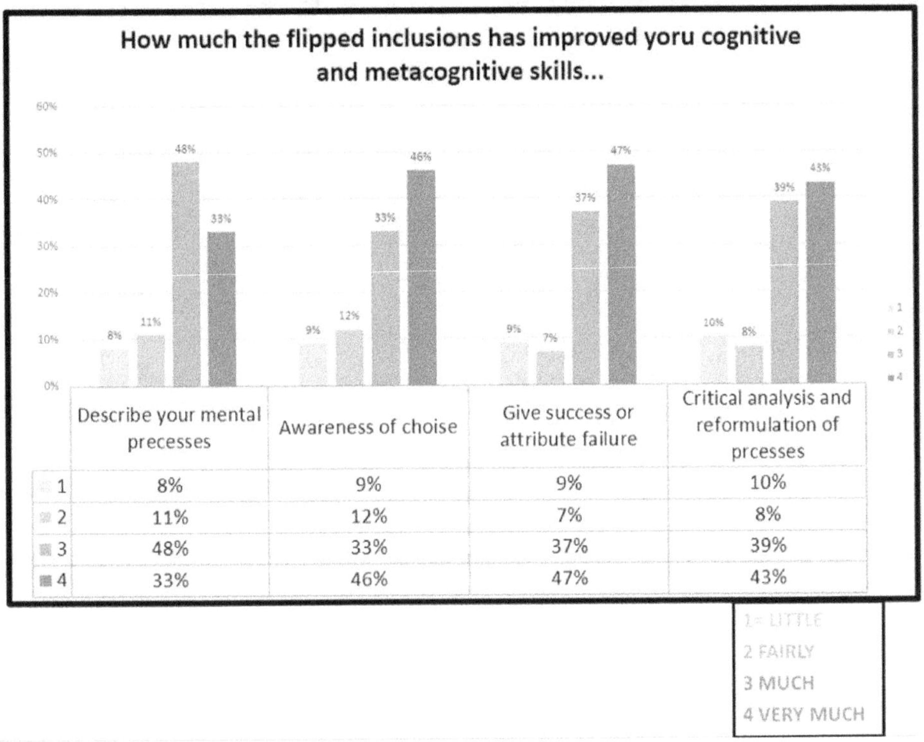

	Describe your mental precesses	Awareness of choise	Give success or attribute failure	Critical analysis and reformulation of prcesses
1	8%	9%	9%	10%
2	11%	12%	7%	8%
3	48%	33%	37%	39%
4	33%	46%	47%	43%

CONCLUSION AND NEW PERSPECTIVES

The research study reported above highlighted the empirical and observable evidence of using a Flipped Inclusion model. The use of simple transformative teaching using such a model, structured with educational planning and improved communication show, promise in promoting better management skills as well as prosocial processes. In fact, the positive trend of the data corroborates the confidence in the applicative validity of an inclusive educational activity, from an augmented classroom. The latter was organized with a recursive cooperation to activate communication and training processes (formal, informal and non-formal), for levels of increasing complexity. The aim was to promote emotional management skills in a pro-social, ecological and systemic perspective. The results of the research seem to indicate that there is the possibility of an effective value of the Flipped Inclusion model on prosocial competences, through a method of co-construction of knowledge, skills and competence, confirming what has already been theorized in numerous disciplinary domains.

Figure 9. Posttest -data analysis on inclusive profiles and inclusive processes

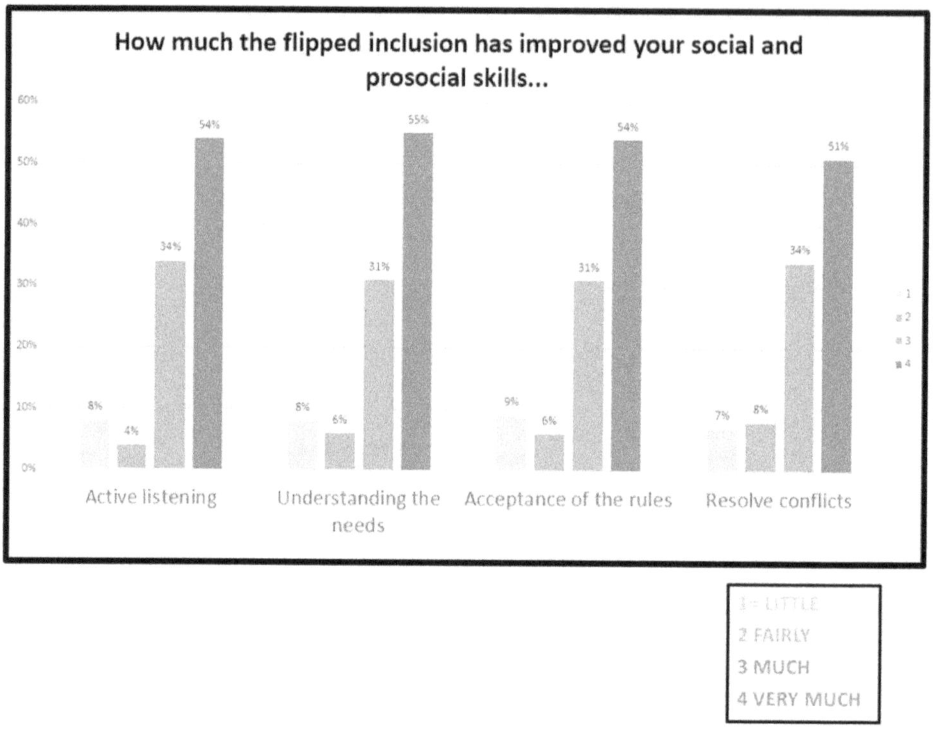

The research surely revealed the need to recognize the prosocial value of a connectionist learning, as an expression of meanings co-constructed by interactive, situated and contextual intelligence (Smolensky, 1988), through the didactic use of real and virtual learning contexts. Therefore, the probabilistic value of the conclusions of the educational research determines the impossibility of reaching the definition of causal inferences, with the exposure of the evaluation of the result to the risks of imprecision and indetermination. These results depend both on the innumerable intervening variables and on the impossibility of a rigorous control of unpredictable dynamic phenomenal events. All this allowed only partially to arrive at a first evaluation of the significance of the research results.

The need for a longitudinal study has emerged and it is necessary to broaden the fields of investigation and experimentation focusing on learning contexts (formal, non-formal and informal, as well as on cross-media instruments, situated and for problems). There is a need to continue experimentation with a wider sampling, also involving different territorial contexts by type, different disciplines and different age groups. Particular pedagogical attention will should be given to potential problematic cases, connected to forms of marginality and differences, in order to promote an

Figure 10. Posttest -data analysis on promotion of inclusive contexts systemic

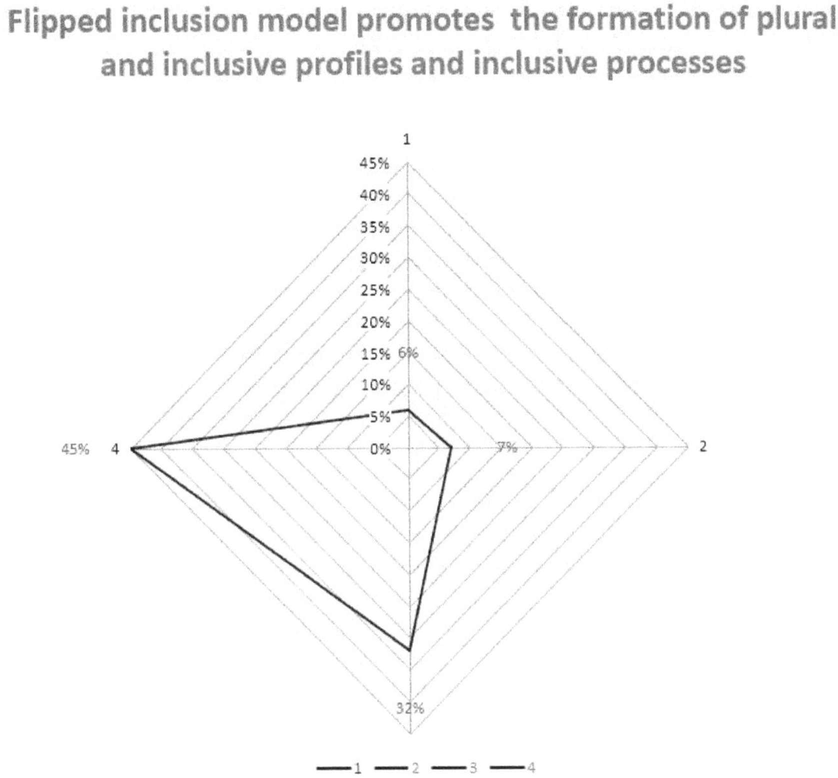

optimization of inclusive pedagogical-didactic interventions. The pervasive impact of new media emerges from the explorative research, also if in a latent way. The cross-mediality represent symbolic resource with which each gives meaning to the social worlds. It is necessary, therefore, to deepen their educational transformative domains. In fact, in the research, the transversal and common elements of the different domains were not explained and a detailed incidence relative to the cross-media informal and non-formal use in the Flipped Inclusion model was not traced.

AUTHORS' NOTE

Tonia De Giuseppe has curated: *Hybrid Society and Dialectical Flipped Inclusion, Between Neo-Generative Systems and Transformative Educational Research*; *Background*; *Solutions and Recommendations*; *Future Research Directions.*

Felice Corona has curated the paragraph written: *Cross-Media and Socio-Psycho-Educational Phenomenologies of the Knowledge Society, Between Investigation Policies and Clinical Taxonomies*; *Introduction*; *Future Research Directions*; *Conclusion*.

REFERENCES

Alberici, A. (2002). *Imparare sempre nella società della conoscenza*. Milano: Mondadori.

American Psychiatric Association. (2013). Diagnostic and statistical manual of mental disorders (5th ed.). Arlington, VA: American Psychiatric Publishing.

Apple, A. L., Billingsley, F., Schwartz, I. S., & Carr, E. G. (2005). Effects of video modeling alone and with self-management on compliment-giving behaviors of children with high-functioning ASD. *Journal of Positive Behavior Interventions*, *7*(1), 33–46. doi:10.1177/10983007050070010401

Baer, D. M., Wolf, M. M., & Risley, T. R. (1968). Some current dimensions of applied behavior analysis. *Journal of Applied Behavior Analysis*, *1*(1), 91–97. doi:10.1901/jaba.1968.1-91 PMID:16795165

Bailey, K. D. (1982). *Methods of social research*. New York: Free Press.

Bandura, A. (1975). *Social Learning & Personality Development*. Holt, Rinehart & Winston.

Bandura, A. (1977). *Social Learning Theory*. Englewood Cliffs, NJ: Prentice Hall.

Bandura, A. (1986). *Social Foundations of Thought and Action*. Englewood Cliffs, NJ: Prentice Hall.

Bandura, A. (1991). Social cognitive theory of moral thought and action. In W. M. Kurtines & J. L. Gewirtz (Eds.), Handbook of moral behavior and development (Vol. 1, pp. 46-102). Florida International University.

Barab, S.A., Scott, B., Siyahhan, S., Goldstone, R., Ingram-Goble, A., Zuiker, SJ., & Warren, S. (2009). Gioco trasformazionale come impalcatura curriculare: utilizzo dei videogiochi per sostenere l'educazione scientifica. *Journal of Science: educazione e tecnologia, 18*(4), 305-320.

Bauman, Z. (2001). *Dentro La Globalizzazione. Conseguenze Sulle Persone*. Roma: Laterza.

Bauman, Z. (2005). *Vita liquida*. Roma, Bari: Laterza.

Bauman, Z. (2010). *La globalización. Consecuencias humanas*. México: Fondo de Cultura Económica.

Bejan, A. (1997). Constructal-theory network of conducting paths for cooling a heat generating volume. *International Journal of Heat and Mass Transfer*, *40*(4), 799–816. doi:10.1016/0017-9310(96)00175-5

Bellini, S., & Akullian, J. (2007). A meta-analysis of video modeling and video self-modeling interventions for children and adolescents with autism spectrum disorders. *Exceptional Children*, *73*(3), 264–287. doi:10.1177/001440290707300301

Benzoni, I. (2004). *Portfolio delle competenze e processi di personalizzazione. BG*. Junior.

Berthoz, A. (2011). *La semplessità*. Torino: Codice.

Bloom, B. S. (1964). *Stability and change in human characteristics*. New York: John Wiley & Sons.

Bocci, F., De Angelis B., Fregola C., Olmetti Peja D., & Zona, U. (2016). *Rizodidattica. Teorie dell'apprendimento e modelli didattici inclusivi*. Lecce-Brescia: Pensa MultimeMedia Bologna: Il Mulino.

Bondì, A. (2013). *Percezione, semiosi e socialità del senso*. Milano: Mimesis.

Bronfenbrenner, U. (1979). *Ecologia dello sviluppo umano*. Bologna: il Mulino.

Bruner, J. (1960). The Process of Education. In Dopo Dewey: il processo di apprendimento nelle due culture. Roma: Armando.

Buggey, T. (2005) Video self-modeling applications with students with autism spectrum disorder in a small private school setting. Sage Journals. doi:10.1177/10883576050200010501

Caprara, G. V., & Cervone, D. (2003). *Personalità: Determinanti, Dinamiche, Potenzialità, 1ª*. Milano: Raffaello Cortina.

Cárdenas, A. (2015). *Lenguaje, analogía y sentido*. Doctorado Institucional en Educación. DIE: Universidad Pedagógica Nacional.

Castells, M. (1997). The Rise of the Network Society. In The Information Age: Economy, Society and Culture (Vol. 1). Cambridge, MA: Blackwell.

Chomsky, N. (2002). *Syntactic structures*. Walter de Gruyter. doi:10.1515/9783110218329

Corona, F. & De Giuseppe, T. (2017b). La Flipped Inclusion, tra impianto teoretico e didattica sperimentale di aula aumentata per una didattica inclusiva. *Pedagogia Più Didattica.*, (3), 2-7.

Cottini, L. (2016). *L' autodeterminazione nelle persone con disabilità. Percorsi educativi per svilupparla.* Trento: Erikson.

Cox, P., Geisen, T., & Green, R. (2011). *Qualitative re-search and social change. European context.* Palgrave Macmillan.

Creswell, J. W. (2003). *Research design: Qualitative, quantitative, and mixed method approaches* (2nd ed.). Thousand Oaks, CA: Sage.

De Giuseppe, T. (2018a). *Flipped Inclusion. L'impianto teoretico tra bisogni emergenti e prospettive epistemologiche. Collana Differenze.* Roma: Aracne.

De Giuseppe, T. (2018b). La media education nell'economia della formazione continua. *Formazione & Insegnamento, 16*(2), 233–247.

De Giuseppe, T., & Corona, F. (2016). Flipped Inclusion per una didattica trasformativa. Convegno. In P. Magnoler, A. M. Notti, & L. Perla (Eds.), *La professionalità degli insegnanti. La ricerca e le pratiche* (pp. 551–570). Lecce: Pensa.

De Giuseppe, T., & Corona, F. (2017). Metodologia Flipped tra sistemica inclusione e prospettive didattico-assertive. *Formazione & Insegnamento, 15*(2), 409–420.

De Giuseppe T., Corona F. (2018). Il complex blended learning di spiral nella prospettiva sistemica della Flipped Inclusion. *Media Education – Studi, ricerche, buone pratiche, 9*(2), 343-356.

de Kerckhove, K. (1997). *Connected intelligence: the arrival of the web society.* Toronto: Somerville House Publ.

de Saussure, F. (1970). *Corso di linguistica generale.* Bari: Laterza.

Dewey, J. (1984). *Democrazia e educazione.* Firenze: La Nuova Italia.

Dewey, J. (1961a). *Come pensiamo. Una riformulazione del rapporto fra il pensiero riflessivo e l'educazione.* Firenze: La Nuova Italia.

Dewey, J. (1961b). *Logica. Teoria dell'indagine.* Torino: Einaudi 1974.

Falloon, G. (2011). Usare avatar e ambienti virtuali nell'apprendimento: Cosa offrono? *Journal of Educational Technology, 41*(2), 108–122.

Foucault, M. (1967). *Le parole e le cose.* Milano: BUR.

Fromm, E. (1994). *Io difendo l'uomo*. Milano: Rusconi.

Fusilli, A. (2016). *Effetto borderline. Soggettivazione e movimenti del desiderio*. Milano: Franco Angeli.

Garb, H. N., Lilienfeld, S. O., & Fowler, K. A. (2008). Psychological assessment and clinical judgement. In B. A. Winstead (Ed.), *Psychopathology: Foundations for a contemporary understanding* (2nd ed.; pp. 103–124). New York, NY: Routledge/ Taylor & Francis Group.

Gee, J. P. (1990). *Linguistica e alfabetizzazione sociale: Ideologia nei discorsi*. Falmer Press.

Gee, J. P. (2005). Learning by Design: Buoni videogiochi come macchine per l'apprendimento. *E-learning*, *2*(1), 5–1.

Gee, J. P. (2008). *Quali videogiochi devono insegnarci sull'apprendimento e l'alfabetizzazione, rivisti e aggiornati*. Basingstoke, UK: Palgrave Macmillan.

Gibson, W. (1994). *Luce Virtuale*. Milano: Mondadori.

Giddens, A. (2000). *Il mondo che cambia. Come la globalizzazione ridisegna la nostra vita*. Bologna: Il Mulino.

Goffman, E. (1974). *Frame analysis. L'organizzazione dell'esperienza*. Roma: Armando.

Goldberg, I. (1996). *Internet addiction disorder*. Retrieved from http://www.urz.uni-heidel-berg.de/Netzdienste/anleitung/ wwwtips/8/addict.htm

Hill, T., & Westbrook, R. (1997). SWOT Analysis: It's time for a product recall. *Long Range Planning*, *30*(1), 46–52. doi:10.1016/S0024-6301(96)00095-7

ISTAT. (2014). *Internet@Italia 2014. L'uso di Internet da parte di cittadini e imprese*. Roma: Fondazione Ugo Bordoni. Retrieved from https://www.istat.it/it/files//2014/12/Cittadini_e_nuove_tecnologie_anno-2014.pdf

Jerald, J., & Block, M. D. (2008). Issues for DSM-V: Internet Addiction. *The American Journal of Psychiatry*, *3*, 306–307.

Jiménez-Murcia, S., Fernández-Aranda, F., Granero, R., Chóliz, M., La Verde, M., & Aguglia, E. (2014). Video game addiction in gambling disorder: Clinical, psychopathological, and personality correlates. *Biomed Reserach International*, *2014*, 1–11. doi:10.1155/2014/31506

Jonassen, D. H. (1995). Operationalizing mental models: Strategies for assessing mental models to support meaningful learning and design supportive learning environments. *Proceeding of the first international Conference Computer Support for Collaborative Learning '95*, 182-186.

Jung, C. G. (1999). *Simboli della trasformazione*. Torino: Bollati Boringieri.

Kaye, S., & Farrell, M. (2014). Disorders associated with excessive use of Internet, computers, smartphones and similar electronic devices. In *Proceedings from the First Act of the World Health Organization Congress: Public health implications of behavioral addictions associated with excessive use of Internet, computers, smart phones and similar electronic devices*. Geneva, Switzerland: WHO Document Production Services.

Kim, B., Park, H., & Baek, Y. (2009). Non solo per divertimento, ma strategie serie: utilizzo di strategie meta-cognitive nell'apprendimento basato sul gioco. *Computer e istruzione, 52,* 800-810.

Kolb, D. A. (2014). *Experiential learning: Experience as the source of learning and development*. Pearson FT Press.

Koster, R. (2004). *A theory of fun for game design* (1st ed.). Phoenix, AZ: Paraglyph Press.

Kuhn, D., Cheney, R., & Weinstock, M. (2000). The development of epistemological understanding. *Cognitive Development, 15*(3), 309–328. doi:10.1016/S0885-2014(00)00030-7

Larson, P.J., & Maag, J. W. (1999). L'assessment funzionale dei comportamenti problema in classe. *Difficoltà di apprendimento, 4*(4), 539 – 556.

Le Breton, D. (2016). *Fuggire da sé. Una tentazione contemporanea*. Milano: Cortina Raffaello.

Leonardi, P. M. (2012). Materiality, Sociomateriality, and Socio-Technical Systems: What Do These Terms Mean? How are They Related? Do We Need Them? In P. M. Leonardi, B. A. Nardi, & J. Kallinikos (Eds.), *Materiality and Organizing: Social Interaction in a Technological World* (pp. 25–48). Oxford, UK: Oxford University Press. doi:10.1093/acprof:oso/9780199664054.003.0002

Lipovetsky, G., & Charles, S. (2006). *Los tiempos hipermodernos*. Barcelona: Anagrama.

Lum, C. M. K. (2006). *Perspectives on Culture, Technology and Communication: The Media Ecology Tradition*. Cresskill, NJ: Hampton Press.

Marchesini, R. (2009). *Il tramonto dell'uomo: la prospettiva post-umanista*. Bari: Dedalo.

Merleau–Ponty, M. (1942). La struttura del comportamento. *Mimesis*.

Merleau-Ponty, M. (1945). *Phénoménologie de la perception*. Milano: Bompiani 2003.

Moore, D. R., Cheng, M. I., & Dainty, A. R. J. (2002). Competence, competency and competencies. Performance assessment in organizations. *Work Study*, *6*(6), 314–319. doi:10.1108/00438020210441876

Neulight, N., Kafai, Y. B., Kao, L., Foley, B., & Galas, C. (2007). Children's participation in a virtual epidemic in the science classroom: Making connections to natural infectious diseases. *Journal of Science Education and Technology*, *16*(1), 47–58. doi:10.100710956-006-9029-z

Nystrom, C. (1973). *Towards a Science of Media Ecology* (PhD dissertation). New York: New York University.

Organizzazione Mondiale della Sanità. (2001). *ICF Classificazione Internazionale del Funzionamento, della Disabilità e della Salute*. Trento: Erikson2004.

Orlikowski, W. J. (1992). The duality of technology: Rethinking the concept of technology in organizations. *Organization Science*, *3*(3), 398–427. doi:10.1287/orsc.3.3.398

Ortiz, L. (2016). *Configuración Sociosemiótica del chat: Una visión multimodal* (Unpublished doctoral dissertation). UPTC.

Parag, K. (2016). *Connectography: Le mappe del futuro ordine mondiale*. Roma: Fazi.

Paulré, B. (2000). De la New economy au capitalisme cognitif. *Multitudes*, *2*(2), 25–42. doi:10.3917/mult.002.0025

Perla, L. (2010). *La didattica dell'implicito. Ciò che l'insegnante non sa*. Brescia: La Scuola.

Postman, N. (1968). Ecologia dei media. L'insegnamento come attività conservatrice. Roma: Armando, 1983.

Postman, N. (1970). The Reformed English Curriculum. In High School 1980. The Shape of Future in American Secondary Education (pp. 160-168). New York: Pitman.

Roche, R. (1995). *La condotta pro-sociale. Terapia del comportamento*. Roma: Bulzoni.

Sibilio, M. (2014). *La didattica semplessa*. Napoli: Liguori.

Simondon, G. (2001). *El modo de existencia de los objetos técnicos*. Buenos Aires: Prometeo libros.

Simondon, G. (2008). *Imagination et Invention (1965-1966)*. Chatou: Éditions de La Transparence

Smolensky, P. (1988). On the Proper Treatment of Connectionism. *Behavioral and Brain Sciences, 11*(1), 1–61. doi:10.1017/S0140525X00052432

Sorzio, P. (2005). *La ricerca qualitativa in educazione*. Roma: Carocci.

Spreng, R.N., McKinnon, M.C., Mar, R.A., & Levine, B.J. (2009). The Toronto Empathy Questionnaire. Scale development and initial validation of a factor-analytic solution to multiple empathy measures. *NCBI Resources, 91*(1), 62.

Stokes, T. F., & Baer, D. M. (1977). On Implicit Technology of Generalization. *Journal of Applied Behavior Analysis, 2*(10), 349–367. doi:10.1901/jaba.1977.10-349 PMID:16795561

Strate, L. (1999). Understanding MEA. *Medias Res, 1*(1), 1–9.

Street, B. V., Bloome, D., & Pahl, K. (2012). Lettera: Apprendere l'empowerment attraverso la formazione nella ricerca in stile etnografico. In M. Grenfell, C. Hardy, & J. Rowsell (Eds.), Lingua, etnografia e istruzione: colmare nuovi studi di alfabetizzazione e Bourdieu (pp. 73-88). New York: Routledge.

Street, B.V., & Lefstein, A. (2007). *Literacy: An Advanced Resource Book. Applied Linguistics.* London: Routledge Applied Linguistics.

Trinchero, R. (2004). *I metodi della ricerca educativa*. Bari: Laterza.

UNESCO. (2005). *Literacy for life. EFA Global Monitoring Report.* Paris, France: Author. Retrieved from http://www.unesco.org/education/gmr_download/chapter1.pdf

UNESCO-Institute for Lifelong Learning. (2009). *Harnessing the power and potential of adult learning and education for a viable future: Belém framework for action*. Hamburg, Germany: Author.

Wagner, D. A., & Srivastava, A. B. L. (1989). *Measuring literacy through household surveys*. New York, NY: United Nations Statistical Office.

Wiggins, G. P. (1993). *Assessing student performance*. San Francisco: Jossey-Bass Publishers.

World Health Organization (WHO). (2015). *Public Health Implications of Excessive Use of the Internet, Computers, Smartphones and Similar Electronic Devices- Meeting report*. Geneva, Switzerland: WHO Press, World Health Organization. Retrieved from https://apps.who.int/iris/bitstream/handle/10665/184264/9789241509367_eng.pdf;sequence=1

Yen, J. Y., Ko, C. H., Yen, C. F., Wu, H. Y., & Yang, M. J. (2007). The comorbid psychiatric symptoms of Internet addiction: Attention deficit and hyperactivity disorder (ADHD), depression, social phobia, and hostility. *The Journal of Adolescent Health*, *41*(1), 93–98. doi:10.1016/j.jadohealth.2007.02.002 PMID:17577539

Zammuner, V. L. (1998). *Tecniche dell'intervista e del questionario*. Bologna: Il Mulino.

Endnote

1) DE GIUSEPPE curated the background, main focus of the chapter, solutions and recommendations, and future research directions. Also, she completed the sections on hybrid society and dialectical flipped inclusion, between neo-generative systems and transformative educational research.

2) CORONA has curated the introduction, future research directions, and conclusion in addition to the cross-media and socio-psycho-educational phenomenologies of the knowledge society, between investigation policies and clinical taxonomies.

Chapter 4
Computational Thinking and Robotics:
What's It All About?

Carol Munn
New Jersey City University, USA

ABSTRACT

This chapter explores a unique framework that expresses freshness and innovation within revolutionary learning experiences. The chapter focuses on the implementation of computational thinking skills as an essential element of a robotics education program that was driven by hands-on activities instilling the notion of imagination through innovative projects. Engineering design applications with robotics created an atmosphere in which students applied abstract mathematics and science concepts. The robot and its technology imploding in the field of education created excitement in the minds of students with renewed, revitalized interest, and curiosity stretching across the areas of science, technology, engineering, and mathematics (STEM). Robotics education along with computational thinking skills are primary keys for unlocking the unlimited creative and innovative potential with engaging hands-on activities. Robots, a fascinating learning instrument, rejuvenate, animate, and revive 21st century skills in tech-savvy language familiar with today's students.

INTRODUCTION

The power of the computer combined with innovation lies with the understanding and engagement in the power of thinking and big ideas implemented in future-driven classrooms. Dating back to 1993, Papert first devises the thinking connection to

DOI: 10.4018/978-1-7998-2914-0.ch004

Computational Thinking (CT) by stating "You can't think seriously about thinking without thinking about thinking about something" (p. 10). In society, a natural human method of understanding is thinking something through, from start to finish. Kordaki and Kakavas (2017) summarized that a specific way of thinking reverts back to the relationship of abstraction and the orderly process of algorithmic problem-solving which contribute either directly or indirectly to the learner to master skills involved with CT. The children become explorers by asking questions, gathering data, analyzing data and creating conclusions and solutions to problems. Taking thinking a bit further with various theoretical thinking philosophies, humans can think, but what does it mean to think? What does it mean to have ideas? What are ideas? How do we teach connecting structures and functions through powerful ideas? Who are the tinkers, planners, thinkers, off-taskers? What relationship is there between thinking and actions? Is there a sequential process? Where does CT fall into the schema?

WHAT IS COMPUTATIONAL THINKING?

Computational Thinking is a thinking process, either by a human or machine, which analyzes a situation or problem in smaller pieces of data or information, allowing for the understanding and creation of a systematic solution to a problem. The process involves thought out strategies involving four key steps which are (a) decomposition; (b) pattern recognition; (c) abstraction; and (d) algorithm design (Computational Thinking in K-12 Education, 2018). **Decomposition** is one of the four steps in the Computational Thinking process focusing on the "breaking down data, processes, or problems into smaller, manageable parts" (Computational Thinking for Educators, n.d.). **Pattern Recognition** is one of the four steps in the Computational Thinking process focusing on "observing patterns, trends, and regularities in data" and creates a visual representation or display of information or data with recognizable similarities within a specific problem (Computational Thinking for Educators, n.d.). **Abstraction** is one of the four steps in the Computational Thinking process. Abstraction acts as the core and beginning step toward the breakdown of a problem by "identifying the general principles that generate these patterns" (Computational Thinking for Educators, n.d.). **Algorithm Design** is one of the four steps in the Computational Thinking process of "developing the step by step instructions for solving...problems" (Computational Thinking for Educators, n.d.).

Infusing and applying CT practical applications and framework leads the way to energized classroom activities crafting and including soft skills and a different mindset for each leaner. The soft skills are: (a) creativity; (b) collaboration; (c) critical thinking; and (d) communication; and (e) teamwork. Beside the contours of solving real-world problems, students' engagement, interactivity, and passion are part of the active learners' mindset, which absorbs new and innovative methods of digital literacy education. "Computational thinking is the new literacy of the 21st Century" (Wing, 2010, p. 3) and defines CT as "the thought processes involved in formulating a problem and expressing its solution(s) in such a way that a computer-human or machine-can effectively carry out" (p. 8). With the direction of CT expanding into other educational fields, Wing (2017) redefined and adjusted her description to include the deeper thought process as "mental activity in formulating a problem to admit a computational solution" and can be carried out by a human or machine (p. 8).

Computational Thinking and 21st Century Skills

With today's vision in education focusing on the 21st Century learner, CT is emerging as a key component in the skillset necessary for the new generation of learners. CT poses a strong ideology based on problem-solving equally conveying an essential position in cross-curricular classroom activities. CT is the backbone of sequential processes which evolve into everyday real-world solutions to problems. Papert (1980) stated that computers process data in well-defined sequential steps. Just like computers handle data instruction, students engaged in problem-solving are using skills which are key elements within the CT framework; the processing and analyzing data in a series of steps. In all likelihood, the emphasis being placed on computing abilities within emerging and global education environment, CT competency is considered a fundamental skill and value in the science, technology, engineering, and mathematics (STEM) education fields (Atmatzidou, & Demetriadis, 2014; Constantinou, & Ioannou, (n.d.); National Research Council, 2011). Using analytic and algorithmic concepts and strategies, CT is related to computer science with the strength to formulate, analyze and solve problems.

Recent studies indicate that CT is no longer a single skill taught in the computer science field (Barr, & Stephenson, 2011; National Research Council, 2011; Shute, Sun & Asbell-Clarke, 2017; Wing, 2011). Many academic fields now incorporate CT because of its value in promoting critical thinking skills, whether being in technology, science, mathematics, language arts, history, art, or other curricular fields. CT is a set of skills which materializes in cross-curricular academic areas. For example, a CT lesson would be in a Foreign Language class where students identify word patterns and structures in Spanish and then compare the results to the translated English version and explain the similarities and differences to a classmate. In computer science, CT lesson manages an unplugged or non-technology hands-on coding activity, which is later compared to the outcome of a plugged or digital web-based coding sequence program.

CT poses new quests and opportunities to draw interest in today's learners in the classroom by implementing different technologies. "Solving problems, designing systems, and understanding human behavior by drawing on the concepts fundamental to computer science" (Wing, 2006, p. 33). One such technology are robots. Earlier on in robotics education, robots were used as a fixture in the stand-alone course deemed only at the high school curriculum. High school robotics was designed as an elective course. The robotics course curriculum was developed to work in conjunction with computer science courses. The basic purpose at the high school level was to have students ready to continue with their education at higher institutions. Now, with the increasing attention on early stages in education, there is an exciting opportunity in today's school level curriculum for robots and CT. As this opportunity takes shape in education, CT and robotics can help shape and support education of children by developing their skills around logic, creativity, and problem-solving.

What is Robotics?

In the world of education, terminology is the base platform to build the curriculum, lessons, and activities. Here are some common and key words describing specific applications and key terms. A **Robot** is a mechanical unit comprised of motors, gears, sensors, circuits, and processing unit. The completed autonomous machine senses the environment, receives commands, and carries out instruction (What Is a Robot?, n.d.). **Robotics** are comprised of design and application of robots. An automated process in which machines will accomplish a range of tasks (Becker et al., 2018). **Robotics Education** brings into schools the robot as part of the integration of science, technology, engineering, and mathematics in the everyday activities in the education field.

Robotics curriculum focuses on hands-on learning, tinkering, and learning-by-doing including problem-solving, sequential applications, and fundamental engineering concepts as innovating and engaging experiences (Scaradozzi, Sorbi, Pedale, Valzano, & Vergine, 2015). CT sets up a "playful learning" experience for engaging students at a younger age group in (Bers, 2008, p. 1). Included in this exploration is a "gateway for helping children learn about mathematical concepts" and revitalizes classroom instruction to include all learners (Bers, 2008, p. 3). Schools are utilizing robotics as a concrete physical tool in hands-on activities within the context of the specific academic domain. Robots provide the needed engaging instrument in transferring human thought patterns into physical solutions for a problem. In addition, robots inspire curiosity and imagination while increasing the methodology of how they learn by slowly increasing the difficulty level but still allowing for excitement in a classroom lesson. CT and robotics can aid in identifying and exploring deeper cognitive thinking techniques within the minds of younger students.

Robotics and CT embrace a special partnership which exposes unity with strong practical hands-on approaches to learning in the classroom. Recently this partnership has been gaining ground as an energizing approach to the natural human instinct of computation providing a theoretical framework and foundation encompassing meaningful and essential practices for the mastery by the students (Weintrop, et al., 2016). Introducing the computer and its programming capabilities Papert (1993) invented LOGO. LOGO refers to a specific programming environment by which a computer image or object such as a "turtle" moves across a monitor (Papert, 1993, p. 11). Papert (1993) believes the process students can be meticulously involved with computers, using to the "LOGO environment" as a baseline; controls the ideology of problem-solving so students reach for materials as "objects to think with" and initiate "powerful ideas" (p. 11 & 76). Constructionism through the work of Papert, regenerates the "big idea" and "powerful ideas" in association with a central idea to be assembled, including diverse and non-structured ways of thinking (Papert, 1993). As seen through the unobstructed view of young children, powerful ideas when sequentially worked out can transform thinking into a real visual experience.

With this realization, the education system is taking on a renewed interest in educational robotics and the power that this tool can unleash not only in high school but also in other grade levels. Focusing on specific CT skills enhanced by robotic and real world problem-solving application can reach students in all grade levels (Weintrop, et al., 2016; Witherspoon, Hugashi, Schunn, Baehr, & Shoop, 2017). Schools are utilizing robotics as a concrete physical tool in hands-on activities within the context of the specific academic domain. Robots provide engaging instruments in transferring human thought patterns into physical solutions to a problem and providing additional classroom connections with CT.

What is Programming?

The algorithmic step in CT involves the strands or step-by step process of creating and following a pattern or process in order to come up with a solution. In this process, programming is essential to the completion. Let's look at some of the ley terms involved in this step of CT. A **Bug (Programming Error)** is an error, flaw or fault in a computer program logic or syntax strand(s) which constitutes an unexpected behavior that can negativity effect or cause the function of the program to quit or stop working (McFadden, 2018). **Code** is the individual or piece of computerized instruction; and when sequentially written with other pieces of instruction form an entire program. **Debugging** is finding and correcting errors within a computer program and fixing them. **Input** is "data entered into a computer" (Christensson, 2006, para. 1). **Output** is "data generated by a computer" (Christensson, 2006, para. 1). **Program** is a logical sequence of instructional codes which a computer can understand and carry out. **Programming** is the entire process of creating a specific algorithm and a set of instructions; for a computerized component to follow such as to complete a task, solve a problem, or demonstrate an action. Programming is comprised of many pieces of code. **Sequence** is a specific order or series of events which is followed, one after the other. Placing all the pieces together, CT, robotics, and programming leads to an exciting and invigorating exploration of creating and solving problems.

Computational Thinking

Jeanette Wing, a pioneer in the field of computer science, promoted CT in 2006 as a fundamental and operational steppingstone in the human analytical skill and thought process as it relates to computer science and all professional fields. These fields such as engineering, medicine, manufacturing, education, finance, health science, and cybersecurity, utilize some form of CT every day. With that ideology, Wing (2006) describes CT in terms of (a) "abstraction"; (b) "decomposition"; (c) "system analysis"; and (d) "modularity" to include more abstract ideology specifications (pp. 33-34). Within her research studies, Wing realized that the focusing just on the mathematical aspects of solving problems was not sufficient, whereas within the CT structure there was an intersection with logical thinking and systems thinking allowing for the mixing of many other thought processes such as "compositional reasoning, pattern matching procedural thinking, and recursive thinking" (p. 1). CT was instrumental in formulating the framework evident by reconstructing, rethinking, and addressing the mental process of inquiry into solving all types of real-world problems (Lee, Martin, Apone, 2014; Win 2010).

Papert (1993) discussed how, as a child, he was curious about the function of gears and their relationship to mathematics. The prospect of the functionality of the gear and its correlation to systems ignited a working computation within mathematical applications. The relation of "object-to-think-with" ideology to complete a sequential sentence or program by visualizing with gears the "if" and "then" process or known as CT. Papert (1993) revealed that students lack the understanding, process, and application(s) in thinking, which inhibits the grasp of "one's person's experience(s)" (Papert, 1993, p. 10). Papert's personal experience with the mechanics of gears lead him to explore the processes and laid out the foundation for CT. With gears in motion, robots and CT can provide the fundamentals to lead to new possibilities for learning, thinking, and growing emotionally as well as cognitively" (Papert, 1993, pp. 17-18). Without this personal experience, thinking is restricted and consequently reduces the beauty of the abstract thought process within each learner (Kordaki & Kakavas, 2017).

The human experience personalizes the use of robots and robotics education through cognitive methods by having students working with their hands and engaging in creative and innovative activities. Robots ignite the passion and provide students with the opportunity to experiment, build, problem-solve, think critically, and work together in STEM activities (Caci, Chiazzese, D'Amcio, 2013; Barker, Nugent, Grandgenett, & Adamchuk, 2012). In mathematics, Francis & Davis (2018) stated that robots take the lead as powerful instruments to advocate mathematics and problem-solving in today's education classroom. Leading the way for this emersion encompasses student understanding of skills such as critical thinking, problem-solving, creativity, and teamwork. Robotics education provide learning environment or playground to promote socioemotional skills of engagement, collaboration, communication, and teamwork (Bers, 2008). Continuing with the constructionist philosophy, hands-on application learning and robots address social skills as collaboration, communication, frustration, and anxiety whereabouts provides an authentic real-world learning experience in a safe-zone environment.

Revolutionizing the educational system standards with robotics and CT skills in the curriculum encourages our youth to expand and wonder beyond the concrete structures through imagination, creation, and development of new challenges. Robots promote a playful learning environment which students can be free to create, re-construct, imagine, obtain various levels of abstraction, and decomposition in STEM activities (Barker, et al., 2012). "Using robotics introduces students to emerging and innovative technological creations, as well as encouraging their participation in the act of making, which, in turn, nurtures them to become active creators rather than consumers of technological products in the future" (Eguchi, 2014, p. 30). Robots and robotics education are tools of exploration in the classroom. These technology tools emphasize and invoke new interests and passions through hands-on activities within any classroom. Some of the opportunities of using robots in a STEM classroom is providing the students with a safe and open, risk-free environment to "stop, question, and think deeply" (Eguchi, 2014, p. 30). Open STEM classrooms focus on CT and robots to create amazing learning situations, opening a door of new innovative ideas. This CT opportunity works with challenging the students' abilities and maintaining a high level of comprehension and understanding.

The youth of today are the newest members of the emerging digital age. These students are sponges for absorbing and retaining information when given the opportunity in the classroom. According to Witherspoon, Hugashi, Schunn, Baehr, & Shoop (2017), CT brings to the classroom exciting and innovative activities which infuse robotics with hands-on application platforms in the science and mathematics curriculum. Within the past few years, the momentum of CT integration within the K-12 education system is slowly extending and infiltrating across the science, technology, engineering, and mathematics (STEM) fields, where the CT skillset gap is quite evident is at the intermediate school level (CSTA, n.d.; Grover & Pea, 2013). CT skills are underdeveloped within the standards of the education system. CT is an essential skill for the 21st Century learner. The aim is to establish a channel for students to enhance their ability to formulate questions, articulate answers, and solve problems. The methodology of which the robotics education is currently being taught in the school system is not effective and, moreover, segregated from the core curriculum (Scaradozzi, Sorbi, Pedale, Valzano, Vergine, 2015).

Through the lens of the child, new visions and insights utilizing hands-on activities set in motion a certain spark for the acceleration and eagerness to learn, discover, absorb knowledge, and experience by expanding the in-depth thought process of the mind. Papert (1993) believes children approach learning differently, by removing the lens which filters their surroundings and exposes a new world untainted and natural, as though one is looking through a naked eye. Children accomplish this through means of exploring their surroundings and "learning without being taught" (Papert, 1993, p. 7). Children developmental stages are important levels to understand when implementing activities which promote understanding and knowledge. Accordingly, "Piagetian learning" is divided into four stages of childhood development: (a) "sensorimotor (birth to two"; (b) "preoperational (two to seven years old)"; (c) "concrete operational (seven to eleven years old)"; and (d) "formal operational (eleven years and older)" (Mcleod, 2018, p. 7). Piagetian learning theory revolves around the "readiness" of the child which are indications of when and at what stage learning concepts are to be introduced in specific stages according to the cognitive development in education (McLeod, 2018, p.10). According to Papert (1993), who worked closely with Piaget, he concludes children's cognitive development at the age of eleven can lead to and constitute more abstract and logical sequences resulting in additional trial and error stages in order to complete the problem. This study directly links to this "formal operational stage," as the students are in sixth grade with an average age of eleven. This stage emerges as enlightening imagery through the unobstructed view of the child and this visualization brings in and paves the way for furthering the need and importance of CT emerging as a fundamental skill in the classroom.

Understanding students' learning approaches, from conception to result, can improve and customize curriculum activities. By "getting things to work" and not "knowing the right answer" exploring starts a natural flow of ideas in science, mathematics, and other areas (Papert, 1993, p. 7). In association with science and math, CT and design thinking focuses on problem-solving, which enables students, through trial and error, to solve practical problems (Shute, Sun, & Asbell-Clark, 2017). This association is the essence for digital learners to "think articulately about learning" and "learning without being taught" (Papert, 1993, p. 7).

Constructionism

As the 21st Century classroom emerges, Papert's theory of constructionism takes on a new light with the union of CT and robotics education. Bilkstein (2018) resonates that the connections created to invoke the thinking and learning as erecting knowledge construction in a way that allows the free flow of creative ideas following through the educational context of developing and learning. There are four pillars to constructionism (a) "learning-by-doing"; (b) "objection recognition-supporting the development of ways of thinking and learning about abstract phenomena"; (c) "powerful ideas"; and (d) "self-reflection" (Bers, 2012, p. 4). These four pillars can be expanded into the cross-cutting theoretical framework, which connects CT and robotics education through diverse hands-on classroom activities.

Children think differently than adults; their minds are open with no constraints or pre-determined assumptions and putting the cognitive skills, creativeness, and behaviors in motion and can lead to discoveries and "big ideas" (Papert, 2000). Along with the hands-on and learning-by-making (constructionism) Papert (1993) references and refines the creativity of and children's natural curiosity becomes an adventure that opens the mental activities leading the way to CT. Interactivity, engagement, and passion navigate the students' of today to the mindfulness of active learners in the new century of digital education. Children innate inquisitiveness for exploring and playing are bursting with borderless ideas immersed in thinking and tinkering. "Seymour focused on how, as a child, he fell in love with gears, and that love for a material changed his self-perception and allowed him to become more interested in learning about math, science and engineering".

The "builder" in a child stems from "Piagetian Learning" which describes children as "doing without being taught," the natural process of curiosity and exploration without boundaries or limitations (Papert, 1993, p. 7). In the transition from concepts and ideas to the products and solutions, providing children with materials to physically build or code present a unique and valuable opportunity to take the lead and search out various resources, including tangible manipulatives to engage in the logical and sequential thinking of CT. With constructivism as a background to constructionism, "constructionism states that children learn better when making their projects, constructing their ideas, and designing their own solutions to problems... providing tools for children to playfully design their own 'objects to think with'" (Bers, 2012 p. 14; Bers, 2008; Kafai & Resnick 1996).

Additionally, constructionism, as it progresses in a classroom, invites children to become inventors, designers, and programmers promote and encourages peer learning with engaging and exploring in engineering, science and other fields of study. All are leading toward CT concepts and skills associating with the construction paradigm of ideas and thoughts to craft human knowledge into the design and evaluation of problems and solutions which is part of everyone's brain (Bers, 2012, p. 14). This connection with CT concepts maintains the paradigm set forth by Papert's as "one learns that the most powerful idea of all is the idea of powerful ideas" (Papert, 1993, p. 76). Creating various opportunities for children to develop their ideas and projects, substantiating the validity of constructionism; computers are power tools combined with other peripherals, such as robots, to provide robust and playful opportunities to foster the designing and constructing ideas.

Accordingly, Papert (1993) stated that once children are involved in this activity, there is an upsurge in engagement and exploration, which lends itself to complex and multifaceted outcomes reflecting their thought process and solutions. This process authenticates the finding from Bers (2012), which children become "creators of interactive content, as opposed to consumers of information" (p.15). Children are absorbed in the action of learning by doing, playing, investigating, and uncovering their thought processes with various applications and platforms regarded as CT and problem-solving (Papert, 1993). Barr, Harrison, & Conery (2011) and Wing (2006) believe students who were exposed in early stages of education to CT skills possess advanced processing skills over a student who has not been exposed earlier on to the skills. The student is the central player in constructionism, applying hands-on activities which revolve around the self-directed learning cycle to support the how and meaningful methods of reshaping and restructuring through the mindful thought process. The process encourages the mind to dig further into the thinking process, analyze, and make decisions based on the gathered information. Critical exploration and encouraging of self-expression create insights on "how children are really seeing the problem" which prompts meaningful exchanges of ideas and thoughts to help in solving issues (Barker, et al., 2012, p. 12).

Social constructions, the human social interactions influence the social setting and environment which children are part of in and out of the classroom. This social engagement directly compounds the "cultural materials" and "social relations" comparable to the skills emphasized when applying CT (Kafai & Resnick, 1996, p. 181). Actively engaged through social constructions provide the opportunity to learn or construct new knowledge from within and through physical demonstration provide the concrete evidence of the obtained knowledge. Working on the basis of constructionism, Kafai & Resnick (1996) allots to the consensus which constructionism is the social playground which sparks the learner in "creating something meaningful to themselves and to others around them" (p. 214).

Educational Considerations

Education curriculum has taken a positive turn by driving attention to the addition of robotics education and CT skills with new and invigorating hands-on, creative classroom activities and lessons. Curriculum initiatives need to move forward and not stay with past stagnate technology applications in the classroom. CT and robotics education can be the forward-thinking initiatives which provide the challenges required for the digitally savvy students. In recent years, robotics education is gaining speed and momentum as becoming an essential part of a school's curriculum (Francis & Davis, 2018). Subsequently, the educational framework has taken on a new configuration for students to engage as "better thinkers" by combining robots and CT skills such as a sequential and abstract perception of a proposed real-world problem within the classroom (Resnick & Robinson, 2017, p. 48). A cohesive bonding of CT and robotics education can effectively redirect the school curriculum to embrace these 21st Century applications for the 21st Century learner and the future global workforce.

There is revitalized spark and energy in restructuring and reevaluating the education curriculum stretching across the STEM fields. A focus on CT and combining the power of robotics are fundamental keys to unlocking the unlimited potential with engaging hands-on activities to deliver innovation, problem-solving, and critical thinking (Alimisis, 2013). Being able to dig deeper and understand from the students' point of view their flow of ideas, from conception to result, can improve and customize curriculum activities in a classroom. By "getting things to work" and not "knowing the right answer" undertakes a connection with exploration in all curriculum areas creating a prominent linkage which can coexist in today's classroom activities (Papert, 1993, p. 7).

The curiosity and sense of innovation motivate the digitally learner and feeds upon the natural sense of exploring new and exciting challenges. A students' passion and interest can start the innovative and creative thinking process moving in many directions with purpose, exploration, determination, and power of imagination which moves students closer to the solution. Besides the idea of problem-solving, the activities which are conducive to the constructionism process are "learning, thinking, playing" all bound to encourage the exploration of new ideas (Kafai & Resnick, 1996, p. 11). Referenced in the literature numerous times are the philosophies of Papert, Piaget, Resnick, and Bers concurring that students learn best by constructing and doing; not by sitting at a desk in rows and lectured by the instructor. Traditional education relies on facts and statistics, while certain practices in constructionism directs computational ideology, technology, and motivation to carry through and provide learners with new options "learning, thinking, and growing emotionally as well as cognitively" (Papert, 1993, p. 17). CT skills, as part of the daily activities in the classroom, can provide an architecture structured on discoveries every day.

In the classroom, digital tools including robots have found their place among the basic classroom tools and instruments. Bers (2008) refers to the tools as an intellectual way of thinking and connecting the "powerful ideas" described by Papert in 1993, as continuum of mixing new and old ideas creating new possibilities to disperse knowledge. Adding to the excitement and inquisitiveness of today's digital learners, the education phenomenon of CT and robots enlivens the regular daily classroom events. Transformation to this digital technology connects knowledge-based activities to various skills such as "critical thinking, problem-solving, creativity, strategizing, and originality" (Bates, 2016, p. 18). From concept to hands-on knowledge-based activities, CT skills restructure learning and applications throughout the students' lifetime to include, "how to find, evaluate, analyze, apply and disseminate information, within a particular context" (Bates, 2016, p. 19).

CT activities create a forward momentum in the classroom and provide outlets for the emerging and innovative ideology for which our students today will need to understand and apply in the real world. The culmination of thinking and sequential writing appears in many curriculum fields, are now taken the front seat directing to innovative approaches in the classroom. According to Resnick & Robinson (2018), there are numerous opportunities which engage and implementing student CT skills in a classroom. The opportunities foster and enrich the Engineering Design Process (EDP). The EDP is the systematic process of steps for the solving of a problem, whereas it can be an ongoing process of trial and error, which leads to a solution or other possibilities. The design is a continuous application of six steps (See Appendix L), which (a) identify the problem; (b) explore; (c) design; (d) create; (e) try it out; and (f) make it better (Engineering Design Process, n.d.). Connecting all the elements and components of CT, Robotics, and EDP and one entity creating a community of life-long learners.

Defining Computational Thinking

Experts in the field still are debating as to what syntax and words which define CT (Wing, 2010; Grover & Pea, 2013). Without direct and concise agreement on a CT definition, researchers are struggling with the ideology, core step, and elements which create the CT philosophy. Papert (1996) hints at a definition, whereas, Wing (2006) defines as "problem-solving, designing systems, and understanding human behavior, by drawing on the concepts fundamental to computer science" (p. 33). Recently, Wing (2017) devised the definition of CT as "computational thinking is the thought processes involved in formulating a problem and expressing its solution(s) in such a way that a computer-human or machine-can effectively carry out" (p. 8).

The level of involvement in CT starts with the abstraction process which Wing (2017) claims is the key component to carrying out and implementing the involved CT skillset and relevant to all students. CT is an essential core element within computer science education (Hendrickson, n.d.). Thus, providing learners with hands-on opportunities to exhibit creativity, demonstrate sequential processes, and collaborating with peers. In association with the sociological pragmatic world view of the CT phenomenon, Resnick & Robinson (2018) and Brennan & Resnick (2012) breakdown CT into key dimensions; not to be confused with CT four pillars conceived by Wing (2006); which transition the traditional classroom into an energetic and hands-on environment that allows for the implementation of the framework. The three key dimensions are (a) computational concepts; (b) computational practices; and (c) computational perspectives (Brennan & Resnick, 2012, p. 1). CT four pillars are (a) "decomposition - breaking down a complex problem into smaller, more manageable parts"; (b) "pattern recognition – looking for similarities"; (c) "abstraction – focusing on the important information only" and (d) "algorithms - developing a step-by-step solution to the problem" (Wing, 2006, pp. 33-34) (see Appendix A). The three dimensions correlate with the CT steps by means of advocating for additional critical thinking processes and sequential analysis opportunities for increase awareness by students in the classroom.

Additionally, Grover & Pea (2013) implies that CT viewed in a collection of computational competencies and "procedural thinking" to be included in all areas of the educational curriculum (p. 38). This relates to the system where the computer science field can be the starting point in from early school years through higher education. The regular flow from year to year in education can address the skillset at the beginning stages of the students' educational career. Computer Science field is categorized as interdisciplinary creates and sustains elaborate playing fields for CT to generate and foster originality and encourage responsiveness through design thinking processes (Hendrickson, n.d.).

Many decades ago, experts realized the value and power of CT and making headway back into the education system (Bers, 2012). With the onset of programming languages of BASIC and LOGO, the power of computation connected with robotics ingenuity contrives to offer appealing activities (Resnick & Robinson; Grover & Pea, 2013). Robotics, along with programming are suitable matches which engage activities and "iterative exploration" of CT core elements for students (Grover & Pea, 2013, p. 40). CT as an essential core element within computer science and other fields in education (Hendrickson, n.d.). Thus, leaning towards and opening up the computer science elements in all curriculum fields.

Robots and their diverse resources can be the driving, engaging resource creating a new force for infusing CT skills within a classroom. Introducing robotics education provide a viable opportunity to pursue within a curriculum. As a learning tool, robots primarily engage students in activities which use visual block-based coding to perform simple or complex tasks (Barker, et al., 2012). Robotics education can provide a focus on the measurement of transferrable CT skills in the intermediate and middle school robotics (Koh, Repenning, Nickerson, Motter, 2013; Lin, 2015). With elements of innovation, creativity, and collaboration robotics education acts as a viable and exciting means of engaging our digitally literate students.

Papert (1993) refers to education, computers, and computer culture as pervading "new opportunities for imagination and creativity" (p. 37) and formulating a climate for CT activities involving computers and robots. Both peripherals have the potential to interlock all these fields with engaging hands-on activities which provide and foster CT in keeping with stimuli of innovation (Alimisis, 2013). Expressed in terms of "all-in-one technological learning tool" CT and robotics programs, enhances the hands-on applications which engage and drive the student thinking process with engaging and enlightening activities which crisscross the entire curriculum (Eguchi, 2014, p. 34).

An earlier attempt to define CT was constructed by Seymour Papert in 1980. Papert identified as the "father of LOGO programming (computerized educational programming system) described CT in terms of a fundamental human experience powerful enough to tie every-day real-life experiences with conceptual, computational procedures (Logo Foundation, n.d.). As the root of CT, LOGO, developed by Bobrow, Feurzeig, Papert, and Solomon in 1987, detailed CT as an operational and fundamental guide to help students in the process of critically thinking within the mathematics forum and solve problems (L'Amoreaux, 2019; Logo Foundation, n.d.; Papert, 1980). The LOGO programming reinforced the shifting of computer applications with a "new conceptual framework for thinking" (Papert, 1980, p. 184). In 1996 Papert expanded on the concept by re-empowering critical thinking and systematic strategies as an accelerant into solving problems.

CT and robotics programming utilize a specialized visual language. The language relies on easily recognized images which initiates a response by sparking an interest which advances into powerful inspiration, problem-solving, and collaboration. Martin (2016) reported, "computational thinking has its core interest in thinking in multiple levels of abstraction and conceptualization…it is being seen at the center of learning in math, science, and engineering" (p. 24). In keeping with the curriculum areas, Hutchison, Nadolny, & Estapa, (2015) blends the area of language arts literacy and programming as an "evolving form of literature" attaining a core set of CT skills (p. 493). Leading to the augmentation of communication outlets between different and diverse learning skills within an environment conducive in the education field.

With the multiple opportunities in education, CT concepts converge on the 21st Century learner, involving authentic hands-on activities with one such activity utilizing the robot. Wolfram (2017) stated, "the role of computational language is to be able to convert how we think about things into something that is computationally understandable" (With, n.d.). Robots possess unique computational language. This language is constructed to initiate and follow through procedures and commands as given by the user. The user is able to decipher and breakdown the elements which make up the program and reassemble to solve a real-world problem in the core of CT. The educational robotics programs referred to as "all-in-one technological learning tool" and intermix the hands-on applications that drive the focus back to the student with engaging and enlightening activities which crisscross the entire curriculum (Eguchi, 2014, p. 34).

Leonard, Mitchell, Barnes-Johnson, Unertl, Outka-Hill, Robinson, & Hester-Croff, (2018) believes the culture set by using CT skills in the education system broadens the problem-soling process and build students' capacity to understand and apply the methods in science and other fields of study. The CT classroom methodology can be instrumental in triggering positive experiences resulting in an understanding of the different components and the interaction and connectivity to generate logical awareness and produce solutions.

Computational Thinking Model

CT segmented into four steps involving a process to establish a solution to a problem. CT is a logical and sequential method starting with a problem then breaking into down into smaller manageable components. Each step is connected or linked in some way within the four pillars of CT - decomposition; pattern recognition; abstraction; and algorithm design (see Appendix A). Thus, formulating a structure for teamwork activities in alliance with CT to break down the problem into the four steps creating a problem formulation which involves logical reasoning and visual notation to entice effective learning.

Each step is specifically defined to work together as a process to complete and solve problems. The decomposition steps are the breaking down of a problem into smaller and more manageable pieces for the students to understand and work with. Pattern recognition is the method or way the student looks at and observes the patterns, similarities, trends in the given information or data. Abstraction is an important step in identifying the basic ideas which create the patterns. The algorithm design is the actual sequential, step-by-step process and instructions which lead to the solution for the problem.

Like, Resnick & Robinson (2017), Wolfram (2017) and Grover & Pea (2013), Ortiz believe that CT is the 21st Century literacy and can be implemented and sustained at all ages and disciplines in education. Core set of skills include algorithm design, sequential process, collaboration, creativity, critical thinking, and innovation all built in with hands-on activities. By linking the four steps to CT model, the authentic learning process lends itself and contributes to the students' hands-on experiences in the classroom and linked to problem-solving in the real world. By repurposing robots as a teaching tool for CT creates vibrant and engaging "learning hubs" and "centers of change" (Teich, 2019, para. 1).

According to Paiget: "discovery learning – the idea that children learn best through doing and actively exploring - was seen as central to the transformation of the primary school curriculum" (McLeod, 2015, p. 10). Robotics encompasses the learning and methodology delivered through hands-on discovery learning and can be differentiated at the same time, paving the way for all students to be able to participate in STEM classroom activities and apply learned skills and knowledge in all aspect of life. Along with activities comes empowerment and understanding of knowledge, being in line with the "big idea" written about by Papert (2000) which is stipulates "the sense of power comes here from being able to solve well-known or self-generated problems, puzzles, and paradoxes from personal knowledge using very general techniques that are all part of what I call 'technological fluency' (p. 8). Involving CT and robots as a basic hands-on activity creates and acts as a catalyst connecting to Papert's Big Idea and allowing for the freedom to creativity think and apply learned knowledge in situations beyond the classroom.

Computational Thinking in K-8 Curriculum

Plaza, et al., (2017) references key processes that are necessary for 21st Century learners. The key processes are: (a) "systems thinking"; (b) "programming mindset"; (c) "active learning"; (d) "mathematics"; (e) "science"; (f) "judgement and decision making"; (g) "good communication"; (h) "technology design"; (i) "complex problem solving"; and (j) "persistence" (p.1). Along with this, the researcher emphasized fun as being an essential piece to the project which includes the participants' understandings and experiences when using robotics in conjunction with the classroom activity. The thought of CT connecting with robots as creative play in the classroom stands out as a methodology which helps boost students find joy, meaning in doing and learning, experimentation, minds on thinking, hypothesis testing, and social interaction.

Bers (2018) breaks down CT into "expression" and "communication" (p.63). CT utilizes sequential programming language, referred to as digital literacy. Taking into consideration that robots do require programming, integrating applications in relation to the mind thinking and social interactions through trial and error problem-solving scenarios in the classroom. Exploring the two areas advocates for additional approaches reverting back to Papert's Big Idea of expression by means of the programming sequences. Reaching for students to evolve as a "thinker" not a "tinkerer". CT as proposed by Bers (2018) creates a unity and connection of expression (Bers) with problem-solving (Papert) creating a new phrase of "computational thinking by expression" (Bers, 2018, p. 60). CT from the standpoint of Bers (2018) characterizes "analytical thinking that share similarities with mathematical thinking (e.g. problem-solving), engineering thinking (designing and evaluation processes), and scientific thinking (systems analysis) (p. 60). Functionally all three "thinking" processes direct link together to form a unique creative engineering design process and CT which enables students to understand and apply knowledge in relation to future endeavors.

Francis & Davis (2018) stated that robots are powerful instruments to advocate mathematics and problem-solving in today's education classroom. Leading the way for this emersion contains multiple student understanding skills such as critical thinking, problem-solving, CT, and creativity. Students are engaged no matter what the results of their solution, correct or incorrect which drive their curiosity and insight into the mathematical applications and coherent spin-offs. This experience reiterates that CT involves multifaceted areas such as arithmetic, coding, and logic representations. For instance, students can use the robot for calculating measurements in distances and locations of objects which leads to innovative and engaging activities which create a partnership between multiple educational subjects' areas.

Korkmaz (2016) study implied that programming with LEGO Mindstorms EV3 is an example of a "complex and abstract process" (p. 1). Leading to the education and application of a coding program involving the robot energizing the human creative and algorithm design to solve problems through mindful thought processes. The study revealed that students understand and grasp the ideas behind logical and sequential mathematical thinking skills including solving real-world problems. The value helps to uncover the tools of creativity and critical thinking in the field of managing information. In addition, the LEGO Mindstorms EV3 design activities maintained a positive involvement and influence in the directing students to problem-solve. The results of this study provide valuable data in which CT activities provide n lasting educational experience based on non-traditional teaching activities, and the EV3 programming was conducive to the problem-solving skills (Korkmaz, 2016).

Envisioning CT in the classroom identifies the Papert's "big idea" in methods and potential use as the power of reasoning. CT in the classroom enables integration of module pieces from different curriculum and transfused into a working educational forum. The students do not merely use computers or robots as a tool but become tool builders (Barr & Stephenson, 2011). CT power in the classroom comes in different forms such as "abstraction, recursion, and iteration, to process and analyze of data, and to create real and virtual artifacts" (Barr & Stephenson, 2011, p. 4). The artifacts can provide insight of the thinking process, comfort level of trial and error, learning by doing, and figuring things out which are key concepts for students to digest and producing a workable solution to a problem. For example, student involved in an activity are more likely to use key CT concepts and the evidence appears as students talking to other students and using CT related vocabulary such as sequences, inputs, outputs, line detection, degrees, angles, move forward, looping, saved value, friction, and motion. Another interesting and important CT skill is observation by peers in the classroom, by way the students get that there is more than one way to come up with a solution, and flexibility ties in with the entire process. These concepts reestablish the constructionism theory and opportunities for students to move forward in their learning.

Additional classroom activities involving CT crisscross the curricula. In science, students test out force and inertia, make predications, record evidence and produce algorithmic solution. In math, students test our certain color patterns in relation to an activity such as a contest or race. In language arts, utilizing the robot as a character in a story and create a different ending by means of coding or programming. In physical education map out a new game or revise a traditional physical activity. These activities involve abstraction, critical thinking, and problem-solving, all part of CT.

Computational Thinking and Robotics

Creative chaos, excitement, non-traditional learning, encouraging new ideas and "active participants in a learning community" intensify when children are actively involved in (Barker, et al., 2012). Robots and programming together form a powerful educational tool which provides immediate feedback to the students. The instant feedback incorporates CT and solving problems without having to wait for results. If there are problems, the participants can address them immediately, which involves motivation for continuing to solve the problem-engaging the interest and experience mimic situations people encounter in the real world. Instant feedback in the classroom is an essential element for collaboration and the ability to talk problems out, in turn, being receptive of other ideas.

Resnick & Robinson (2017) relates LEGO Mindstorms robotics "…remixing is aligned with our core values of openness and creativity" (p. 103). Throughout the book, Resnick illustrates scenarios in which students are involved in hands-on activities from building, coding, and executing the designed program on a robotic unit to fit the classroom activity. The association between building the LEGO EV3 robot and coding and programming of the action constitutes an alliance rich in creativity and CT illuminating a powerful voice in the world of education. The robotics education platform represents a real, concrete, physical, and tangible option which allows students to comprehend the learning processes such as abstraction, construction, and evaluation (Resnick & Robinson, 2017).

With the onset of programming languages of BASIC and LOGO, the power of computation connected with robotics ingenuity contrives to offer appealing activities (Resnick & Robinson, 2017; Grover & Pea, 2013). Robotics, along with programming, are suitable matches which engage activities and "iterative exploration" of CT core elements for students (Grover & Pea, 2013, p. 40). Correspondingly, many decades ago, experts realized the value and power of CT and making headway back into the education system (Bers, 2012). Thus, leaning towards and opening up the computer science elements in all curriculum fields.

Robots and their diverse resources can be driving, engaging resource, creating a new force for infusing CT skills within a classroom. Introducing robotics is a viable avenue to pursue a curriculum area. A research study conducted by Chaudhary, Agrawal, & Sureka (2016) provided data based on the infusion of robotics and programming, characteristics consistent with the technology-based groundwork consist of lesson and activities which foster teamwork, problem-solving, creativity, collaboration, and project management. The students' progress aligned within the direction of the four steps of CT.

As a learning tool, robots primarily engage students in activities which use visual block-based coding to perform simple or complex tasks (Barker, et al., 2012). Robotics education can provide a focus on the measurement of transferrable CT skills in the intermediate and middle school robotics (Koh, Repenning, Nickerson, Motter, 2013; Lin, 2015). With elements of innovation, creativity, and collaboration, robotics education acts as a viable and exciting means of engaging our digitally literate students.

The robot is the "machinery" to motivate children and sustain a love learning in a classroom. Robots have been dispersed in classrooms as educational tools locally and globally, ranging from Pre-K to higher level education (Datteri, Zecca, Laudisa, & Castiglioni, 2013). A robot is a fascinating piece of technology with its circuits, wires, and mechanisms. One such robot produced by LEGO creates a current motivational vehicle when infused within the technology and science activities (Norton & McRobbie, 2015). The robot-based curriculum has the potential to intertwine multiple core content fields with engaging hands-on activities which provide and foster CT in keeping with stimuli of innovation (Alimisis, 2013).

Positive Technology Development (PTD)

Positive Technological Development (PTD) described as a framework for linking individual characteristics with technology-related mannerisms that are situated within a field (Bers, 2016). The six characteristics are categorized as intrapersonal or interpersonal and are essential in the positive development of a student. The six behaviors which can lead to positive results are: (a) content creation, (b) creativity; (c) choice of conduct; (d) communication; (e) collaboration; and (f) community building (Bers, 2019, pp. 11-12). Aligning these behaviors with CT, creates the mental tools to master the multiple steps involved when given a real-world problem to solve using hands-on activities.

PTD constitutes an area of CT which draws on the student's metacognitive functions in their use of specific vocabulary directed at the processing of information throughout a hands-on activity, especially robotics and programming languages. Bers (2016) states "robotics kits can be wonderful playgrounds for young children because they encourage problem solving and logical thinking, creativity, and love of learning, through playful exploration" (p. 29). Robotics involves student risk-taking, approached as a fun and engaging activity the positive aura can seamlessly attribute in many aspects of CT in the classroom. Risk-taking develops a sense of confidence which fosters and evolves into an experience for the student as "hard fun" involving creativity and cleverness to solving problems and crafting a range of CT skills (Kay, 2003).

PTD and constructionism combine to activate tactics to contribute to higher-level thinking, including hands-on activities and learning by doing, constitutes best practices for multi-intelligence and crisscrosses into other cognitive skills (Bers, 2012). Content creation and creativity (two of the four C's) fall in line with constructionism (Bers, 2012). The C's are essential areas of the PTD framework to link, develop, and stretch thinking within the human mind. This multi-dimensional learning applications characterizes a positive reinforcing comfort zone where students can experiment, choose, collaborate, and create on their terms by promoting a metacognitive approach to learning. PTD delivers and creates new technological aspirations in placing the emphasis on positive behaviors which are the outcome of exciting and engaging hands-on activities in the classroom. Using prompts to stimulate in-depth thinking and develop learning environments which foster growth mindset of the 21-st century learner. CT & PTD interconnect in the open-ended and critical thinking skills based on a real-world problem. The playground, as Bers (2012) surrounds the students with multiple outlets or choices involving hands-on activities devoted to the tools and materials of technology, through unplugged or plugged solutions.

On multi-levels surrounded by streams of active involvement and cognitive applications, Piaget, Papert, Bers, and Resnick are drawn to the connections which spark the imagination, excitement, intuitiveness, innovation, and creativity driven by the individuality of the learner. These connections are the functioning sparks which initiate the curiosity which sets gears in motion. "His [Papert] constructionist theory of learning, emphasizing collaboration, iteration, and agency, has formed the foundation of our work in the Tinkering Studio and has deeply influenced the entire maker movement" (Ryan, 2018).

CT AND ROBOTICS: FRAMEWORK OF A FUTURE DRIVEN CLASSROOM

The opportunity, spontaneity, and unpredictability in absorbing critical 21st Century skills are evident within the limitless CT framework available in today's classroom to enable the power of imagination and curiosity of children to investigate and create the unknown. This framework advances the restructuring of today's education system to integrate CT into the classroom and provide students with the opportunity to interact with a familiar and disruptive tool, the robot. Creating controlled chaos invoking the energy of disruption as it may seem, students are fascinated and completely mesmerized with the opportunity to learn and understand containing spur-of the moment, unpredictable, and random actions provided by robotics.

There is revitalized energy in restructuring and reevaluating the education curriculum stretching across the STEM fields. A focus on CT and combining the power of robotics are fundamental keys to unlocking the unlimited potential with engaging hands-on activities to deliver innovation, problem-solving, and critical thinking (Alimisis, 2013). Being able to dig deeper and understand from the students' point of view their flow of ideas, from conception to result, can improve and customize curriculum activities in a classroom. By envisioning through the pure lens of a child and visualizing by "getting things to work" and not "knowing the right answer" undertakes a connection with exploration in all curriculum areas creating a prominent linkage which can coexist in today's classroom activities (Papert, 1993, p. 7).

Beliefs and experiences influence instructional paradigms. Whatever the community of academia buzz words are diversity, disruption, emerging, the cycle of innovative curriculum is exciting and provides evolutionary theories and practices driving the future of the classroom and education. Since the world does not stop and there are constant new and updating, restructuring and connecting the curriculum areas for the 21st Century learner and can be refocused to integrate robotics education effectively in the classroom in all subject content areas.

CT through the lens of a child lends itself to a playground of creativity, imagination, and the wonder of play. With the addition a playful instrument of the robot, the future driven classroom is alive with engaging, exciting, and limitless learning opportunities for all students. The curiosity and sense of innovation motivate the digitally learner and feeds upon the natural sense of exploring new and exciting ideas and challenges. A students' passion and interest spark innovative and creative thinking process with purpose, exploration, determination, tinkering, and imagination driving students closer to the solution. Seen through the naked eye of the child, no boundaries exist with limitless opportunities to move like synchronized gears in motion, making connections, developing critical thinking skills, and igniting curiosity to go beyond and venture into the inquisitive thought patterns that promote 21st Century skills.

REFERENCES

Alimisis, D. (2013). Educational robotics: Open questions and new challenges. *Themes in Science & Technology Education, 6*(1), 63-71. Retrieved from: http://earthlab.uoi.gr/theste/index.php/theste/article/view/119/85

Atmatzidou, S., & Demetriadis, S. (2014). *How to support students' computational thinking skills in educational robotics activities*. Paper presented at the 4th International Workshop Teaching Robotics, Teaching with Robotics & 5th International Conference Robotics in Education, Padova, Italy.

Barker, B. S., Nugent, G., Grandgenett, N., & Adamchuk, V. I. (2012). *Robots in K-12 education: A new technology for learning*. Hershey, PA: Information Science Reference. doi:10.4018/978-1-4666-0182-6

Barr, D., Harrison, J., & Conery, L. (2011). Computational thinking: A digital age skill for everyone. *Learning and Leading with Technology, 38*(6), 20–23.

Barr, V., & Stephenson, C. (2011). Bringing computational thinking to K-12: What is involved and what is the role of the computer science education community? *ACM Inroads, 2*(1), 48–54. doi:10.1145/1929887.1929905

Bates, T. (2016). *Teaching in a digital age: guidelines for designing teaching and learning*. SFU Document Solutions.

Becker, S., Brown, M., Dahlstrom, E., Davis, A., DePaul, K., Diaz, V., & Pomerantz, J. (2018). *NMC Horizon Report: 2018 Higher Education Edition.* Louisville, CO: Educause.

Bers, M. (2008). *Blocks to robots: Learning with technology in the early childhood classroom.* New York: Teachers College Press.

Bers, M. U. (2010). The tangible robotics program: Applied computational thinking for young children. *Early Childhood Research & Practice, 12*(2).

Bers, M. U. (2012). *Designing digital experiences for positive youth development: From playpen to playground.* New York: Oxford University Press. doi:10.1093/acprof:oso/9780199757022.001.0001

Bers, M. U. (2018). *Coding as a playground: programming and computational thinking in the early childhood classroom.* New York, NY: Routledge.

Blikstein, P. (2018). *Seymour Papert's legacy: Thinking about learning and learning about thinking.* Retrieved from: https://tltl.stanford.edu/content/seymour-papert-s-legacy-thinking-about-learning-and-learning-about-thinking

Brennan, K., & Resnick, M. (2012). *New frameworks for studying and assessing the development of computational thinking.* Paper presented at the 2012 Annual Meeting at the American Educational Research Association, Vancouver, Canada.

Christensson, P. (2006). *Input Definition.* Retrieved September 21, 2019 from https://techterms.com

Christensson, P. (2006). *Output Definition.* Retrieved September 21, 2019 from https://techterms.com

Computational Thinking for Educators - - Unit 1 – Introducing Computational Thinking. (n.d.). Retrieved from https://computationalthinkingcourse.withgoogle.com/unit

Computational Thinking in K-12 Education. (2018). Retrieved June 1, 2019 from https://cspathshala.org/2017/10/25/computational-thinking-curriculum/

Constantinou, V. & Ioannou, A. (n.d.). Development of computational thinking skills through educational robotics. *Ceur-Ws*, 2193.

CSTA. (n.d.). *About CSTA standards.* Retrieved May 13, 2019 from https://www.csteachers.org/page/about-csta-s-k-12-nbsp-standards

Datteri, E., Zecca, L., Laudisa, F., & Castiglioni, M. (2013). Learning to explain: The role of educational robots in science education. *Themes in Science and Technology Education*, *6*, 29–38.

Eguchi, A. (2014). *Robotics as a learning tool for educational transformation.* Paper presented at the Teaching with Robotics & 5th International Conference Robotics in Education. Retrieved from http://www.terecop.eu/TRTWR-RIE2014/files/00_WFr1/00_WFr1_04.pdf

Eguchi, A. (2016). Computational thinking with educational robotics. *Site,* 79-84.

Francis, K., & Davis, B. (2018). *Coding Robots as a Source of Instantiations for Arithmetic.* Digital Experiences in Mathematics Education. doi:10.100740751-018-0042-7

Grover, S. (2018). *The 5th 'C' of 21st century skills? try computational thinking (not coding).* Retrieved on March 1, 2018 from: https://www.edsurge.com/news/2018-02-25-the-5th-c-of-21st-century-skills-try-computational-thinking-not-coding?

Grover, S., & Pea, R. (2013). Computational thinking in K–12: A review of the state of the field. *Educational Researcher*, *42*(1), 38–43.

Hendrickson, K. (n.d.). *Computational Thinking*. Retrieved May 11, 2019 from https://k12cs.org/computational-thinking

Kafai, Y. B., & Resnick, M. (1996). *Constructionism in practice: Designing, thinking, and learning in a digital world*. Mahwah, NJ: Erlbaum.

Kay, A. (2003). Interview with Alan Kay. *Computers in Entertainment, 1*(1), 8. doi:10.1145/950566.950580

Koh, K., Repenning, A., Nickerson, H., & Motter, P. (2013). Will it stick? Exploring the sustainability of computational thinking education through game design conference. *Proceeding of the 44th ACM technical symposium on computer science education*, 597-602. doi:10.1145/2445196.2445372

Kordaki, M., & Kakavas, P. (2017). *Digital storytelling as an effective framework for the development of computational thinking skills*. Paper presented at the International Conference on Education and New Learning Technologies, Greece. 10.21125/edulearn.2017.2435

Korkmaz, O. (2016). The effect of scratch-and Lego Mindstorms Ev3-based programming activities on academic achievement, problem-solving skills and Logical Mathematical thinking skills of students. *Malaysian Online Journal of Educational Sciences, 4*(3).

L'Amoreaux, C. (2019). *Who is Seymour Papert (and why does it matter)?* Retrieved June 7, 2019 from https://www.claudialamoreaux.com/who-is-seymour-papert-and-why-does-it-matter/

Lee, I., Martin, F., & Apone, K. (2014). Integrating computational thinking across the K–8 curriculum. *ACM Inroads, 5*(4), 64–71. doi:10.1145/2684721.2684736

Leonard, J., Mitchell, M., Barnes-Johnson, J., Unertl, A., Outka-Hill, J., Robinson, R., & Hester-Croff, C. (2018). Preparing teachers to engage rural students in computational thinking through robotics, game design, and culturally responsive teaching. *Journal of Teacher Education, 69*(4), 386–407. doi:10.1177/0022487117732317

Lin, V. (2015). *Computational thinking and technology toys* (Honors Computer Science Thesis Collection). Available from Wellesley College. (307). Retrieved on June 28, 2019 from https://repository.wellesley.edu/thesiscollection/307

Logo Foundation. (n.d.). Retrieved June 7, 2019 from https://el.media.mit.edu/logo-foundation/

Martin, S. A. (2016). *A descriptive and explorative case study of a scratch programming experience involving the creation of a lunar Simulation/Model with grade six learners.* Retrieved from https://prism.ucalgary.ca/bitstream/handle/11023/2851/ucalgary_2016_martin_stephen.pdf?sequence=7&isAllowed=y

McFadden, C. (2018, September 24). *The origin of the term 'computer bug'.* Retrieved September 21, 2019 from https://interestingengineering.com/the-origin-of-the-term-computer-bug

McLeod, S. A. (2018, June 6). *Jean Piaget's theory of cognitive development.* Retrieved October 9, 2019 from https://www.simplypsychology.org/piaget.html

Morgan, D. L. (1997). *Qualitative research methods: Focus groups as qualitative research.* Thousand Oaks, CA: SAGE Publications, Inc.; doi:10.4135/9781412984287

Morgan, D. L. (1999). *The focus group guidebook.* Thousand Oaks, CA: SAGE.

National Research Council. (2011). *Successful K-12 STEM education: Identifying effective approaches in science, technology, engineering, and mathematics.* Washington, DC: National Academies Press.

Papert, S. (1980). *Mindstorms: Children, computers, and powerful ideas* (1st ed.). New York: Basic Books.

Papert, S. (1993). *Mindstorms: Children, computers, and powerful ideas* (2nd ed.). New York: Basic Books.

Papert, S. (1996). An exploration in the space of mathematics educations. *International Journal of Computers for Mathematical Learning*, 95–123.

Papert, S. (2000). What's the big idea? Toward a pedagogy of idea power. *IBM Systems Journal, 29*(3 & 4), 720–729. doi:10.1147j.393.0720

Patton, M. Q. (2015). *Qualitative research & evaluation methods: Integrating theory and practice*. Los Angeles, CA: SAGE.

Piaget, J. (1954). *The construction of reality in the child.* New York, NY: Basic Books. doi:10.1037/11168-000

Resnick, M. (2018). *Computational Fluency – Mitchel Resnick – Medium.* Retrieved from https://medium.com/@mres/computational-fluency-776143c8d725

Resnick, M., & Robinson, K. (2018). *Lifelong kindergarten: Cultivating creativity through projects, passion, peers, and play*. Cambridge, MA: The MIT Press.

Ryan. (2018, July 31). Tinkering about thinking about Seymour [Blog post]. Retrieved from https://www.exploratorium.edu/tinkering/blog/2017/02/03/tinkering-about-thinking-about-seymour

Scaradozzi, D., Sorbi, L., Pedale, A., Valzano, M., & Vergine, C. (2015). Teaching robotics at the primary school: An innovative approach. *Procedia: Social and Behavioral Sciences*, *174*, 3838–3846. doi:10.1016/j.sbspro.2015.01.1122

Shute, V., Sun, C., & Asbell-Clarke, J. (2017). Demystifying computational thinking. *Educational Research Review*, *22*, 1–18. doi:10.1016/j.edurev.2017.09.003

Teich, A. G. (2019). *Repurposing school libraries as vibrant hubs and centers of change*. Retrieved September 21, 2019 from https://www.techlearning.com/resources/repurposing-school-libraries-as-vibrant-hubs-and-centers-of-change

Weintrop, D., Beheshti, E., Horn, M., Orton, K., Jona, K., Trouille, L., & Wilensky, U. (2016). Defining computational thinking for mathematics and science classrooms. *Journal of Science Education and Technology*, *25*(1), 127–147. doi:10.100710956-015-9581-5

Wing, J. M. (2006). Computational thinking. *Communications of the ACM*, *49*(3), 33–35. doi:10.1145/1118178.1118215

Wing, J. M. (2006). Computational thinking viewpoint. *Communications of the ACM*, *49*(3), 33. doi:10.1145/1118178.1118215

Wing, J. M. (2017). Computational thinking's influence on research and education for all. *Italian Journal of Educational Technology*, *25*(2), 7–14. doi:10.17471/2499-4324/922

With, A. C. (n.d.). *Mining the Computational Universe*. Retrieved from https://www.edge.org/conversation/stephen_wolfram-mining-the-computational-universe

Witherspoon, E. B., Hugashi, R., Schunn, C., Baehr, E., & Shoop, R. (2017). Developing computational thinking through a virtual robotics programming curriculum. *ACM Transactions on Computing Education*, *32*(1), 1–20. doi:10.1145/3104982

Chapter 5
eService-Learning:
Bridging Online Graduate Students' Sense of Belonging With Community Engagement

Mandi M. Laurie
The University of Texas at Tyler, USA

ABSTRACT

The service-learning pedagogy has flourished in popularity and student outcomes, and with the online student population expanding, institutions of higher learning must develop programs with similar content across both face-to-face and distance-learning applications. Institutions are using eService-Learning (eS-L) programs to address this divide. Commonly reported service-learning outcomes include student's sense of belonging and community engagement, and this chapter aims to present a case with similar outcomes for graduate-level students enrolled in a distance learning course with a substantive service-learning component.

INTRODUCTION

With the increased focus on student outcomes, the service-learning pedagogy is gaining traction nationwide. Online education is one of the fastest growth areas across higher education, leading institutions to create educational programs with content consistent across both face-to-face and distance-learning applications (Allen, Seaman, Poulin, & Straut, 2016; Berry, 2019). One method utilized by universities and educational institutions to address to divide is eService-Learning (eS-L). The service-learning pedagogy has expanded its reach to be inclusive of online and

DOI: 10.4018/978-1-7998-2914-0.ch005

distance education students; this form of service-learning is entitled eService-Learning or electronic service-learning (eS-L). A review of the literature indicates student's sense of belonging and community engagement are commonly reported service-learning outcomes. Therefore, the present chapter aims to present the progression of service-learning's entry into academia through modernized e-Service-Learning, and a case study utilizing similar outcomes in unique population: graduate-level students, enrolled in a distance-learning course, with a substantive service-learning component.

Service-learning has been implemented across the world to allow students the benefit of serving their communities and has been proven to have multiple positive experiences for students (Walker & Walker, 2018). Defined recently as an educational experience leading to academic course credit which allows students to participate in activities that benefit the community (Warren-Gordon & Graff, 2018), service-learning aids in student personal and academic growth. Research indicates service-learning has many benefits for student participants: aiding in the development of critical thinking skills; gaining a higher level of confidence; increased organizational engagement; increased institutional engagement; improved professional socialization; and enhanced professional skills (Garcia-Romero, Sanchez-Busques & Lalueza-Sazatornil, 2018; Lu & Duan, 2018; Pak, 2018; Rincon & Castillo-Montoya, 2018). Kwenani and Yu (2019) posit service-learning projects create a greater sense of belonging and increased international student engagement, which in turn help break down barriers in the international student's new prospective community (Warren-Gordon & Graff, 2018). Despite the plethora of benefits of social learning, many characteristics of service-learning projects must be met before students can experience those benefits.

BACKGROUND

Service-learning has its roots in Dewey's (1933) Program for Educational and Social Reform: experiential learning, reflection, and reciprocal learning (Champagne, 2006). Dewey posited (1933) experiential learning would meet societal needs while also encouraging students to connect these issues with classroom learning for a deeper understanding of the material. He believed traditional classroom environments failed to create this student engagement within their local communities. Dewey (1933) argued that in order to truly understand something, one must spend ample time reflecting upon the topic. The final piece of Dewey's trifecta was reciprocal learning: learning was multi-directional and involved students, teachers, and the organization in which acts of service occurred.

Sigmon (1975; 1979) built his three principles of Service-Learning upon Dewey's (1933) Program for Educational and Social Reform and Greenleaf's (1977) model of Servant Leadership. Working through Greenleaf's (1977) Servant Leadership tenets, Sigmon (1975; 1979) developed his principles of Service-Learning:

Table 1. Sigmon's (1975) Three Principles of Service-Learning

Principle One	Those being served control the service(s) provided
Principle Two	Those being served become better able to serve and be served by their own actions
Principle Three	Those who serve are also learners and have significant control over what is expected to be learned.

Sigmon (1977) posited that Greenleaf's definition of service required the party receiving the service define what needs to be done. The person providing the service should meet the service-receivers most exigent needs first. When providing service, it should help the individual or organization grow in such a manner that they themselves can become the servant. And finally, service should benefit those in society that have the least ability and power to serve themselves. According to Sigmon (1975; 1979), the end result of Dewey's (1933) Program for Educational and Social Reform and Greenleaf's (1977) Servant Leadership model was in *learning objectives*. These learning objections are the defining factor of the service-learning model, not the service itself. Thus, service activities are situational, and it is the context that provides the setting for learning to occur.

Service-Learning in Higher Education

Building upon Dewey (1977) and Sigmon (1979), Bringle and Hatcher (1995) developed the definition on service-learning most widely used today: "course-based, credit-bearing educational experience in which students (a) participate in an organized service activity that meets identified community needs and (b) reflect on the service activity in such a way as to gain further understanding of course content, a broader appreciation of the discipline, and an enhanced sense of civic responsibility." Notice that Bringle and Hatcher (1995) emphasized that these activities will earn institutional credits, this is in direct contrast to service activities one may engage in where there are no educational evaluations to certify that academic learning has occurred. Learning may very well happen during extracurricular service, but this is not evaluated (graded), academic learning. Perez (2000) expands this idea to say service-learning in higher education is the juxtaposition of experiential learning and community service.

Academic institutions have begun to embrace a civic minded climate through a "scholarship of engagement" (Boyer, 1990, p.15). The traditional academic model of higher education has begun to expand to include experiential education and service-learning to create stronger ties between students and their local communities (Batchelder & Root, 1994; Bringle & Hatcher, 2000; Eyler, Giles & Braxton, 1997; Hesser, 1995; McWhorter, Delello, & Roberts, 2016; Poon, Chan, & Zhou, 2011). Furthermore, there is an ever-expanding body of literature supporting the effectiveness of service-learning in the undergraduate population through student outcomes (Batchelder & Root, 1994; Boyer, 1990; Eyler & Giles, 1999; Hesser, 1995; Holland, 1997). Undergraduate student outcomes are frequently reported but there appears to be a marked lack of studies reporting on outcomes of the Graduate student. A meta-analysis of 62 service-learning studies, which included 11,837 students, reported only a small amount of literature included the graduate student population (Celio, Durlak, & Dymnicki, 2011). Another meta-analysis of service-learning included 65 studies, of which only eight included graduate students (Celio, Durlak, & Dymnicki, 2011). To address the gap in literature where this population is concerned, the case-study presented in the present chapter includes the outcomes of graduate students participating in service-learning activities.

Moreover, Boyer's (1990) scholarship of engagement creates a relationship between theory and practice which encourages learners to apply obtained knowledge in order to move between the realms of theory and practice. Scholarship of engagement is rooted in collaborative work environments, the same collaborative style required for service-learning to occur (Boyer, 1990; Boyer, 1997; Bringle & Hatcher, 2000; Holland, 1997). Linking the academic learning that occurs in the classroom setting to the exogenous environment forces students to develop much needed critical thinking skills (Keh-Wen & Kuan-Chou, 2013). Civic-minded educational opportunities allow student learners to get involved with their local communities, apply learned material to problem solving, and make meaningful connections to course materials (Bok, 2006; Bringle & Hatcher, 1995; 1996; Eyler & Giles, 1999; Godfrey, Illes, & Berry, 2005; Holland, 1997; Markus, Howard, & King, 1993; Sax & Astin, 1997).

In addition to the aforementioned advantages, students benefit from service-learning academically (Astin, Vogelgesang, Ikeda & Yee, 2000; Brandstad, 2015; Bringle & Hatcher, 2000; Strait, 2004). During an extensive multi-site, mixed-methods study, Astin, Vogelgesang, Ikeda and Yee (2000) found support for service-learning across academic outcome measures: higher GPA's, improved writing skills, and development of critical thinking skills. Brandstad (2015) further advances the idea of high-level academic outcomes through service-learning.

A compelling argument is made which suggests when linked to course objectives, students are better equipped for *knowledge application*: the ability to transfer knowledge across settings (Brandstad, 2015). This relates directly to Boyer's (1990) "Scholarship of Application" where new knowledge is applied in new ways. Other researchers (Dienhart, Maruyama, Snyder, Furco, McKay, Hirt & Huesman, 2016) suggested a linkage between service-learning and a greater understanding of course materials by student learners, though it could be argued this is a continuation of knowledge application skills. Moreover, students that participate in service-learning carry the effects with them into real-world scenarios. Astin, Vogelgesang, Ikeda and Yee (2000) proposed that service-learning not only increases student self-efficacy, but in a multi-site mixed-methods study, they found service-learning created a value shift among students. Those who participated in service-learning activities exhibited a new found commitment to community engagement, understanding of racial disparities, and cross-racial understanding. This body of literature supports service-learning' s effectiveness in the undergraduate population (Batchelder & Root, 1994; Boyer, 1990; Eyler & Giles, 1999; Hesser, 1995; Holland, 1997), but what about online students? How is service-learning beneficial to them?

Service-Learning Must Cross the Digital Divide

The National Student Clearing House (2019) reported that post-secondary enrollment is down for the eighth consecutive year, with the Spring 2019 semester showing a 1.7% decrease over the previous spring. There was one high performer in the realm of higher education, online-based, distance-learning courses (Allen, Seaman, Poulin, & Straut, 2016; Berry, 2019). To put it into perspective, in the Fall 2014 semester, 5.8 million college students enrolled in at least one online-based, 100% distance-learning course at US-based postsecondary institutions (Allen, Seaman, Poulin, & Straut, 2016). There was a 3.9% increase in the enrollment rate in online-based, 100% distance-learning courses from the number reported in 2014, the most recent report revealed that 28% of college students were enrolled in at least one online-based, 100% distance-learning course (Allen, Seaman, Poulin, & Straut, 2016).

In summary, more than one-fourth of all students enrolled in college coursework are taking at least one online course. Twenty years ago, online courses were in their infancy and critics argued that the inherent lack of critical discourse between students and the faculty in the distance-learning instructional methodology hindered the development of student critical thinking skills (Bullen, 1998; Burge, 1988; Garrison, 1993; Lauzon, 1992). Laurillard (1993) argued critical discourse is a requirement for reflection and deeper understanding on course content.

There are still many challenges in online instruction, but reflection and deeper understanding are not among them. The largest challenge is technology (Yoo &

Huang, 2013), as some instructional designers are not developing course content in a manner appropriate to online instruction (Devlin & McKay, 2016). Though, evidence suggests less than half (44%) of the faculty surveyed were not working with an instructional designer to revise or create their online courses for student-friendly courses (Lederman, 2018). Another challenge for online instruction is the apparent lack of critical discourse between student and faculty, which was warned by early critics of online instruction (Ilgaz & Gülbahar, 2015), although there are ways to mitigate the issue.

In addition, attrition is routinely higher in distance-learning environments than face-to face classes, with reports indicating a 20% discrepancy among the two groups (Angelino, Williams, & Natvig, 2007; Carr, 2000; Greenland & Moore, 2014; Moody, 2004; Stone et al., 2016). Terry (2001) posited that graduate-level business classes had an even worse fate: online graduate-level courses in "business statistics and finance have attrition rates 20 – 25% higher than face-to-face." (Angelino, Williams, & Natvig, 2007, p. 6). When faced with high attrition levels, the university may consider making changes to their program to ensure that students are willing and able to complete their coursework (Nash, 2005). This may include changes to delivery methodology, curriculum design, engagement activities, or student services (Carnevale, 2000; Nash, 2005). This is especially important in the distance education sphere where those attrition rates are the highest (Angelino, Williams, & Natvig, 2007; Carr, 2000; Moody, 2004; Terry, 2001).

Service-learning activities in face-to-face higher education courses reportedly lead to: higher grades; stronger critical thinking skills; increased writing skills; greater student community involvement; potential workplace experiences; active problem solving; reduced attrition, a greater understanding of course materials by student learners; greater knowledge application; increased student self-efficacy; greater meaningful connections to course materials; and even affects perception of ethical issues (Astin et al., 2000; Author, 2016; Brandstad, 2015; Dienhart et al., 2016; Keh-Wen & Kuan-Chou, 2013;). Thus, the benefits of service-learning (S-L) are numerous, but the question posed, can distant learning students benefit from incorporating this teaching strategy into online courses?

With purpose and intention, faculty can incorporate S-L activities into online course work, but unfortunately, online students are not routinely offered this opportunity (Bourelle, 2014; Guthrie & McCracken 2010; Salam, Iskandar, Ibrahim, & Farooq, 2019; Strait & Nordyke, 2015; Thurston 2005; Waldner, Widener, & McGorry, 2012). Referred to as Electronic Service-Learning (eService-Learning; e-SL) when offered in an online environment, like its face-to-face counterpart, eS-L can be used to develop relationships between higher education students, educational institutions, and community partners (Bourelle, 2014). Dailey-Hebert, Donelli-Sallee, and DiPadova (2008) reported the mistaken belief that S-L pedagogy was thought to be contradictory with an online learning environment which ultimately resulted in the underutilization of this dynamic teaching tool.

The use of S-L in face-to-face learning environments has grown exponentially over the years, and as the use of online postsecondary learning environments also rise, it can be expected that eService-Learning (eS-L) will also increase in popularity. In online courses, eS-L may be offered utilizing on-site or online service activities (Guthrie & McCracken, 2010; Strait & Sauer, 2004; Waldner et al., 2012). McWhorter, Delello, & Roberts (2016) asserted that the eS-L pedagogy creates a more personalized learning environment in which learners have a greater sense of connection to course material through the application of educational experience to real-world problems in a host organization.

eService-Learning: Student Engagement Bridges Sense of Belonging

Online students frequently report not feeling like part of their university system (Delahunty, Verenikina, & Jones, 2014; Masika & Jones, 2016; Thomas, Herbert, & Teras, 2014; Waldner, Widener, & McGorry, 2012), but how can the university system encourage a sense of community amongst distance students? Instructors are learning to utilize eS-L and experiential learning methods to increase study engagement (Angelino, Williams, & Natvig, 2007), but specifically how are they addressing this issue? Masika and Jones (2016) contend that are three modes of belonging: engagement, alignment, and imagination. Student belonging requires ongoing engagement activities which entail sense-making of learned materials that in turn lead to a sustained identity.

Alignment, as defined by Masika and Jones (2016) is the synchronization of student academic endeavors across learner group activities with learner output that culminates in a shared outcome. Finally, imagination is the outcome of the overall experience, such that the learner will develop a schema based upon the results of the individual sense-making that develops into a sustained identity (Masika & Jones, 2016). It is ultimately the individual student's participation, actually doing, those group-format, relational activities (the group-work, talking out ideas and constructs, utilizing course artifacts, and utilizing critical thinking skills) that lead to learner connections of material and the institutional community on a whole (Masika & Jones, 2016). According to Wenger (1998), it is the intertwined mechanisms that combine relationships between people, goings-on, and the external world which intersect and change in the peripheral over time that facilitate negotiation, learning, meaning-making and ultimately group identity. Furthermore, "because learning transforms who we are and what we can do, it is an experience of identity" (Wenger 1998, p. 215); therefore, learning becomes an integral part of identity development (Smith et al., 2004) and it is through the development of identity that a sense of belonging is forged (Masika & Jones, 2016). Since this phenomenon is still relevantly new in the academy, there is a need to compare results amongst graduate students' service-learning sense of belonging and institutional community engagement, which this chapter aims to do (Carpenter, 2011).

eService-Learning: Markers for Success

While eService-Learning can be applied across any academic field, there are caveats that should be followed when attempting to integrate academic learning and community engagement. Astin et al. (2000) suggested that the success of a Service-Learning/eService-Learning project lies within reflection. Students should be given the opportunity to process the service experience. Further, students should discuss the outcomes of their service experiences amongst each other, as well as relaying learning to the instructor. Early Service-Learning research referred to this activity as reflection (Dewey, 1933; Sigmon, 1979). Service-Learning/eService-Learning reflection may take on the guise of written response to the service experience in a journal or a more formal paper, classroom or forum post to encourage student discussions, or discussion with instructor. Reflection is quite literally the student learner reflecting upon the Service-Learning/eService-Learning experience in order to create connections between the content of the academic course and the resulting service-learning experience (Astin et al., 2000). It is the sensemaking that occurs as the student learner reflects inward upon what the Service-Learning/eService-Learning experience meant to them; what they learned and how the experience affected them.

According to Astin et al. (2000), the student learners' level of interest in the subject matter is the deciding factor into whether the service-learning experience will be a positive one for both the student learner and the organization receiving the service. This level of interest will determine whether the student determines the service experience to be a learning event. It is because of this that Astin et al. (2000) recommend that all service-learning course offerings be confined to the student learners' major field.

In order to encourage student learner success and adequate participation in the service-learning activity, students should be informed of the service-learning project well in advance, preferably on the first day of class (McLaughlin, 2010; Milton and Otty, 2018; Kwenani & Yu, 2019). Student learners should be informed of the importance of the service-learning project, including what the student should gain from the project, as this will increase buy-in on behalf of the student learner (Kwenani and Yu, 2019; McLaughlin, 2010; Milton & Otty, 2018). Students should be fully informed of all necessary aspects and expectations of the service-learning project, as the service-learning project will require significant preparation on the part of the student learner (Milton & Otty, 2018). Another important facet of student buy-in involves autonomy. It is important that the faculty member gives the student learner autonomous ownership over their service projects, whereby students are afforded the opportunity to self-select their service-learning organization (McLaughlin, 2010; Milton & Otty, 2018). By selecting their preferred organization, the student will be encouraged to find an organization in which they feel a sense of belonging, a sense of community, thereby increasing the likelihood that the student learner will engage at a higher level with the organization receiving the service (Milton & Otty, 2018).

To increase the impact of the learning events that occurred in the service-learning organization, the student learner should be expected to engage in classroom discussion whereby they recount their experiences to their peers and instructor (Culhane, Niewolny, Clark, & Misyak, 2018). Class discussion is one of the most significant markers of a positive service-learning experience among student learners, as it allows learners to relay their experiences as a means of increasing self-awareness (Astin et al., 2000; Milton & Otty, 2018; Wagner & Pigza, 2016). Recounting the student learners' experience allows the student the opportunity to assess whether their actions and behaviors, which occurred during the service-learning experience, were in agreement with the student's original reflection, perceptions and understanding of events (Wagner & Pigza, 2016).

By developing the practice of reflection, the student learner begins to develop the habit of reflecting upon the day's activities and attempting to create a schema whereby an overlay of learned course work coincides with the events that transpired (Milton & Otty, 2018). The ability of a student learner to turn an eye inward and self-reflect in addition to reflecting upon service-learning experiences for the sole purposes of sense-making and determining whether the results of their learning activity is congruent with the overall expressed impact of the service-learning event. Wagner and Pigza (2016) suggested that allowing students to engage in class discussion and write their reflections of their experiences is crucial for students to have a greater sense of impact from their service-learning project (Culhane et al., 2018; Mezirow, 1998). It should also be noted that students perceive a greater sense of learning when their experience is challenging (Ngai, Chan, & Kwan, 2018).

In addition to reflection activities, how often the instructor creates a connection between the service-learning experience and the academic subject matter will determine whether the academic subject matter behaves as a sort of enhancer to the service-learning experience on its own merit, or whether it is the service-learning experience that leads to understanding of the materials regarding the academic subject matter. There is a body of literature which indicates more positive outcomes when the faculty and/or school are also highly engaged with the service-learning project as well (Arellano & Jones, 2018). For example, when students feel that they have support from faculty and a partnership exists between the students and professor, students will have a more positive outcome with their project (Leedahl, Brasher, Estus, Breck, Dennis & Clark, 2019).

Purpose Statement

The purpose of the present study was to explore graduate student participants' perceptions of their experiences in an eService-Learning assignment in a nonprofit organization.

Research Questions

Three research questions guided this study:

- **Research Question 1:** What were the benefits of eService-Learning for the graduate students perceived personal significance and learning?
- **Research Question 2:** What were the benefits of eService-Learning for the nonprofit organization?
- **Research Question 3:** What were the benefits of eService-Learning for the university?

METHODOLOGY

A qualitative case study utilizing archival data collected over a two-year period was selected to best examine graduate students' experiences with service-learning projects (Merriam & Tisdell, 2016). The research team chose to examine the archival data as a case study because the data was gleaned from graduate students enrolled in an online Master's degree program in Human Resource Development. Graduate students enrolled in a graduate-level course entitled Leadership and Ethics in Business were instructed to complete 20 hours of volunteering at nonprofit organization. They were instructed to select a 501(c)3 nonprofit of their choice but warned that the organizations would be reviewed by the instructor before approval would be granted. Students who had difficulties finding an organization that fit the assignment parameters were offered an alternative assignment: transcribing data local historical records. The records were obtained from a historical society located in the community in which the university was located and could be used on an hour-per-hour basis until the entire 20 hours were completed. Data analyzed in the case-study were collected from a final written project report reflecting upon the entire eService-Learning experience submitted by the graduate students.

Prior to actually serving in a 501(c)3 nonprofit organization, the graduate students completed a "Pre-Work" Proposal packet for the nonprofit organization which included a thorough online investigation of the organization and documentation of a meeting (conducted either in-person or over the phone) with the volunteer coordinator at the nonprofit. The meeting was required to define the scope of work the students would be performing, draft a schedule of their volunteer hours and, complete a release form signed by both the graduate student and the volunteer coordinator. In addition, students were required to provide the volunteer coordinator verification from the faculty who approved them to perform the eS-L at the chosen organization prior to beginning their 20 required service-learning hours. Students were permitted to include pictures from the organization such as photos of brochures, students performing eS-L at the organization, or other graphics that the nonprofit allowed them to utilize. Students could also include photocopies of communication between the student and volunteer coordinator of the organization. If the student completed any volunteering hours via historical indexing, an additional log sheet was also included.

Data Analysis

A total of 61 service-learning reflection reports were collected from graduate students enrolled in the leadership and ethics in business course during the Fall 2016 and Fall 2017 semesters. The researchers concentrated their efforts analyzing the final service-learning written report submitted by the graduate students. The reflection reports routinely contained supplemental items submitted alongside the report to review for additional insight such as: student time log with reflective comments, the proposal document, pictures. Students were permitted to include photographs or images from the organization such as pictures of brochures, student performing eS-L at the organization, or other graphics that the nonprofit allowed them to utilize. Students also included photocopies of communication between the student and volunteer coordinator of the organization.

It was noted that one of the case-study researchers involved with data analysis was a graduate student at the time the data was collected and completed an eS-L project during both semesters. In order to prevent bias, her data was not included in the analysis, rather it was used to offer an "insider perspective" (Dwyer & Buckle, 2009) since she had personally participated in eS-L experiences when she was a graduate student.

The three researchers independently examined an equal number of documents from each of the two semesters where data was gathered and they established intercoder agreement (Creswell, 2015). The research team followed Creswell's (2015) guide on analysis of raw data form codes specifically, reading the documents to have a general summary of their information and adding a code label to sections of the documents (Creswell, 2015). Then, the researchers combined redundant codes to create themes. They conducted analysis until data saturation was reached and then met together to compare the 11 themes. At that time, they were able to collapse the 11 themes into 7 broader themes.

A conceptual map was created to visually compare the findings. It is noted that primarily, the student's final report of the account was analyzed. However, if information was missing from the final report, other documents were also examined. For example, missing in one final report was the reason why a student chose the eService-Learning organization. This information was found within the Pre-Work proposal packet that supplemented the data in the report.

Findings

Several findings from the case study were noted. The most prominent was community engagement through service-learning as it was a consistent theme amongst the themes. However, it is noted that there were differences in student perceived outcomes when they volunteered for 20 hours at a non-profit organization as compared to those students who had a split-assignment that involved 10 hours on-site at a nonprofit organization and 10 hours working from home on archiving historical documents. It appeared that those who were immersed in one assignment on-site at a nonprofit, appeared overall to take more ownership in their volunteer duties and were more inclined to continuing to volunteer even after their service-learning assignment concluded. In addition, their reflective reports indicated that the students with the single assignment on-site also had the opportunity to gain greater depth into the inner workings of their chosen non-profit. They were also found to have more accessibility to non-profit staff and executive leadership so the longevity of their assignment appeared to contribute to their community engagement and learning. Likewise, increased time in one setting increases the credibility of the study.

Graduate students involved with only one nonprofit were found to have more suggestions and insight on future changes which may benefit their chosen non-profit organization having spent more time on-site (prolonged engagement) and seeing the workings of the organization than were students with a split-assignment. They were also found to be more invested in selecting the non-profit organization and related the selection to their personal life on a more substantive level, as compared to the students who had a split-assignment. Lincoln and Guba (1985) noted that prolonged engagement, persistent observation, and triangulation were three valuable techniques for establishing credibility in a qualitative study to have confidence in the findings.

The majority of graduate students with a service-learning assignment reported being the first student to volunteer at that organization from the [University]. This benefits the educational institution by providing evidence of brand development

Table 2. Themes, Working Definitions, and Examples of Quotations in Case

Themes	Working Definitions	Examples of Quotations in Case
Non-profit's awareness of the University	The organization's knowledge/awareness of university	"I was happy to serve as the first volunteer for the [university] and I believe I served as a great example for the university and my classmates"-(SL005F16, p. 4). "This organization has never had volunteers from [University] prior to myself"-(SL007F16, p.4)
Connection to personal life	Why volunteer chose the organization	"I come from nothing and was raised by my brothers in a small shack in south Texas. I wore hand-me-downs, showered in a small room with a water hose that we fit through the crack in the door, and would sometimes visit the city dump to look for clothes that might fit a young girl like myself." (SL010F16, p. 4). "I am not typically involved in the food distribution aspect of the Food Bank of Central despite my employment there. I am not involved because I manage another program" (SL011F16, p. 3).
Significance (as an outcome)	Includes personal significance, organizational significance and appreciation of the non-profit	"This service-learning assignment helped me realize how privileged we are as individuals, lifelong learners, and contributors to society. I reflected on how often certain individuals are forgotten in society and how often individuals are judged through no fault of their own" (SL011F16, p. 10). "This experience has helped me to see that the community needs more safe havens for women and that my purpose is to build and open one of those places" (SL005F16, p. 6).
Insight/change for the future	Any suggestions that the student thought would be beneficial for the organization	"I realized I must continue to be a positive role model and lead by example, "If leaders do not actively serve as role models for the organization's core values, then those values become nothing more than lip service" (SL009F16, p. 5). "This is by far, one of the best things I've had to do in 2016. It's taught me compassion and understanding, and I hope I can take that with me as I develop my career in Human Resources. Also, I plan on volunteering long after this course has ended, and have already received permission from my employer and the wonderful staff at Dress for Success" (SL010F16, p. 5).
Leadership	Leadership characteristics of the leaders of the organizations	Staff indicted non-profits philosophy, "...that's it's important to have many people from all walks of life because clients want to be able to relate to someone when they walk into the building" (SL010F16, p. 7).
Volunteer Duties	A list of duties given to the student/volunteer	"Assisted Ms. Tiffany with set up for the dates of classes, transportation, helped makeup artist, photographers, bring clothes, host and accepting tickets" (SL007F16, p. 4). "I relieved the House Manager, Annette, from duty so that she could attend her weekly prayer and bible study meeting that was about two and a half hours long" (SL005F16, p. 5).
Volunteers' Perspective	Inferences made by the student possible based on their own experiences/ world view	"Somehow my work and my health had taken away an activity and a community that was and is vital to my sense of identity, especially when working in HR can be so draining, I think it is essential that I maintain this connection not that I have found it again" (SL006F16, p. 4). "Being there with them brought me back to a place in my life where hope was all I had. It was such a scary place and as I was leaving home I began reflecting on how I could minister to these women" (SL005F16, p. 5).

by creating *brand awareness,* or the notion that consumers are familiar with the brand (*i.e.*, name of the university) to aid recruitment and funding; *differentiating themselves from their competitors* (i.e. having students providing service-learning that enhances their reputation that others in their market may not be doing); *student engagement*, one that provides rich individualized educational experiences in unique and compelling ways [University] Strategic Plan (2018), and *community engagement* by establishing collaborations with community partners and nonprofit organizations to address regional needs for the institution while improving the communities where the students live.

Students found being on-site for their e-SL experience gave them motivation to talk with the executive leadership in the organization where they were employed and some were even able to get paid release from their traditional duties to perform volunteer hours during work hours while still on their employer's payroll.

Graduate students in a leadership and ethics course were able to spend more time with the non-profit's staff and executive leadership, which resulted in either growth in the services that the organization offered or additional funding for the mission of the organization. These students also found creative ways to infuse their technical and soft skills to help the organization grow their services offered to the community. In one case, a student volunteered to teach Caucasian foster parents how to manage their African American foster children's hair. Another student built stage props for a community theatre with a $0 budget. The eService-Learning students in this study were able to accomplish a number of their [university] goals through e-SL by performing nonpaid work tasks at nonprofit organizations such as: religious organizations, foster care and child adoption organizations, family services and women's empowerment organizations, workforce rehabilitation and job training organizations, and development and fundraising organizations, among many others. A number of the students articulated a desire to continue their volunteer work following the conclusion of the service-learning assignment.

DISCUSSION

The findings in this study revealed prolonged engagement at one setting for eService-Learning students had a number of benefits as already noted. This has implications for faculty members who want to increase the engagement of online students to connect them to their community that can lead to powerful experiential learning through eService-Learning. It appears to be meaningful (Chalofsky, 2010) and possibly even transformational (Mezirow, 1997), that duration and sustained efforts are success factors.

Waldner, Widener and McGorry (2012) noted that they experienced a number of colleagues who were abandoning service-learning because courses were being moved from face-to-face instructional arrangements to online learning and learning management systems (LMSs) where collaboration would primarily be through asynchronous discussion boards in small groups or whole class. As they pointed out, "online learning is a facilitator rather than a barrier to service-learning....[and] holds potential to transform both service-learning and online learning by freeing service-learning from geographical constraints, and by equipping online learning with a tool to promote engagement" (p. 123).

For many online students, there is little or no interaction with peers and faculty members in their distance education. Many even live in other cities, states, or countries, far from their university. Angelino, Williams, and Natvig (2007) noted that increasing engagement in online courses often focuses on improving reading or other instructional material to make it more exciting or interactive such as adding media (*i.e.*, YouTube videos or outside resources), or improving students' interaction with course content, their peers in the course, and the instructor.

In contrast, eService-Learning provides an entirely new dimension for online students: interacting with workers and leaders of a nonprofit and serving in their community. It takes them away from their computer, and into their community to foster life-long partnerships that can create meaning through experiential learning. Students are afforded a sense of satisfaction that they are improving the conditions of people and resources where they live. Furthermore, students are connecting with content in new way which creates meaningful coursework.

Online students found personal significance in their choice of organization (self-directed learning (Merriam & Bierema, 2014) with some students having a direct experience (*i.e.*, adopting a child) while others had an indirect experience (i.e. having friends who have adopted a child). Students noted they appreciated their chosen non-profit's presence in their community. A majority of the students shared their personal journey in their reasoning for choosing their perspective non-profit. Some students also had personal needs and shared that they used the services that they were assigned for their volunteering duties, bringing the students' learning full circle as both a facilitator of services and a receiver of those services as they were also in need. The online students shared their humble beginnings, gratefulness for their current life status and understanding of their "voice" in making their communities a better place through volunteering and making a difference.

FUTURE RESEARCH DIRECTIONS

The present study was conducted using archival data collected over a two-semester time period. One area of future research would be a longitudinal data collection: upon graduation or separation from the university student participants would be asked a series of survey questions regarding their time spent in their eService-Learning projects. Of particular interest would be whether the students continued their efforts of community outreach in any form.

Future research should also include exit data from the nonprofit leaders to gain their perspective on eS-L students volunteering at their organization. It is likely that this project could be improved by gaining new insights from the volunteer coordinators and the leadership at the nonprofits where students served. Another study could encompass examining the "brand recognition" in the areas where two or more students served in nonprofits. It is possible over time, that utilizing eService-Learning could actually spread the good will of the university to other untapped areas for future students or strategic partners.

Implications of Research

The present study is an important first step towards understanding the temporal effect of service-learning of learners enrolled in online courses. Through service-learning, students are taught to enter their communities and make a change. Therefore, it is important that organizations incorporate the best service-learning methods. In the present study, personally significant learning locations were gathered from locations approved by the faculty member who designed the project. Additionally, the length of time associated with a service-learning project and subsequent student involvement or personally perceived outcomes were assessed. Organizations should review the findings of the previous research when developing their own service-learning opportunities. The present study offers a beginning reference on the time needed for positive learner perceived outcomes, as well as the requirement of learner selected learning locations.

CONCLUSION

This study demonstrates that eService-Learning for online students can provide a meaningful project to engage online graduate students in their community, experience course objectives (such as seeing organizational processes in action) and, engage in meaningful volunteerism with a purpose. It takes graduate students away from their computer and into their community to foster partnerships that can create personal satisfaction for each student that are improving conditions of people and resources where they live and connecting with learning content in an entirely different way through meaningful coursework.

ACKNOWLEDGMENT

I would like to acknowledge Ashley Wilson and Natasha Ross for their help with early drafts of the reported case study. I would also like to acknowledge Dr. Rochell McWhorter for her data collection, assistance with data analysis, and writing on an earlier draft of this chapter.

REFERENCES

Allen, I. E., Seaman, J., Poulin, R., & Straut, T. T. (2016). *Online report card: Tracking line education in the United States*. Babson Park, MA: Babson Survey Research Group and Quahog Research Group LLC.

Anderson, T. (2004). Toward a theory of online learning. In T. Anderson & F. Elloumi (Eds.), *Theory and Practice of Online Learning* (pp. 33–60). Athabasca University.

Angelino, L. M., Williams, F. K., & Natvig, D. (2007). Strategies to engage online students and reduce attrition rates. *Journal of Educators Online*, *4*(2). doi:10.9743/JEO.2007.2.1

Arellano, I., & Jones, S. J. (2018). Exploration of University Faculty Perceptions and Experiences of Service-Learning as Engaged Scholarship of Teaching and Learning. *The Journal of Scholarship of Teaching and Learning*, *18*(4). doi:10.14434/josotl.v18i4.23178

Astin, A. W., Vogelgesang, L. J., Ikeda, E. K., & Yee, J. A. (2000). *How service learning affects students*. Academic Press.

Batchelder, T. H., & Root, S. (1994). Effects of an undergraduate program to integrate academic learning and service: Cognitive, prosocial cognitive, and identity outcomes. *Journal of Adolescence*, *17*(4), 341–355. doi:10.1006/jado.1994.1031

Berry, S. (2019). Professional development for online faculty: Instructors' perspectives on cultivating technical, pedagogical and content knowledge in a distance program. *Journal of Computing in Higher Education*, *31*(1), 121–136. doi:10.100712528-018-9194-0

Bok, D. (2006). *Our underachieving colleges: A candid look at how much students learn and why they should be learning more*. Princeton University Press.

Bourelle, T. (2014). Adapting service-learning into the online technical communication classroom: A framework and model. *Technical Communication Quarterly, 23*(4), 247–264. doi:10.1080/10572252.2014.941782

Boyer, E. (1996). The scholarship of engagement. *Journal of Public Service and Outreach, 1*(1), 11–20. Retrieved from http://openjournals.libs.uga.edu/index.php/jheoe/article/view/253/238

Boyer, E. L. (1990). Scholarship reconsidered: Priorities ofthe professoriate. Princeton, NJ: The Carnegie Foundation for the Advancement of Teaching.

Boyer, E. L. (1997). The scholarship of engagement. *Journal of Public Service and Outreach, 1*(1), 11-20.

Brandstad, J. (2015). *A service-learning and transfer-oriented approach to teaching developmental reading and writing students*. Retrieved from https://bwe.ccny.cuny.edu/Brandstad_Service_Learning.htm

Bringle, R. G., & Hatcher, J. A. (1996). Implementing service-learning in higher education. *The Journal of Higher Education, 67*(2), 221–239. doi:10.2307/2943981

Bullen, M. (1998). Participation and critical thinking in online university distance education. *Journal of Distance Education, 13*(2), 1–32.

Burge, L. (1988). Beyond andragogy: Some explorations for distance learning design. *International Journal of E-Learning & Distance Education/Revue internationale du e-learning et la formation à distance, 3*(1), 5-23.

Carnevale, D. (2000). 'Boot camp' helps new online students at Boise State U. *The Chronicle of Higher Education, 46*(24), A48.

Carpenter, H. (2011). How we could measure community impact of nonprofit graduate students' service-learning projects: Lessons from the literature. *Journal of Public Affairs Education, 17*(1), 115–132. doi:10.1080/15236803.2011.12001630

Carr, S. (2000). As distance education comes of age, the challenge is keeping the students. *The Chronicle of Higher Education, 46*(23), A39–A41.

Celio, C. I., Durlak, J., & Dymnicki, A. (2011). A meta-analysis of the impact of service-learning on students. *Journal of Experiential Education, 34*(2), 164–181. doi:10.1177/105382591103400205

Chalofsky, N. E. (2010). *Meaningful workplaces: Reframing how and where we work*. John Wiley & Sons.

Champagne, N. (2006). Service-learning: Its origin, evolution, and connection to health education. *American Journal of Health Education*, *37*(2), 97–102. doi:10.1080/19325037.2006.10598885

Creswell, J. W. (2015). *30 essential skills for the qualitative researcher*. Sage Publications.

Culhane, J., Niewolny, K., Clark, S., & Misyak, S. (2018). Exploring the Intersections of Interdisciplinary Teaching, Experiential Learning, and Community Engagement: A Case Study of Service Learning in Practice. *International Journal on Teaching and Learning in Higher Education*, *30*(3), 412–422.

Dailey-Hebert, A., Donelli-Salee, E., & DiPadova, L. N. (2008). *Service E-Learning: Educating for citizenship*. Charlotte, NC: Information Age Publishing.

Delahunty, J., Verenikina, I., & Jones, P. (2014). Socio-emotional connections: Identity, belonging and learning in online interactions. A literature review. *Technology, Pedagogy and Education*, *23*(2), 243–265. doi:10.1080/1475939X.2013.813405

Devlin, M., & McKay, J. (2016). Teaching students using technology: Facilitating success for students from low socioeconomic status backgrounds in Australian universities. *Australasian Journal of Educational Technology*, *32*(1). doi:10.14742/ajet.2053

Devlin, M., & McKay, J. (2018). Teaching inclusively online in a massified university system. *Widening Participation and Lifelong Learning: the Journal of the Institute for Access Studies and the European Access Network*, *20*(1), 146–166. doi:10.5456/WPLL.20.1.146

Dewey, J. (1933). *How we think: A restatement of the relation of reflective thinking in the educative process*. Chicago: Henry Regnery.

Diaz, D. P., & Bontenbal, K. F. (2001). Learner preferences: Developing a learner-centered environment in the online or mediated classroom. *Ed at a Distance, 15*(8). Retrieved April 11, 2006, from http://www.usdla.org/html/journal/AUG01_Issue/article03.html

Dienhart, C., Maruyama, G., Snyder, M., Furco, A., McKay, M. S., Hirt, L., & Huesman, R. (2016). The impacts of mandatory service on students in service-learning classes. *The Journal of Social Psychology*, *156*(3), 305–309. doi:10.1080/00224545.2015.1111856 PMID:27064181

Dwyer, S. C., & Buckle, J. L. (2009). The space between: On being an insider-outsider in qualitative research. *International Journal of Qualitative Methods*, *8*(1), 54–63. doi:10.1177/160940690900800105

Eyler, J., Giles, D. E. Jr, & Braxton, J. (1997). The impact of service-learning on college students. *Michigan Journal of Community Service Learning*, *4*, 5–15.

García-Romero, D., Sànchez-Busqués, S., & Lalueza-Sazatornil, J. L. (2018). *Exploring the Value of Service Learning: Students' assessments of Personal.* Procedural and Content Learning.

Garrison, D. R. (1993). A cognitive constructivist view of distance education: An analysis of teaching-learning assumptions. *Distance Education*, *14*(2), 199–211. doi:10.1080/0158791930140204

Godfrey, P. C., Illes, L. M., & Berry, G. R. (2005). Creating breadth in business education through service-learning. *Academy of Management Learning & Education*, *4*(3), 309–323. doi:10.5465/amle.2005.18122420

Greenland, S., & Moore, C. (2014). Patterns of online student enrolment and attrition in Australian open access online education: A preliminary case study. *Open Praxis*, *6*(1), 45–54. doi:10.5944/openpraxis.6.1.95

Greenleaf, R. K. (1977). *Servant Leadership: A Journey into the Nature of Legitimate Power and Greatness.* New York, NY: Paulist Press.

Guthrie, K. L., & McCracken, H. (2010). Teaching and learning social justice through online service-learning courses. *The International Review of Research in Open and Distributed Learning*, *11*(3), 78–94. doi:10.19173/irrodl.v11i3.894

Harasim, L. (1999). A framework for online learning: The Virtual-U. *Computer*, *32*(9), 44–49. doi:10.1109/2.789750

Hesser, G. (1995). Faculty Assessment of Student Learning: Outcomes Attributed to Service-Learning and Evidence of Changes in Faculty Attitudes About Experiential Education. *Michigan Journal of Community Service Learning*, *2*, 33–42.

Holland, B. (1997). Analyzing institutional commitment to service: A model of key organizational factors. *Michigan Journal of Community Service Learning*, *4*(1), 30–41.

Hughes, J. A. (2004). Supporting the online learner. In T. Anderson & F. Elloumi (Eds.), *Theory and Practice of Online Learning* (pp. 367–384). Athabasca University.

Hui, W., Hu, P. H., Clark, T. H., Tam, K. Y., & Milton, J. (2008). Technology-assisted learning: A longitudinal field study of knowledge category, learning effectiveness and satisfaction in language learning. *Journal of Computer Assisted Learning, 24*(3), 245–259. doi:10.1111/j.1365-2729.2007.00257.x

Ilgaz, H., & Gülbahar, Y. (2015). A snapshot of online learners: e-Readiness, e-Satisfaction and expectations. *The International Review of Research in Open and Distributed Learning, 16*(2). doi:10.19173/irrodl.v16i2.2117

Johnson, N. (2012). *The Institutional Costs of Student Attrition*. Delta Cost Project at American Institutes for Research. Retrieved from https://deltacostproject.org/sites/default/files/products/Delta-Cost-Attrition-Research-Paper.pdf

Keh-Wen, C., & Kuan-Chou, C. (2013). Designing service-learning: Project in systems analysis and design course. *Academy of Educational Leadership Journal, 17*(2), 47–60.

Kwenani, D. F., & Yu, X. (2019). Maximizing international students' service-learning and community engagement experience: A case study of student voices on the benefits and barriers. *Journal of Higher Education Outreach & Engagement, 22*, 29–52. Retrieved from http://openjournals.libs.uga.edu/index.php/jheoe/article/view/2153

Laurillard, D. (1993). *Rethinking University Teaching: Rethinking University Teaching: a Framework for the Effective Use of Educational Technology*. Routledge.

Lauzon, A. C. (1992). Integrating computer-based instruction with computer conferencing: An evaluation of a model for designing online education. *American Journal of Distance Education, 6*(2), 32–46. doi:10.1080/08923649209526784

Lederman, D. (2018). Who is studying online (and where). *Inside Higher Ed*. Retrieved from https://www.insidehighered.com/digital-learning/article/2018/01/05/new-us-data-show-continued-growth-college-students-studying

Leedahl, S. N., Brasher, M. S., Estus, E., Breck, B. M., Dennis, C. B., & Clark, S. C. (2019). Implementing an interdisciplinary intergenerational program using the Cyber Seniors® reverse mentoring model within higher education. *Gerontology & Geriatrics Education, 40*(1), 71–89. doi:10.1080/02701960.2018.1428574 PMID:29364782

Lincoln, Y. S., & Guba, E. G. (1989). Ethics: The failure of positivist science. *The Review of Higher Education, 12*(3), 221–240. doi:10.1353/rhe.1989.0017

Lu, D., & Duan, J. (2018, July). Systematic Thinking on University Service-learning. In *2018 3rd International Conference on Education, Sports, Arts and Management Engineering (ICESAME 2018)*. Atlantis Press. 10.2991/icesame-18.2018.62

Markus, G. B., Howard, J. P. F., & King, D. C. (1993). Integrating community service and classroom instruction enhances learning: Results from an experiment. *Educational Evaluation and Policy Analysis*, *15*, 410–419.

Masika, R., & Jones, J. (2016). Building student belonging and engagement: Insights into higher education students' experiences of participating and learning together. *Teaching in Higher Education*, *21*(2), 138–150. doi:10.1080/13562517.2015.1122585

McLaughlin, E. (2010). The" real-world" experience: Students' perspectives on service-learning projects. *American Journal of Business Education*, *3*(7), 109–118. doi:10.19030/ajbe.v3i7.463

McWhorter, R., Delello, J. A., & Roberts, P. B. (2016). *Giving back: Exploring service-learning in an online learning environment*. Academic Press.

Merriam, S. B., & Bierema, L. L. (2013). *Adult learning: Linking theory and practice*. San Francisco, CA: Jossey-Bass Higher and Adult Education.

Merriam, S. B., & Tisdell, E. J. (2016). Designing your study and selecting a sample. *Qualitative research: A guide to design and implementation*, 73-104.

Mezirow, J. (1981). A critical theory of adult learning and education. *Adult Education*, *32*(1), 3–24. doi:10.1177/074171368103200101

Milton, L. E., & Otty, R. (2018). Innovations in occupational therapy education: The centralized service-learning model. *Journal of Occupational Therapy Education*, *2*(1), 8. doi:10.26681/jote.2018.020108

Minich, E. (1996). Using student feedback to improve distance education. Florida Community College, Jacksonville, FL. (ERIC No. ED 397893)

Moody, J. (2004). Distance education: Why are the attrition rates so high? *The Quarterly Review of Distance Education*, *5*(3), 205–210.

Nash, R. (2005, Winter). Course completion rates among distance learners: Identifying possible methods to improve retention. *Online Journal of Distance Learning Administration*, *8*(4). Retrieved from http://www.westga.edu/%7Edistance/ojdla/winter84

Ngai, G., Chan, S. C., & Kwan, K. P. (2018). Challenge, Meaning, Interest, and Preparation: Critical Success Factors Influencing Student Learning Outcomes from Service- Learning. *Journal of Higher Education Outreach & Engagement*, *22*(4), 55–80.

Norman, D. A., & Spohrer, J. C. (1996). Learner-centered education. *Communications of the ACM*, *39*(4), 24–27. doi:10.1145/227210.227215

Pak, C. S. (2018). Linking service-learning with sense of belonging: A culturally relevant pedagogy for heritage students of Spanish. *Journal of Hispanic Higher Education*, *17*(1), 76–95. doi:10.1177/1538192716630028

Pan, M. L., Lawrence, L., Creswell, J. W., Hacker, D., Merriam, S., Cone, J. D., & Creswell, J. W. (2007). *Preparing literature reviews*. Los Angeles, CA: Pyrczak Publishing.

Perez, S. (2000). *Assessing service learning using pragmatic principles of education: A Texas charter school case study*. Academic Press.

Poon, P., Chan, T. S., & Zhou, L. (2011). Implementation of service-learning in business education: Issues and challenges. *Journal of Teaching in International Business*, *22*(3), 185–192. doi:10.1080/08975930.2011.653746

Rincón, B., & Castillo-Montoya, M. (2018). Learning Assessment in Student Affairs through Service-Learning. *Research & Practice in Assessment*, *13*, 39–50.

Rovai, A. (2002). Building sense of community at a distance. *International Review of Research in Open and Distance Learning*, *4*(1), 1–9.

Salam, M., Iskandar, D. N. A., Ibrahim, D. H. A., & Farooq, M. S. (2019). Service learning in higher education: A systematic literature review. *Asia Pacific Education Review*, 1–21.

Sax, L. J., & Astin, A. W. (1997). The benefits of service: Evidence from undergraduates. *The Educational Record*, *78*(3/4), 25–32.

Schneider, M. (2010). *Finishing the first lap: The cost of first year student attrition in America's four year colleges and universities*. American Institutes for Research. Retrieved from https://www.air.org/sites/default/files/AIR_Schneider_Finishing_the_First_Lap_Oct101.pdf

Sigmon, R. L. (1975). *Service-learning: A new path for the transition from adolescence to adulthood*. Western Carolina University Journal of Education.

Sigmon, R. L. (1979). Service-learning: Three principles. *Synergist*, *8*(1), 9–11.

Smith, B. L., MacGregor, J., Mathews, R., & Gabelnick, F. (2004). *Learning Communities: Reforming Undergraduate Education*. Jossey-Bass.

Stone, C., O'Shea, S., May, J., Delahunty, J., & Partington, Z. (2016). Opportunity through online learning: Experiences of first-in-family students in online open-entry higher education. Academic Press.

Strait, J., & Sauer, T. (2004). Constructing experiential learning for online courses: The birth of e-service. *EDUCAUSE Quarterly*, *1*, 62–65.

Terry, N. (2001). Assessing enrollment and attrition rates for the online MBA. *T.H.E. Journal*, *28*(7), 64–68.

Thomas, L., Herbert, J., & Teras, M. (2014). A sense of belonging to enhance participation, success and retention in online programs. *The International Journal of the First Year in Higher Education*, *5*(2), 69–80. doi:10.5204/intjfyhe.v5i2.233

Thurston, A. (2005). Building online learning communities. *Technology, Pedagogy and Education, 14*(3), 353-369.

Tinto, V. (1975). Dropout from higher education: A theoretical synthesis of recent research. *Review of Educational Research*, *45*(1), 89–125. doi:10.3102/00346543045001089

Towles, D. E., Ellis, J. R., & Spencer, J. (1993, May). *Student persistence in a distance education program: The effect of faculty-initiated contact.* Paper at the Annual Forum of the Association for Institutional Research, Chicago, IL. (ERIC No. ED 360931)

Volery, T., & Lord, D. (2000). Critical success factors in online education. *International Journal of Educational Management*, *14*(5), 216–223. doi:10.1108/09513540010344731

Vygotsky, L. S. (1978). *Mind in society, the development of higher psychological processes*. Cambridge, MA: Harvard University Press.

Wagner, E. D., & McCombs, B. L. (1995). Learner centered psychological principles in practice: Designs for distance education. *Educational Technology*, *35*(2), 32–35.

Wagner, W., & Pigza, J. M. (2016). The intersectionality of leadership and service-learning: A 21st-century perspective. *New Directions for Student Leadership*, *2016*(150), 11–22. doi:10.1002/yd.20167 PMID:27150901

Waldner, L. S., Widener, M. C., & McGorry, S. Y. (2012). E-service learning: The evolution of service-learning to engage a growing online student population. *Journal of Higher Education Outreach & Engagement*, *16*(2), 123–150.

Walker, W., & Walker, C. (2018). Do Course-Based Service Learning Programs Offer More Benefits to Students than Non-Course-Based Programs?" The Research Says' Yes'. *Journal of Education and Learning*, *7*(6), 197–202. doi:10.5539/jel.v7n6p197

Warren-Gordon, K., & Graff, C. S. (2018). Critical Service-Learning as a Vehicle for Change in Higher Education Courses. Change. *The Magazine of Higher Learning*, *50*(6), 20–23. doi:10.1080/00091383.2018.1540817

Wehlage, G. G., Rutter, R. A., & Smith, G. A. (1989). *Reducing the risk: Schools as communities of support*. New York: The Falmer Press.

Wenger, E. (1998). Communities of practice: Learning as a social system. *The Systems Thinker*, *9*(5), 2–3.

Wojciechowski, A., & Palmer, L. B. (2005). Individual student characteristics: Can any be predictors of success in online classes? *Online Journal of Distance Learning Administration*, *8*(2). Retrieved from http://www.westga.edu/~distance/ojdla/summer82/wojciechowski82.htm

Yoo, S. J., & Huang, W. D. (2013). Engaging online adult learners in higher education: Motivational factors impacted by gender, age, and prior experiences. *The Journal of Continuing Higher Education*, *61*(3), 151–164. doi:10.1080/07377363.2013.836823

Zaharias, P. (2002). On-line Learning Courses: A review and usability attributes. In M. Driscoll & T. Reeves (Eds.), *Proceedings of E-Learn 2002--World Conference on E-Learning in Corporate, Government, Healthcare, and Higher Education* (pp. 1056-1062). Montreal, Canada: Association for the Advancement of Computing in Education (AACE). Retrieved from https://www.learntechlib.org/primary/p/15341/

ADDITIONAL READING

Dolgon, C., Mitchell, T. D., & Eatman, T. K. (Eds.). (2017). *The Cambridge handbook of service learning and community engagement*. Cambridge University Press. doi:10.1017/9781316650011

Furco, A. (2015). *EService-learning: Creating experiential learning and civic engagement through online and hybrid courses*. Stylus Publishing, LLC.

Herrera, E. (2017). The Impact of Service-Learning on Undergraduate and Graduate Students.

Hill, C. (2012). Four keys to successful service learning in online courses. *Faculty Focus: Distance Education Report, 14*(24), 1–5.

Purcell, J. W. (2017). Community-Engaged Pedagogy in the Virtual Classroom: Integrating eService-Learning Into Online Leadership Education. *The Journal of Leadership Studies, 11*(1), 65–70. doi:10.1002/jls.21515

Shumer, R. (2015). Invited retrospective: The future of service-learning and community engagement research is mixed. In V. Jagla, A. Furco, & J. Strait (Eds.), *Servicelearning pedagogy: How does it measure up?* (pp. 365–382). Charlotte, NC: Information Age Publishing, Inc.

Stefaniak, J. E. (2015). The implementation of service-learning in graduate instructional design coursework. *Journal of Computing in Higher Education, 27*(1), 2–9. doi:10.100712528-015-9092-7

Warren, J. L. (2012). Does service-learning increase student learning? A meta-analysis. *Michigan Journal of Community Service Learning, 18*(2), 56–61.

You, Z., & Rud, A. G. (2010). A model of Dewey's moral imagination for service-learning: Theoretical explorations and implications for practice in higher education. *Education and Culture, 26*(2), 36–51.

KEY TERMS AND DEFINITIONS

Alignment: The synchronization of student academic endeavors across learner group activities with learner output that culminates in a shared outcome.

Engagement: Student belonging requires ongoing engagement activities which entail sense making of learned materials that in turn lead to a sustained identity.

eService Learning: An educational experience for online or distance students which leads to academic course credit, allowing students to participate in activities that benefit the community.

Experiential Learning: The process of learning through experience, with learning occurring after reflecting upon learned materials.

Institutional Community Engagement: Universities offer their resources (in the case of eService-Learning, students and their learned skills) to address issues and problems in the community.

Knowledge Application: The ability to transfer knowledge across settings, to apply learned material to real-life applications.

Reciprocal Learning: The process in which the teacher becomes the learner and the learner becomes the teacher.

Reflection: The student learner reflecting upon the Service-Learning/eService-Learning experience to build connections between academic content and the eService-Learning experience.

Scholarship of Engagement: Scholarship of Engagement creates a relationship between theory and practice which encourages learners to apply knowledge in order to move between the realms of theory and practice.

Service Learning: An educational experience leading to academic course credit which allows students to participate in activities that benefit the community.

Chapter 6
Digital Learning Management Systems Case Study:
Instructors' Perspective

Abdeleh Bassam Al Amoush
University of New England, Australia

Kamaljeet Sandhu
https://orcid.org/0000-0003-4624-6834
University of New England, Australia

ABSTRACT

Digital learning management systems (DLMSs) are an important and necessary educational tool for learning activities. Universities are using the Digital LMS tools that are unique to their environment. Different universities have different DLMS tools that are software driven and allow the users (management, instructors, and students) to use them for regular tasks. This research discusses the qualitative data collected using a case study methodology. In this research, a case study design was selected for the qualitative methodology and semi-structured interviews were employed as the data collection method. This case study is based on a deanship of a university implementing a digital learning management system in the country of Jordan. The research provides an analysis of the interviews to gain insights into instructors' perspectives regarding the factors influencing the implementation of the digital learning management system in a higher education sector.

DOI: 10.4018/978-1-7998-2914-0.ch006

INTRODUCTION

Digital Learning Management Systems (DLMSs) are widely used at universities for streamlining teaching and learning tasks, as they effectively manage learning tools and resources for user participants (students and teachers) (Kats, 2010; Unwin et al., 2010). A phenomenal revolution has occurred in the university education over the past decade as a result of digital learning management systems, as they have been found to significantly benefit students' understanding and academic progress (Alnsour, Muhsen, Dababnah, Eljinini, & Barhoum, 2011). DLMS applications are multifunctional, enabling students to have ubiquitous access to systems, watch lectures, download course materials and upload assignments online (AlQudah, 2014).

The latter provides convenience to students as never before and is conducive to study. To have efficient functioning of a DLMS, it is necessary to enable well-established links between data that is indexed by search engines, and web bots that connect and bind information together (Babić, 2012). DLMS empowers users and offers a greatly enhanced user experience overall (McIntosh & Torres, 2014; Phillipo & Krongard, 2012).

A DLMS greatly simplifies learning and deliverables, such as tests, assignments and other forms of assessment, with the use of multimedia applications that incorporate team learning through dynamic, real time, documented collaboration. Some examples include the use of chat, file transfer and assessment applications. A DLMS creates new platforms for the active learner and results in greatly expanding the learning curve. To fully understand the nature and implications of different DLMSs, it is vital to explore exactly what the DLMS context is (Ahmad, Chinade, Gambaki, Ibrahim, & Ala, 2012). This study is based on the most recent and related literature explaining various scenarios where DLMSs address learning issues in the digital environment in a way that was not possible in the previous confines of print logics. Table 1 below shows definition of the latent variable for this study.

Table 1. Definition of the latent variable.

Latent variable	Definitions
IT infrastructure	The availability of technology in the country in general, and in universities in particular, such Internet connections, computers and DLMS tools (AlQudah, 2014).
Jordanian Culture	The unique culture in Jordan is a very important factor that influences the use of DLMSs in different ways, and this research will focus on this factor through perceived use and perceived ease of use.
Perceived ease of use (PEOU)	Perceived ease of use refers to *the degree to which a person believes that using a particular system would be free of effort* (Davis, 1989, p. 320).
Perceived usefulness (PU)	the degree to which the user believes that using the DLMS would improve their learning performance (Davis, 1989).
Acceptance of an DLMS	The perceived usefulness and perceived ease of use of a DLMS under a given behaviour for continued use of DLMSs in Jordanian universities.

The universities are an educational institutions with a good reputation and each university has several branches across Jordan. The university management recognises the importance of IT for educational purposes; in its initial stages, it created an Information Technology Centre for supporting and train the users on DLMS. The deanship provides IT and digital learning services for both staff and students. In describing the mission of the deanship, the dean said:

Our mission is seeking an increase and a development of electronic dealing services to the beneficiaries which contribute to the development of all tasks related to academic, administrative and e-learning.

The university management seeks, through related parties, to achieve the following goals:

1. Provide all possible means of modern machine technologies (servers- information centre-operation centre-computer workshops-television digital learning management system such as Blackboard or Moodle) and necessary operating systems.
2. Guarantee the application of information systems.
3. Automate (machines of procedures and models) all information tasks.
4. Take advantage of networks affiliated with the university, such as email services, banner system and television transmission of lectures and activities.
5. Provide an environment suitable for spreading digital learning management system and applications and related training.
6. Achieve high levels of information security from the source to the users.
7. Safety of procedures and policies that guarantee the provision of electronic services.

STUDY OF FACTORS FOR THE ACCEPTANCE OF DLMS AT JORDANIAN UNIVERSITIES

The Technology Acceptance Model (TAM) (Davis, 1989; Venkatesh, 1999) has usually been used to examine the acceptance of a new technology. In this study, a research model is based on the TAM to investigate the acceptance of using DLMSs at Jordanian universities. The component of technical support will be incorporated through the TAM model, and serves as an extension to TAM. This is conducted in order to measure the acceptance of using DLMSs at Jordanian universities. The research model explains the system usage of applying DLMSs at Jordanian universities. It consists of concepts that will be refined in relation to the Digital

Learning Management System (DLMS) and include technical support, perceived usefulness (PU) and perceived ease of use (PEOU) (Venkatesh & Davis, 2000).

Information technology infrastructure (IT infrastructure) and the Jordanian culture, which are the main element factors that directly affect the acceptance of DLMS in Jordan universities, followed by perceived usefulness and perceived ease of use of the DLMS. After that will be tested and discussed the acceptance of using DLMS and make sure it will continue use of DLMS in Jordanian universities.

Information Technology (IT) Infrastructure for DLMSs at Jordanian Universities

IT infrastructure for DLMSs in Jordan refers to the availability of technology in the country in general, and in universities in particular, such Internet connections, computers and DLMS tools (AlQudah, 2014). Many developing countries are still struggling with lack of IT infrastructure, and this research will highlight this factor as being one of the main factors influencing the acceptance of the DLMS in Jordan universities. According to Samak and Tawfik (2017), the Jordanian Ministry of Education made a five-year plan to establish IT infrastructure for public schools. This plan will be useful to identify the amount, methods and types of infrastructure used in the schools.

According to Khwaldeh, Al-Hadid, and Masa'deh (2017), the infrastructure for DLMS in Jordanian universities needs to improve and suit the demands of users (students, instructors and administration users). The study highlighted some users' infrastructure needs, such as deaf and the blind users, which means there is a gap in the IT infrastructure that needs to be identified and improved to be usable for all users (students and stuff).

According to (Wimpenny, Adefila, & DeWinter, 2018), Jordanian academic staff plan around the existing technology. As some interviewers stated, *access to technology appears to be limited in certain organisations, and there is also evidence of unequal provision around the country. This could therefore justifiably explain why relatively few respondents (10.8%) feel that they can draw from a range of activities to meet their learner needs* (Wimpenny et al., 2018, pp. 33-48).

Donation from developed countries and organisations are being used to help upgrade the IT infrastructure for DLMSs in Jordanian universities from the old infrastructure to a new platform infrastructure; however, because of other obstacles like refugees from the border country and Jordan's lack of financial resources, the improvement is slow, which means that there is a lack of infrastructure, especially in the education sectors.

IT infrastructure for DLMSs in Jordanian universities positively affects the acceptance of DLMS through perceived use and perceived ease of use. Two hypotheses (H) were formulated regarding IT infrastructure for DLMSs as follows:

H1: IT infrastructure for DLMSs has a positive effect on perceived usefulness for the DLMS; and
H3: IT infrastructure for DLMSs has a positive effect on perceived ease of use for the DLMS.

Culture in Jordan

Cultural factors affect the use of DLMS tools in Jordanian universities. The unique culture in Jordan is a very important factor that influences the use of DLMSs in different ways, and this research will focus on this factor through perceived use and perceived ease of use (Al-Shaikh, 2003). According to Cho and Berge (2002), the most influential factors regarding the adoption and deployment of a DLMS are the culture and norms of those working in higher education faculties. Therefore, pedagogical issues and the lack of technology professional training are affecting the acceptance of a DLMS.

Many researchers highlight culture as one of the most significant factors affecting the acceptance of using DLMSs in Jordanian universities. A limitation in the studies done in Jordan universities is that they do not reflect the real number of users, and to be accurate, the sample size should represent all types of users (Abdullateef, Elias, Mohamed, Zaidan, & Zaidan, 2016; M. Al-Shboul, 2013; Al-Zoubi & Ali, 2019; Al Musawi, Ambusaidi, Al-Balushi, & Al-Balushi, 2015; Atoum, Otoom, & Ali, 2017; Hofstede, 1983; Khwaldeh et al., 2017). On the other hand, *emotions play a significant part in students' learning experiences within complex educational environments* (Arguel, Lockyer, Kennedy, Lodge, & Pachman, 2019, pp. 44-56).

According to M. Al-Shboul (2013, pp. 13-28), *We need to spread the culture of using e-Learning technology to enhance the quality of learning.* The university management advised that more focus on user training is needed and the university culture needs to be more open to new technology to improve the quality of teaching and learning. With reference to Abbad, Morris, and De Nahlik (2009, p. 19) *Cultures that are more focused on oral traditions may be less engaged with e-learning.* Management should work more on user training so the users can accept and enjoy the new technology and look forward to new updates. According to Varis and Al-Agtash (2008, p. 74) *As a relevant factor in the dissemination of values and in the formation of society's thinking, the media are responsible for disseminating local culture and the events that occur in the world.* The media today is a very important player in local social life and improving the use of any new technology. It's the key player in changing user culture and improving positively any new technology.

In this research, a significant relationship exists between the influence of culture and the perceived usefulness and perceived ease of use of adopting a DLMS in Jordanian universities. There is a need to confirm this prediction with respect to the location (Jordanian universities) through this research. The relationship between cultural influence and the use of a DLMS is hypothesised in this research as follows:

H2: Culture has a positive effect on perceived ease of use for the DLMS; and
H4: Culture has a positive effect on perceived usefulness for the DLMS.

Perceived Usefulness of a DLMS

The perceived usefulness of a DLMS is defined as the degree to which the user believes that using the DLMS would improve their learning performance (Davis, 1989). It is *the degree to which a person believes that using a particular system could enhance their job performance* and *it is the extent to which an individual believes that using the system enhances his/her performance* (Saadé & Bahli, 2005, pp. 217-240). According to Davis (1989, p. 320) *People tend to use or not use an application to the extent they believe it will help them perform their job better. We refer to this first variable as perceived usefulness.*

In the TAM, Davis (1989) suggests two determinants of computer usage: PU (perceived usefulness) and PEOU (perceived ease of use). Later, other investigators and researchers extended the TAM model to combine additional variables that could account for extra variance in computer usage technology (Gefen & Straub, 1997; Venkatesh & Davis, 2000). The advantage paradigm from behavioural decision theory is relevant to perceived usefulness and perceived ease of use (Beach & Mitchell, 1978).

The benefits of the hypotheses of IT infrastructure and culture are to encourage users (students, instructors and management) to accept any new technology and work on it to gain all the advantages and improve the quality of learning. The IT infrastructure and culture in Jordan are very important factors affecting the acceptance of any new technology to accept, including the use of DLMSs in Jordanian universities (M. Al-Shboul, 2011; M. A. Al-Shboul & Alsmadi, 2010; Alnsour et al., 2011; AlQudah, 2014).

This research will highlight the perceived usefulness on DLMS tools from the IT infrastructure and the culture in Jordan context to show the attention of acceptance using DLMS in Jordan universities. The relationship between perceived usefulness and the acceptance of DLMS to adopt DLMS is hypothesised as follows:

H5: Perceived usefulness in accepting a DLMS has a positive effect on acceptance of the DLMS.

Perceived Ease of Use of a DLMS

The perceived ease of use of a DLMS is defined as the degree to which the user believes that using the DLMS will be free of effort (Davis, 1989). It can be described as *the degree to which a person believes that using a particular system is free of effort. Previous research has demonstrated that individuals are more likely to use a new technology if they perceive that it is easy to use* (Saadé & Bahli, 2005, pp. 101-132).

Davis (1989) shows that ease of use had a direct effect on perceived usefulness. Many studies on the TAM also provided strong experimental support for a positive relationship between perceived usefulness and perceived ease of use. The TAM shows that perceived ease of use and perceived usefulness have a direct effect on acceptance of using a new technology (Adams, Nelson, & Todd, 1992; Szajna, 1996; Venkatesh & Davis, 2000). Venkatesh (1999) discovered that *facilitating conditions and external control served as anchors that users employ to inform perceived ease of use about information technology* as cited in (Abbad et al., 2009, p. 167). According to Williams (2002), an important factor in determining the acceptance of technology such as a DLMS is the availability of 'technical support'.

Both IT infrastructure and culture are very important factors that affect ease of use of a DLMS. According to M. Al-Shboul (2013) IT infrastructure and culture are key players in enhancing the quality of learning. IT infrastructure should be able to support and accept any technology and software tools without obstacles. Poor infrastructure will cost any organisation money and time in activating the new tools or new technology software.

It is hypothesised that IT infrastructure and culture affect the perceived ease of use a DLMS positively and negatively. Good infrastructure makes any new software easy to apply and use. With regard to culture, knowledge of the culture where a new software will be applied will allow the new technology to be easy for the uses to use for the users to use (students, instructors and managements) (Al-Dmour, 2014; AlQudah, 2014).

Perceived ease of use of a DLMS shows the intention a user has to make use of a technology under a given behaviour. It was predicted that users' behavioural intention to adopt DLMS has an encouraging influence on acceptance of the DLMS.

H6: Perceived ease of use of a DLMS has a positive effect on acceptance for the DLMS.

Acceptance and Continued Use of DLMS

The Technology Acceptance Model (TAM) *has become well-established as a robust, powerful, and parsimonious model for predicting user acceptance* (Venkatesh & Davis, 2000, p. 19). A large number of researchers have adopted the TAM to test and examine the acceptance of new technologies like personal computers (Igbaria, Zinatelli, Cragg, & Cavaye, 1997).

Davis (1989, p. 335) indicates that: *the possibility of dysfunctional impacts generated by information technology emphasizes that user acceptance is not a universal goal and is actually undesirable in cases where systems fail to provide true performance gains. Although there has been a growing pessimism in the field about the ability to identify measures that are robustly linked to user acceptance, the view taken here is much more optimistic. User reactions to computers are complex and multifaceted. But if the field continues to systematically investigate fundamental mechanisms driving user behaviour, cultivating better and well measures and critically examining alternative theoretical models, sustainable progress is within reach.*

To accept a new system, hypotheses for both perceived usefulness and perceived ease of use should be examined (Davis, 1989). The model reflects these hypotheses to measure the acceptance of new software for any organisation, especially Jordanian universities, and it will then be possible to see if it is appropriate to use this new software in these universities. The model can then be used for all universities in Jordan. The outcome will influence the continued use of DLMS software in Jordanian universities in general (Almarabeh, 2014).

Acceptance of a DLMS shows the perceived usefulness and perceived ease of use of a DLMS under a given behaviour for continued use of DLMSs in Jordanian universities.

H7: Acceptance of a DLMS has a positive effect on continued use of the DLMS.

CONCEPTUAL CONSTRUCTS AND THE RESEARCH MODEL

The research utilises the Technology Acceptance Model (TAM) as a framework to address all the relevant issues around the DLMS context. In this research, the proposed model based on the extension of the Technology Acceptance Model (TAM) (Davis, 1989) that has been applied, tested and examined for the discovery of new information and understanding of the Digital Learning Management System model (DLMS). The following is a review and definition of this framework with critical perspectives.

Figure 1. Conceptual framework for digital learning management system model at Jordan universities (DLMS)

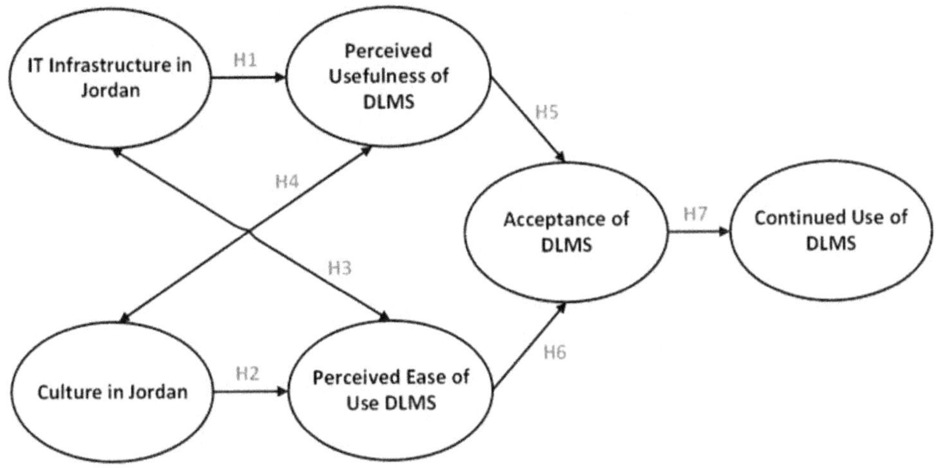

A number of issues can affect users' decision to accept DLMS tools of choice, and Jordan is one of the developing countries that experience problems such as lack of network connectivity, slow system response and social cultural issues. In this research, the many factors that affect acceptance and rate of adoption are used to understand how they influence the diffusion rate of DLMS tools as innovative instructional delivery tools. Therefore, we investigate behavioural intent to use DLMS tools by employees, based on attitude, relative advantage, complexity and compatibility (Mkhize, Mtsweni, & Buthelezi, 2016).

RESEARCH APPROACH AND METHODOLOGY

The approach of this research considered the need for this research in order to find the values related to the research and also to constitute ethical behaviour in this research (Ridder, 2017). Adopting this research approach demands a clear concept of how the research approach has been presented (Mills, Harrison, Franklin, & Birks, 2017; Yin, 2011). The researchers observed limitations in the process that effects the research in general, and the outcome in particular, therefore, it is very important to realise the effects of such developments and preserve control over the research (Creswell, 2014; Yin, 1981).

A semi-structured interviews method was adopted and built in order to find qualitative data in this research (Rowley, 2002; Yin, 2012). The interviews method can be critical, positivist or interpretive, depending on the researcher's philosophical assumptions. In this research, the researcher adopted a positivist interview method for the qualitative data phase (Gupta & Rao, 2018; Yin, 2015).

The interviews conducted among the research participants within Jordanian universities (six universities) to extract information. The researcher collected qualitative data from interviews. It has been adopted to obtain qualitative data in order to develop and refine concepts identified from the literature for inclusion in the conceptual model and to examine the acceptance of the Jordanian Digital Learning Management System Model (DLMS) at Jordanian universities (Creswell, 2013; Yin, 1981).

The research conducted in the context of Jordanian universities. It was particularly focused on six main universities in Jordan: the University of Jordan (UJ) (www.ju.edu.jo), which is the oldest government university in Jordan, Princess Sumaya University for Technology (PSUT) (www.psut.edu.jo), which is a mixed university system (government and private), and Philadelphia University (PU) (www.philadelphia.edu.jo), which is one of the private universities in Jordan. Jordan University of Science and Technology (JUST) (www.just.edu.jo), which is one of the public universities in Jordan. The last university is The Applied Science Private University (ASU) (www.asu.edu.jo), which is one of the private universities in Jordan. These six different types of universities represent the major universities in Jordan. According to (Rank, 2019), table 2 below shows some important details of these universities from which the research data was collected.

Table 2. Research context: Jordanian universities

University	World Rank	Country Rank	Type	Enrollment Range Students	location
UJ	1643	1	non-profit public	35,000-39,999	Large city Amman
PSUT	6452	11	non-profit private	3,000-3,999	Amman
JUST	2315	2	non-profit public	20,000-24,999	Irbid
PU	2528	3	profit private	5,000-5,999	North Amman
ASU	7342	16	profit private	8,000-8,999	Amman
AABU	5362	9	non-profit public	15,000-19,999	Mafraq

Note. UJ = University of Jordan; PSUT = Princess Sumaya University of Technology; JUST = Jordan University of Science and Technology; PU = Philadelphia University; ASU = Applied Science Private University; AABU = Al al-Bayt University (Rank, 2019).

Instructors' Perceptions on the Digital Learning Management System

This section explores instructors' perceptions on the factors affecting the acceptance and continuance usage of digital learning management systems in Jordanian universities on the basis of qualitative data collected using the case study methodology. This is significant as instructors give more attention to discuss digital learning management systems from a pedagogical perspective, as opposed to IT staff, who are more concerned with the technological features of the system. The following are the instructors' perceptions on the IT infrastructure, Jordanian culture, perceived usefulness, perceived ease of use, acceptance of DLMS and continuance usage of the digital learning management system.

It was clear from the instructors' interviewees that the digital learning management system encourages cooperative learning among instructors themselves and/or between instructors, administrators and students. This cooperative practice shows to extend users (instructors', administrators' and students') prospects and add value, by giving them to look at same topic from another perspectives. Instructor 3 agreed, stating:

The essential way my experience of the digital system has evolved my learning is by encouraging cooperative practices. When instructors and students cooperate among themselves and/or between groups, it delivers more possibility to consider learning materials and study topics from different perspectives and to find more learning styles.

Instructor also painted that digital learning management system provides a better and new communication experience with users. Instructors believed that the digital systems open more places for bi-directional communication and collaboration with their users (instructors, administrators and students), as users can rate different recommended courses based on their experiences, which is helpful for all users they use the digital learning management system. With digital learning management systems, instructors should be more confident about the efficacy of the topic to their special study area. There is an advantage for instructors as they become more expert and aware when they keep using the digital learning management system continuously. For instance, instructor 3 pointed:

The digital system offers a better experience in relation to the communication pathways between users (instructors, administrators and students). It is more meaningful as the instructors can make recommendations to use learning topics and the students can provide feedback, and even provide their own recommendations if they want. This assistances to involve the students in the education process.

Instructors' Perceptions on the IT infrastructure of using the Digital Learning Management System

From the interviews, it was clear that the digital learning management system is still not fully utilised by users (instructors and students). According to Instructor 2:

We recognized that the majority of users still do not realise the full capacity of the digital learning management system.

Instructor 3 agreed with Instructor 2, saying:

Still we need spacious efforts to get instructors and students use the digital learning management system frequently.

Nevertheless, although the digital learning management system is in its early stages of adoption, some users (instructors and students) demonstrated sound knowledge of the system and held positive perceptions about the importance of implementing the system. The interviewees stressed the importance of adopting innovative systems to deliver high quality educational services to users (instructors and students). For instance, Instructors 3 stated:

The university management is trying to increase its information technology infrastructure to provide better education quality to the users. To be competitive, you must have a good quality of IT infrastructure and we always try hard to make the IT services accessible for everyone, and I think the university management is supporting us to present innovative service to both users (instructors and students).

Instructor 2, who is an e-learning executive, concurred, and highlighted the importance of the digital learning management system for connecting instructors and students in a virtual community infrastructure, saying:

The university uses the digital learning management system to help students, instructors, etc. choose and use quality content of IT infrastructure. It is being used to create an e-university environment whereby instructors and students can be better connected with each other and also the global learning community infrastructure.

Similarly, instructors 3 showed that he had knowledge of the features of the digital learning management system. He mentioned some of these features of DLMS, and linked them to students' learning performance:

The digital learning management system combines content-based filtering and learner ratings to provide recommendations on learning materials. It gives me an easier way to upload documents related to the lecture like guidelines. The aim at the university is to increase the students' learning performance, and the system used at the university I think it combines some very useful platforms.

Moreover, digital learning management system provides students with personalised content based on their needs, which enables them to make choices without prior experience of the alternatives, and it is noticeably required to reduce the information overload. Instructors 4 and 5 explicitly noted this, stating:

The university is using the learning system to provide instructors and students with better quality digital learning platforms. These types of systems help to personalise the information retrieval and management process. The university is aiming to deliver a better infrastructure tool to help personalise the education experience.

Nonetheless, the implementation of the digital learning management system was challenging and required much effort in IT infrastructure. The concept itself is still new in the education domain. Additionally, university management and IT staff were unfamiliar with the digital learning management system, which makes them perceive it as a complex technology. Instructor 1 clarified:

I would like to mention that the implementation process of the digital learning management system was not an easy task. We faced many challenges including the old IT infrastructure, the unfamiliarity of our staff with the system and the complexity of the concept itself.

Moreover, instructors indicated that the university management needs to provide a mandatory adequate training programmes on the topic especially for the new instructors' and students'. In this regard, Instructors 2, 3 and 5 mentioned:

We believe it's really high important to train students and instructors on the digital learning management system to fully utilise its valuable services, and it should be a mandatory for both users (instructors and students).

Besides that, some users expected more from the system, and indicted that the implemented digital learning management system needs improvement. They recommended extra features to improve to the current system. Instructor 2, 4 and 6 provided examples of useful features:

We think the university management could improve the way of the digital learning management system is used. The way technology is advancing at the moment, there are always new systems to embed within the IT infrastructure framework, such as examination systems, digital learning management systems and learner directory platforms. We would like to see the university continue to grow the system.

Instructors' Perceptions on the Jordanian Cultural of Using the Digital Learning Management System

Cultural factors affect the use of DLMS tools in Jordanian universities. The unique culture in Jordan is a very important factor that influences the use of DLMSs in different ways, and this research focuses on this factor through perceived use and perceived ease of use. According to Cho and Berge (2002), the most influential factors regarding the adoption and deployment of a DLMS are the culture and norms of those working in higher education faculties. Therefore, pedagogical issues and the lack of technology professional training are affecting the acceptance of a DLMS.

It was reported, that the university management were aware of the cultural issues. However, as technical people, it is evident that their focus was on the system quality, and little attention has been paid to the Jordanian culture. Interviewees mentioned culture and its dimensions several times during interviews, from different perspectives. For example, Instructor 1 mentioned Jordanian culture and how the culture affect using digital learning management system at Jordanian universities when he said:

Jordanian culture affect using digital learning management system (DLMS) tool in both side (positive and negative), and not only for DLMS tool, it can affect any new application to use. The culture is a main factor effect using DLMS tool in the university, and it determines the acceptance of using DLMS tool.

Instructor 4 mentioned cultural from another perspective, when he noted the relationship between culture and age:

Most of the old fashion instructors they lost desire to use a new software and that maybe because the age, or the way that they learned. On the other hand, the fresh instructors are so excited to learn and use new software's especially if that software helps them in their daily works.

As mentioned earlier in this section, the main focus of interviewees was on the acceptance of using digital learning management system in higher education sector at Jordanian universities. The implementation of the digital learning management system has some difficulties and required much effort. As for accessibility, Instructor 5 stated:

I do not know if the cultural implications of using a digital learning management system software is a problem or that the problem is that some users are rickety to use digital learning management system, it's very difficult to say. In my opinion, there is no problem from students wise to use the DLMS software but it should be driven by management, like the assessments, material, assignments and exams can be online. The problem here is the cheating, and that's not for specific culture only, it can be anywhere. That's why the main exams are still paper based and we should keep it like that from my experience. However, if there is a system can avoid the cheating, it will be easier to make it online instead of paper base. For me, I just do the quizzes online only.

Instructor 4 indicated that the management was aware of the lack of using digital learning management system, and reported that the digital learning management system is reliable, available and accessible, without any major problems:

From my opinion, I don't think the culture is effecting using digital learning management system, because in these days most of the students can use the social media to share their information, and they prefer to use social media instead of other system.

Instructor 7 also highlighted the role of culture in predicting the usage of the digital learning management system. Most of new and novel IT technologies are developed and designed in developed countries. However, it was evident in the literature that differences in cultures systematically influence whether and for what reasons users adopt—or fail to adopt—new technologies (Al-Hujran, Al-Debei, Chatfield, & Migdadi, 2015; Steers, Meyer, & Sanchez-Runde, 2008). The culture must be considered when designing and developing new systems. Instructor 7 said:

Most of the digital learning management systems are developed and designed in the Western world. Developers and designers usually forget to take the differences in culture into consideration. This is become as a serious matter that needs to be taken into consideration when designing and implementing emerging technologies in our country.

Students' Perceptions on Perceived Usefulness of the Digital Learning Management System

Perceived Usefulness (PU) affect directly the acceptance of technology. Indeed, if the user perceives the digital learning management system to be useful in terms of increase or improvement in work productivity, then the user will have a favourable perception of accepting the digital learning management system. Institutions of higher education around the globe implemented digital learning management systems as a channel to effective deliver educational material to users. The use of digital learning management systems can be considered as a gain to develop teaching and learning processes (Hsbollah & Idris, 2009). In this research, it was revealed that perceived usefulness of the digital learning management system was known by instructors' staff. Most interviewees admitted sufficient knowledge of the perceived usefulness of the system. In this context, Instructor 1 believed that the digital learning management system would help both users (instructors and students), saying:

I know that a lot of instructors find the system useful so I suppose it works for them. If I was a student, I would like the fact that I have a centralised platform with all the information I need to locate people or course materials. I believe that the students think it helps them.

Instructor 4 agreed with his colleague, saying:

It is not easy to argue against the fact that technology is fundamental for learning and teaching in the new world. Student needs to be able to manage information and learning materials properly and instructor wants to be able to apply the latest teaching techniques. Most of universities have digital learning platforms, and digital systems are becoming more and more important.

Moreover, Instructor 5 indicates the value of the digital learning management system compared with the traditional learning systems, commenting:

Digital learning management system makes our daily work easier, instead of printing and talk to each one in the classes, it gives us more flexibility to communication with other. We can do our daily works by using DLMS software processes in efficient time (save time). We can save the lecture time instead of paper work.

According to the interviews, instructors were also fully aware of the personalisation features of the digital learning management system, such as collaborative filtering and rating. Instructor 3 noted that:

From my side, I use digital learning management system software in my daily base and there are many benefit for me as an instructor. I uploaded all the assignments, guidelines, materials and study plans, and I uploaded videos for some classes.

Similarly, Instructor 7 agreed on the role of the digital learning management system for personalising the educational materials, stating:

These types of software's help to personalise the information retrieval and management process. The university is aiming to offer a tool to help personalise the education experience.

The instructor also advised of the usefulness of the digital system to build a collaborative learning environment. Instructor 3 mentioned:

A digital learning management system is becoming an important tool. The more that the system supports users to identify interesting content, the more it will play a key role in learning and teaching. As the filtering systems become more elaborate for identifying content and generating collaborative learning, instructors' will use them more and more.

Instructors' Perceptions on Perceived Ease of Use of the Digital Learning Management System

Perceived ease of use concurrently pretended a strong direct effect on behavioural to use any new technologies, and previous research has found a positive relationship between PEOU and continuance usage. Thus, it is essential for designer and developer to design and develop an easy to use tool, which would lead to better user engagement and motivation. Although the system reduces users' efforts by showing them personalised material according to their needs, instructor presented real concern about the complexity of the digital learning management system. For instance, instructor 4 stated:

The feedback from students regarding the DLMS, I can tell the system is complicated for the new user and they need orientation workshops and training to assist them to react with the system. This exemplifies a real concern for me.

To bypass this essential issue, he proposed that the institute needs to provide orientation workshops and training programmes especially for the new users' on the system. Instructor 3, 5 & 6 agreed on the complexity of the digital learning management system for new users', they seeing it as a new system that was rarely used by users':

The digital learning management system is new technology adopted in the high education field and many users' may not be familiar with it yet. It's not easy to ask users' to use advanced technology without adequate knowledge.

Instructor 7 highlighted some features in using DLMS in daily base. Saving time and effort are an important feature for all the system users'. It makes them (users') happy and comfortable to use it in any time. Instructor 7 stated:

... This tool is easy to use, and it makes my daily job easier. When I uploaded the lecture and the assignments to DLMS tool, this will make interaction easier to access their need, so they don't have to visit me in the office and asking me for a copy of anything like assignments, and that makes me and my students happy to use in any time.

Instructors' Perceptions on Acceptance and Continuance Usage of the Digital Learning Management System

It was obvious that the usage regularity of the digital learning management system was the ultimate goal of university management. During the interviews, instructors frequently commented that the value of implementing new system will not be known without usage regularity of the system by users' (instructors and students). Some Jordanian universities invested extensive finances into deploying the digital system to improve users' (instructors and students) education process. Nevertheless, higher university management assumed that the return on this investment would be the effective use of the digital system. Instructor 1 stated:

In theory, the value of the digital system comes from the usage regularity of the system by both users' (educators and learners). In another words, the system that becomes enhanced the more people are using it because this means there is more data go into the system to direct the recommendation. If users (instructors) can use platforms to store and deliver information, and if users (students) can access what they need to for the course they are needing, then it has got to make a contribution to more effective learning.

University management indicated that they understand that the digital learning management system needs to store more contents and learning materials to work efficiently and effectively. If instructors and students do not effectively involve with the system, the usefulness of the system will be restricted. Thus, university management thought often about how to rise the usage regularity by both instructors and students to fully utilise the valuable facilities provided by the digital system. Instructor 2 noted:

I consider it's highly significant to train instructors and students on the digital learning management system to fully utilise its valuable facilities. We stated that the majority of users still do not understand the full capacity of the digital system. The challenge is to get more instructors and students to use the system regularly in order to recognise its benefits for efficient and effective learning.

To rise the usage levels of the digital learning management system, users (instructors and students) must perceive its ease of use, usefulness and quality. Improved user experience by training and workshops was also an important factor. As regards the user experience and the quality, Instructor 4 stated:

The existing digital system model is good, but there is always need to improve outcome quality and user experience to rise usage knowledge levels.

SOLUTIONS AND RECOMMENDATIONS

DLMS's are managed by the administrators to restrict access to some or all documents stored (depends on the permissions for each user). However, in some cases the administrators can override the safeguard to access information in order to provide the best services. It is particularly important that Jordanian universities use one learning system for culturally and linguistically diverse. This also requires everyone to have access to smart phones, laptop and internet connections. On the other hand, it is essential for the users (instructors, administrators and students) to also embrace, access and manage the privacy of their profile's to enhance their practice and provide positive outcome to the daily works.

The challenge is to strike a balance between the qualities of using and managing the online digital learning management system by the users is very important in partnership with each other's. Following recommendations can be taken to improve using of digital learning management system as well as e-learning so that users can use 'DLMS' to empower them for their goal benefit in best and professional way;

- Promoting using digital learning management system is essential, as DLMS facilitates and motivates users to self-manage and empower themselves that determine their daily processes with other users or with themselves.

- It is essential for users to not just have access, know how to use their digital learning system records, but foremost how to use the basics computer skills, as a minimum requirement. In each university at Jordan has a specialised computer centre can be run one semester digital/computer literacy units to teach how to use the computer and internet search engine skills to users with low-computer literacy and non-English speaking background.
- All key stakeholders such as the Ministry of Higher Education at Jordan, Jordanian university management (public and private) must work together for the adoption, safety, security, privacy of the digital learning management system, for beneficial usage of electronic educational system to empower the users to manage themselves and to improve using DLMS.
- It's obvious that DLMS must be improved and changed to suit users. Many users have complained about the complexity of use the digital learning management system due to the difficulty of accessing the information. Further, they propose improving the DLMS system through improving the notifications system, so they can access and know any updates on the DLMS instead of using unprofessional tools such as social media between each other.

FUTURE RESEARCH DIRECTIONS

Future research could collect interview data with the instructors and the administrators to test the validity of the qualitative findings, followed by quantitative survey with the users' (instructors', students' and administrators'). Studies can also survey users satisfaction with the use of 'DLMS' by purposeful sampling of diverse population belonging to different demographic groups, education, gender characteristics and rural urban setting, due to digital divide. Such as youth, middle age. The hypothesised relationship between digital literacy, improves communication and enhances using DLMA can be explored and empirically tested with different groups of people in developed countries such as Australia, USA, UK...

CONCLUSION

To sum up, university management fully understand that digital system use and continued use are significant factors for the fruitful implementing of the digital learning management system. Users should use the digital system efficiently and effectively to get the fully value and advantages of system. Nevertheless, to rise up usage levels, users (instructors, administrators and students) need to realise the advantages of using the digital system. This is believed to be a respectable indication, as university management perception was that increased ease of use, usefulness and quality of the digital system would lead to advanced usage by instructors and students.

Case study (qualitative analysis) showed that university management's main objective in deploying the digital learning management system was to deliver better quality education to users (instructors, administrators and students). University management is fully recognise the expected benefits of such a system. Depend on interviews, instructors clearly showed the value of using the digital learning management system in terms of suggesting personalised contents and learning materials according to users' need and preferences, building a digital collaborative learning environment, saving money and time, which may stimulate collaboration among users (instructors, administrators and students) and conclusively evolve higher learning efficiency and effectiveness of users.

Universities management notified a number of challenges facing the application and use of the digital learning management system, which are the limited use of the digital system by users (instructors, administrators and students), some complexity of the digital system and the inelastic to satisfy and support users' needs.

Interviewers (instructors) showed a serious intention to raise the usage level of the digital system as to raise the significance of the system for users (instructors, administrators and students). Users' believed that the system significance will not be recognised without efficiently and effectively using it. Nevertheless, to encourage the acceptance process of DMLS, each university management is required to conduct regular orientation programmes and awareness campaigns especially for the fresh users, to clarify the expected features of the digital system. Furthermore, the university management should offer tutorials and training to increase the user experience with the digital system.

In summary, the university management recognised that the developed and increased of IT infrastructure, cultural, ease of use and usefulness would lead to better acceptance and usage by instructors, administrators and students. Hence, it is substantial for the university management to think seriously about developing and implementing a strategy and action plan to meet these strategic goals.

REFERENCES

Abbad, M. M., Morris, D., & De Nahlik, C. (2009). Looking under the bonnet: Factors affecting student adoption of e-learning systems in Jordan. *The International Review of Research in Open and Distributed Learning*, *10*(2). doi:10.19173/irrodl.v10i2.596

Abdullateef, B. N., Elias, N. F., Mohamed, H., Zaidan, A., & Zaidan, B. (2016). An evaluation and selection problems of OSS-LMS packages. *SpringerPlus*, *5*(1), 1. doi:10.118640064-016-1828-y PMID:27064567

Adams, D. A., Nelson, R. R., & Todd, P. A. (1992). Perceived usefulness, ease of use, and usage of information technology: A replication. *Management Information Systems Quarterly*, *16*(2), 227–247. doi:10.2307/249577

Ahmad, S. A., Chinade, U. B., Gambaki, A. M., Ibrahim, S., & Ala, N. A. (2012). The need for MOODLE as a learning management system in Nigerian Universities: Digesting University Utara Malaysia Learning Zone as a case study. *Academic Research International*, *2*(3), 444.

Al Amoush, A. B., & Sandhu, K. (2019). *A New Model for Acceptance of Analytics in Learning Management Systems at Jordanian Universities (JLMS). In Modern Technologies for Teaching and Learning in Socio-Humanitarian Disciplines* (pp. 138–161). IGI Global.

Al-Dmour, R. (2014). *An integration model for identifying the determinants of the adoption and implementation level of HRIS applications and Its effectiveness in business organisations in Jordan* (Doctor of Philosophy). Brunel University. Retrieved from http://bura.brunel.ac.uk/handle/2438/9222

Al-Hujran, O., Al-Debei, M. M., Chatfield, A., & Migdadi, M. (2015). The imperative of influencing citizen attitude toward e-government adoption and use. *Computers in Human Behavior*, *53*, 189–203. doi:10.1016/j.chb.2015.06.025

Al Musawi, A., Ambusaidi, A., Al-Balushi, S., & Al-Balushi, K. (2015). Effectiveness of E-Lab Use in Science Teaching at the Omani Schools. *TOJET: The Turkish Online Journal of Educational Technology, 14*(1).

Al-Shaikh, F. N. (2003). The practical reality theory and business ethics in non-Western context: Evidence from Jordan. *Journal of Management Development*, *22*(8), 679–693. doi:10.1108/02621710310487855

Al-Shboul, M. (2011). Potential Use of Course Management Systems in Higher Education Institutions in Jordan. *Online Submission*, *8*(2), 220–232.

Al-Shboul, M. (2013). The Level of E-Learning Integration at the University of Jordan: Challenges and Opportunities. *International Education Studies*, *6*(4). doi:10.5539/ies.v6n4p93

Al-Shboul, M. A., & Alsmadi, I. (2010). Challenges of Utilizing E-Learning Systems in Public Universities in Jordan. *iJET, 5*(1), 4-10.

Al-Zoubi, S. I., & Ali, M. (2019). *The Bearing of Culture upon Intention to Utilize D-learning Amongst Jordanian University Students: Modernizing with Emerging Technologies.* Paper presented at the International Conference for Emerging Technologies in Computing. 10.1007/978-3-030-23943-5_21

Almarabeh, T. (2014). Students' Perceptions of E-learning at the University of Jordan. *iJET, 9*(3), 31-35.

Alnsour, A., Muhsen, Z., Dababnah, M., Eljinini, M. A., & Barhoum, K. A. (2011). Implementing Moodle as a Tool to develop the Isra University e-learn system. *IJCSNS International Journal of Computer Science and Network Security, 11*, 120–124.

AlQudah, A. A. (2014). Accepting Moodle by academic staff at the University of Jordan: Applying and extending TAM in technical support factors. *European Scientific Journal, 10*(18).

Arguel, A., Lockyer, L., Kennedy, G., Lodge, J. M., & Pachman, M. (2019). Seeking optimal confusion: A review on epistemic emotion management in interactive digital learning environments. *Interactive Learning Environments, 27*(2), 200–210. doi:10.1080/10494820.2018.1457544

Atoum, I., Otoom, A., & Ali, A. A. (2017). Holistic Cyber Security Implementation Frameworks: A Case Study of Jordan. *International Journal of Information, Business and Management, 9*(1), 108.

Babić, S. (2012). Factors that influence academic teacher's acceptance of e-learning technology in blended learning environment. *E-Learning-Organizational Infrastructure and Tools for Specific Areas*, 3-18.

Beach, L. R., & Mitchell, T. R. (1978). A contingency model for the selection of decision strategies. *Academy of Management Review, 3*(3), 439–449. doi:10.5465/amr.1978.4305717

Cho, S. K., & Berge, Z. L. (2002). Overcoming barriers to distance training and education. *USDLA Journal, 16*(1), 16–34.

Creswell, J. W. (2013). *Research design: Qualitative, quantitative, and mixed methods approaches.* Sage Publications.

Creswell, J. W. (2014). *A concise introduction to mixed methods research.* Sage Publications.

Davis, F. D. (1989). Perceived usefulness, perceived ease of use, and user acceptance of information technology. *Management Information Systems Quarterly, 13*(3), 319–340. doi:10.2307/249008

Gefen, D., & Straub, D. W. (1997). Gender differences in the perception and use of e-mail: An extension to the technology acceptance model. *Management Information Systems Quarterly*, *21*(4), 389–400. doi:10.2307/249720

Gupta, S. K., & Rao, D. (2018). A case study on contemporary water management innovation (CWMI) in sustainable agriculture: Farmers joined hands for mutual survival and prosperity: The case of Kummarvandlapalli. *IJCS*, *6*(6), 2283–2287.

Hofstede, G. (1983). National cultures in four dimensions: A research-based theory of cultural differences among nations. *International Studies of Management & Organization*, *13*(1-2), 46–74. doi:10.1080/00208825.1983.11656358

Hsbollah, H. M., & Idris, K. M. (2009). E-learning adoption: The role of relative advantages, trialability and academic specialisation. *Campus-Wide Information Systems*, *26*(1), 54–70. doi:10.1108/10650740910921564

Igbaria, M., Zinatelli, N., Cragg, P., & Cavaye, A. L. (1997). Personal computing acceptance factors in small firms: A structural equation model. *Management Information Systems Quarterly*, *21*(3), 279–305. doi:10.2307/249498

Kats, Y. (2010). *Learning Management System Technologies and Software Solutions for Online Teaching: Tools and Applications: Tools and Applications*. IGI Global.

Khwaldeh, S. M., Al-Hadid, I., Masa'deh, R. e., & Alrowwad, A. (2017). The Association between E-Services Web Portals Information Quality and ICT Competence in the Jordanian Universities. *Asian Social Science*, *13*(3), 156. doi:10.5539/ass.v13n3p156

McIntosh, J., & Torres, M. I. (2014). *Learning Management Systems: Revisiting Learning Management Systems at the University Level*. The University of Tampa.

Mills, J., Harrison, H., Franklin, R., & Birks, M. (2017). *Case study research: Foundations and methodological orientations*. Paper presented at the Forum Qualitative Sozialforschung/Forum: Qualitative Social Research.

Mkhize, P., Mtsweni, E. S., & Buthelezi, P. (2016). Diffusion of Innovations Approach to the Evaluation of Learning Management System Usage in an Open Distance Learning Institution. *The International Review of Research in Open and Distributed Learning*, *17*(3). doi:10.19173/irrodl.v17i3.2191

Mutlag, A. A., Ghani, M. K. A., Arunkumar, N., Mohamed, M. A., & Mohd, O. (2019). Enabling technologies for fog computing in healthcare IoT systems. *Future Generation Computer Systems*, *90*, 62–78. doi:10.1016/j.future.2018.07.049

Phillipo, J., & Krongard, S. (2012). *Learning Management System*. LMS.

Rank, U. U. (2019). *Jordanian University Ranking 2019*. Retrieved from https://www.4icu.org/jo/

Ridder, H.-G. (2017). The theory contribution of case study research designs. *Business Research, 10*(2), 281–305. doi:10.100740685-017-0045-z

Rowley, J. (2002). Using case studies in research. *Management Research News, 25*(1), 16–27. doi:10.1108/01409170210782990

Saadé, R., & Bahli, B. (2005). The impact of cognitive absorption on perceived usefulness and perceived ease of use in on-line learning: An extension of the technology acceptance model. *Information & Management, 42*(2), 317–327. doi:10.1016/j.im.2003.12.013

Samak, A., & Tawfik, Z. (2017). *Exploration of Jordanian English Language Teachers' Attitudes*. Skills, and Access as Indicator of Information and Communication Technology Integration in Jordan.

Steers, R. M., Meyer, A. D., & Sanchez-Runde, C. J. (2008). National culture and the adoption of new technologies. *Journal of World Business, 43*(3), 255–260. doi:10.1016/j.jwb.2008.03.007

Szajna, B. (1996). Empirical evaluation of the revised technology acceptance model. *Management Science, 42*(1), 85–92. doi:10.1287/mnsc.42.1.85

Unwin, T., Kleessen, B., Hollow, D., Williams, J. B., Oloo, L. M., Alwala, J., . . . Muianga, X. (2010). Digital learning management systems in Africa: myths and realities. *Open Learning: The Journal of Open, Distance and e-Learning, 25*(1), 5-23.

Varis, T., & Al-Agtash, S. (2008). *Ubiquitous ICT for sustainable education and cultural literacy*. Citeseer.

Venkatesh, V. (1999). Creation of favorable user perceptions: Exploring the role of intrinsic motivation. *Management Information Systems Quarterly, 23*(2), 239–260. doi:10.2307/249753

Venkatesh, V., & Davis, F. D. (2000). A theoretical extension of the technology acceptance model: Four longitudinal field studies. *Management Science, 46*(2), 186–204. doi:10.1287/mnsc.46.2.186.11926

Williams, P. (2002). The learning web: The development, implementation and evaluation of internet-based undergraduate materials for the teaching of key skills. *Active Learning in Higher Education, 3*(1), 40–53. doi:10.1177/1469787402003001004

Wimpenny, K., Adefila, A., & DeWinter, A. (2018). *Jordan Opportunity for Virtual Innovative Teaching and Learning: A Needs Analysis Report Contextualising the State of the Art in International Online Teaching and Learning, with Particular Attention to the Jordanian Case*. Academic Press.

Yin, R. K. (1981). The case study crisis: Some answers. *Administrative Science Quarterly*, *26*(1), 58–65. doi:10.2307/2392599

Yin, R. K. (2011). *Applications of case study research*. Sage.

Yin, R. K. (2012). A (very) brief refresher on the case study method. *Applications of Case Study Research*, 3-20.

Yin, R. K. (2015). *Qualitative research from start to finish*. Guilford Publications.

KEY TERMS AND DEFINITIONS

Acceptance of a DLMS: The perceived usefulness and perceived ease of use of a DLMS under a given behaviour for continued use of DLMSs in Jordanian universities.

DLMS: Digital Learning Management Systems (DLMSs) are a necessary educational tools for learning activities (Al Amoush & Sandhu, 2019).

Information Technology: The technology involving the development, maintenance, and use of computer systems, software, and networks for the processing and distribution of data (Mutlag, Ghani, Arunkumar, Mohamed, & Mohd, 2019).

IT Infrastructure: The availability of technology in the country in general, and in universities in particular, such Internet connections, computers and DLMS tools (AlQudah, 2014).

Jordanian Culture: The unique culture in Jordan is a very important factor that influences the use of DLMSs in different ways, and this research will focus on this factor through perceived use.

Perceived Ease of Use: Perceived ease of use refers to the degree to which a person believes that using a particular system would be free of effort (Davis, 1989, p. 320).

Perceived Usefulness: The degree to which the user believes that using the DLMS would improve their learning performance (Davis, 1989).

Technology Acceptance Model (TAM): Has become well-established as a robust, powerful, and parsimonious model for predicting user acceptance (Venkatesh & Davis, 2000, p. 19).

Section 2

Disruptive and Emerging Technology: Social and Ethical Considerations

The second section of the book reminds us of the importance that emerging and disruptive technologies have on individual lives in education and the workplace. Social and ethical considerations such as issues of privacy, invasiveness, disruptive innovations, and the impact of technology on loneliness are addressed.

Chapter 7
Web and Education as Disruptors of Traditional Education and Development of Future Students and Workers

Marie-Line Germain
Western Carolina University, USA

ABSTRACT

Over the past 30 years, the internet has evolved from being the web of content to being the web of thoughts and the web of things in business, communication, entertainment, and education. To stay competitive, higher education institutions have had to train students on the wide range of skills and experiences and to move to digital platforms to better meet the needs of students, employees, and organizations. This chapter provides an overview of the development of online education, the 1.0 to 5.0 phases of web development, and how the field of education has adapted to these phases. Particular emphasis is placed on the use of mobile learning such as MOOCs, course collaboration software, and how smartphones can be used in courses to interact with peers and faculty. This chapter then presents a case study illustrating how online courses can successfully integrate Web 4.0 and 5.0 technology. It concludes by discussing the benefits and challenges of adopting some disruptive technologies and on how educational institutions can meet the needs of the next generation of students.

DOI: 10.4018/978-1-7998-2914-0.ch007

INTRODUCTION

Higher education has been instrumental in the growth of and participation in our global economy. To be a competitive workforce, students must find opportunities to acquire or update their knowledge and skills beyond the experience on traditional college campuses. This requires expanding access to formal and informal learning experiences and learning resources. Technology can increase relevancy by aligning both content and learning approaches with the immediate and long-term needs and interests of students.

According to the 2019 Babson Survey Research Group, the United States are "mature" online learning markets (Babson Survey Research Group, 2019). The average annual growth rate of online enrollments in the United States between 2003 and 2009 was nearly 20% in higher learning institutions (Allen & Seaman, 2009). According to Allen and Seaman (2010), the growth of the online student body has exceeded the growth of on-site students for the past decade, with a 21% increase in online enrollment versus a less than 2% increase in on-site enrollment from 2008 to 2009. In 2010, 63% of all traditional schools agreed that online education was critical to their future class offerings. Online enrollments have expanded at a rate of 10 to 15% per year over the last five years and constitute around 14% of all post-secondary teaching for credit. Additionally, distance education students are a growing segment of the overall student population--so much so that in the fall of 2013, they comprised 12.5% of all U.S. higher education students. A 2014 survey by the Babson Survey Research Group reveals that online education is no longer an institutional accessory (Babson Survey Research Group, 2014).

Today, almost all American colleges and universities offer online courses. Online courses can be found in almost all subject areas and about 70% percent of post-secondary institutions see online learning as very or extremely important for their future plans (Bates, 2018).

This chapter presents an overview of the evolution of the web and education from the 1990s till today and reviews how online education has changed the higher education landscape. It then presents a case study illustrating a successful integration of experiential learning in an online course. The chapter concludes by suggesting how this evolution may impact the way we will learn and work in the future.

ONLINE EDUCATION AND MOOCS: FROM A TIMID START TO A PARADIGM SHIFT

Online education is typically offered in two formats: online courses (credit-bearing or not), and Massive Open Online Courses (MOOCs), for which certificates can be earned at no cost or for a small fee. Some free MOOCs such as Coursera have a paid "verified certificate" option (Online Course Report, 2019). Even when they are fee-based, MOOCs remain inexpensive and are accessible to almost any budget. As of 2019, the top five MOOC providers by registered users include Coursera (23 million users), edX (10 million users), XuetangX (6 million users), FutureLearn (5.3 million users), and Udacity (4 million users).

Public, For-Profit, and Ivy League Universities

Distance education in public and for-profit universities did not start with the birth of the internet.

Since the 19th century, universities in the U.S. and Great Britain have offered distance education in the form of correspondence courses. In tandem with technological progress, distance learning began to incorporate radio broadcasts, TV programs, and audio- and videocassettes. In the U.S., the University of Chicago became the first university to offer "correspondence" courses in 1892. The first "virtual college", Coastline Community College, was founded in the mid-1970s. It had no campus and all the courses were broadcast. Several universities started online programs in the 1980s as more people had access to the Internet. Founded in 1989, the University of Phoenix became the largest and most popular online university in the U.S. In 1997, the Interactive Learning Network (ILN) was created. It was released to multiple schools as the first eLearning platform used at universities such as Yale, Cornell, and the University of Pittsburgh. Also, Blackboard Inc., a content management system, was founded. Today, Blackboard remains the most commonly used online course management platform at many universities. In the 2000s, an increasing number of U.S. universities began incorporating online courses in their curriculum and started offering degrees entirely online.

Unlike public and private universities, Ivy League universities had a late start in online education. In the 2000s, however, as the negative headwinds facing online education began to shift, elite universities started to show an interest in digital learning. Some of them realized that if they were not getting into the online space, others would and they would miss on an opportunity for growth. In 2012, Stanford, Harvard and MIT entered the digital space by announcing that they would offer online courses for a large number of individuals at no cost. These Massive Open Online Courses (MOOCs) used new technology startups Coursera and edX. Several other top-tier schools, including the University of North Carolina, Georgetown University, and Berkeley, moved some of their graduate programs online. Meanwhile, the company Starbucks announced a partnership with Arizona State University to offer its benefits-eligible employees full tuition reimbursement to complete their degrees online.

Within five years, thousands of students who would never have considered taking an online course from for-profit institutions began taking one for the first time from top-tier universities. These selective universities, along with large organizations, legitimized virtual education.

As of August 2019, all eight Ivy League universities offer online certificates, courses or degrees programs (Littlefield, 2017). Brown University, Columbia University, Cornell University, Dartmouth College, Harvard University, Princeton University, the University of Pennsylvania, and Yale University are offering MOOCs (See Table 1).

Table 1. Ivy league universities' online courses and MOOCs

Ivy League University	Traditional Online Courses	MOOC	Platform Name
Brown University	Yes. Pre-college and college courses	Yes	BrownX
Columbia University	Yes	Yes	ColumbiaX cvn.columbia.edu
Cornell University	No. Certificates only		eCornell
Dartmouth College	Yes	Yes	DartmouthX
Harvard University	Yes	Yes	Harvard Extension School
Princeton University	No	Yes	Princeton Online (Partnerships with Coursera, edX, NovoEd, and Kadenze)
The University of Pennsylvania	Yes	Yes	The Penn Online Learning Initiative (partnership with Coursera and edX) Wharton Online
Yale University	Yes	Yes	Open Yale Courses (partnership with Coursera)

MOOCs are Massive Open Online Courses aimed at unlimited participation worldwide and open access via the web. Unlike regular college or university courses, MOOCs can attract thousands of enrollees around the world. By using online tools, MOOCs aim to provide a richer learning environment compared to traditional ones. There are now thousands of MOOCs available worldwide from several hundred colleges, universities, and other institutions of higher learning (Germain, 2019). They can come in the form of active course sessions with participant interaction, or as archived content for self-paced study. They are facilitated by lecturers and the sessions may feature high profile guest speakers. The majority of top courses come from two platforms: Coursera and edX. According to a 2016 report from MOOCs aggregator Class Central, an estimated 58 million students worldwide are enrolled in 6,850 courses offered by more than 700 universities (Shah, 2016).

To deliver MOOCs, MIT developed the MITx platform, which was renamed edX when it partnered with Harvard. To date, more than four million students have enrolled in Coursera MOOCs. Both Udacity and edX have enrolled over a million students in their MOOCs. Udacity also partnered with San Jose State University to offer for-credit courses. Although not free, these courses are low cost and blended MOOCs material are developed by on-campus professors and teaching assistants.

The MOOC course with the highest known enrollment is co-taught by Dr. Barbara Oakley, a professor of engineering at the University of Oakland, and Dr. Terrence Sejnowski, a professor at the Salk Institute for Biological Studies. The course, "Learning How to Learn: Powerful mental tools to help you master tough subjects", covers a number of learning techniques that are utilized in a wide array of fields, such as medicine, business, and education. This course, which uses the Coursera platform, has garnered over one million enrollees. It is only the second MOOC to achieve this benchmark in history (Online Course Report, 2019). According to Sebastian Thrun, in 50 years, there might only be 10 distinct institutions offering higher education courses (Marks, 2012).

MOOCs present several advantages over traditional online courses. First, they have the potential to reduce costs and expand access to higher learning (Selingo, 2014a). Also, they cover a wide variety of subjects, from philosophy, history to mathematics (Goldy-Brown, 2017), and they tend to be short-termed. Most MOOCs can be completed in 10 to 20 hours, at the learners' own pace, around their own schedule, using their own assimilation style (Condruz-Bacescu, 2015; Sokolova, 2014). This is particularly attractive to individuals and employees who want to sharpen up specific skills without having to commit to completing a degree. MOOCs address the need for what human resource professionals term "on-demand learning." Condruz-Bacescu (2015) posit that employee performance and productivity can improve by up to 55% after completion of online training.

Also, MOOCs erase geographical barriers. The learner does not have to live in the country where the course is offered. In fact, an increasing number of them are offered in languages other than English, such as Chinese, French, and Arabic (Sokolova, 2014). Subtitles are particularly helpful to individuals with hearing or visual impairments (Goldy-Brown, 2017) or for those who are not fluent in the language of instruction. Finally, MOOCs are efficient and economical for both the learners and the educational institutions offering them. Learners do not have to commit to taking more than one course at a time and they do not incur any travel expenses. On their part, educational institutions do not need to rent and maintain on-ground facilities to hold MOOCs (Condruz-Bacescu, 2015) and the cost to the institution or organization is minimal.

MOOCS do present some disadvantages. In his 2014 essay for *The New York Times*' "Education Life" adapted from his book, Jeffrey Selingo explains how MOOCs have largely failed to fulfill their promise to expand access. Selingo (2014a) explains that the average consumer of MOOCs is not the remote village dweller hungry for education or the U.S. college dropout looking for a second chance. The typical MOOC learner is a young white American male with a bachelor's degree and a full-time job and with advanced degrees. MOOCs often serve a professional development function. Furthermore, large MOOCs are perceived as impersonal since they rarely provide personalized courseware or attention from an instructor (Srikanth, 2017). Other disadvantages include the difficulty of tracking students' assignments and involvement. In fact, MOOC dropout rates are notoriously high --much greater than in on-campus, traditional education (Condruz-Bacescu, 2015). The dropout rate can be as high as 90%, especially when they are offered at no cost. Students may be highly motivated at the start of a course but their involvement may fade as the course becomes increasingly difficult. Despite these limitations, MOOCs remain an attractive educational and training resource. An increasing number of organizations use MOOCs to train their employees and over 77% of employers have online professional development programs.

How Universities have Adapted to Wed and Education Changes

Over the past thirty years, the internet has evolved from being the *Web of Content* to being the *Web of Thoughts and Web of Things* in business, communication, and entertainment. This section details the 1.0 to 5.0 phases of web development and how the field of education has adapted to these phases since the 1990s and the exponential development of online courses. Table 2 shows how the web development phases have affected the development in the field of education.

1990-2000. Web 1.0

The first iteration of the web, Web 1.0, was created in the mid-1990s. Before 1999, the Internet was "Read-Only" web. That is, it allowed people to search for information and read it anytime. There was little to no user interaction or content generation. Web 1.0 was innovative in that it removed the geographic restrictions associated with brick-and-mortar businesses, and universities. Millions of static websites such as the online stores Amazon, Ebay, and Priceline were created.

Table 2. Web and education developments

Version	Web	Education
1.0	• Read-Only web • 45 million users (1996) • Focused on companies • Homepages • Owning content • Britannica Online • HTML, portals • Web forms • Page views • Advertising	Centuries of experience with memorization
2.0	• Read-Write-Publish • More than one billion global users (2006) • Focused on Communities • Blogs • Share content • Wikipedia • Web applications • Tagging	Internet-enabled learning
3.0	• Read-write-execute / portable personal web • Focused on the individual • Live stream • Consolidating dynamic content • The semantic web • Widgets, darg, and drop mashups • User behavior • User engagement • Advertainment • MOOC	Consuming & Producing knowledge
4.0	• Mobile Web	• Empowering education to produce innovation. • Course software and Apps for iPad and smartphones
5.0	• Open, Linked and Intelligent Web: the Emotional Web / Sensory-emotive web	Learning tools interacting with students using emotions and facial recognition

From Web and Education 1.0 to Web and Education 5.0

Education 1.0: Centuries of Experience with Memorization

Schools are still using the Education 1.0 model (Gerstein, 2019) focus on an essentialist-based curriculum with related ways of teaching and testing. Classes are teacher-oriented and their focus is on achievement test scores, which are the primary means of evaluating progress. Like Web 1.0, Education 1.0 is mainly a one-way process. In Education 1.0, students sit in rows and are taught in large numbers. They learn passively. They are largely consumers of information resources (Keats & Schmidt, 2007). This model of education rings true with schools that are in an age of standardization, accountability, No Child Left Behind, Race-to-the-Top, and Common Core Curriculum Standards. As such, Education 1.0 is similar to Web 1.0.

2000-2010. Web 2.0

1999 marked the beginning of a "Read-Write-Publish web" era. Unlike Web 1.0, Web 2.0 allowed active interaction of users. Even non-technical users could actively interact with other web users and contribute to the web using different blogs. Publishing content could be done with a few clicks. This interaction and contribution dramatically changed the web landscape. It empowered the common user with new online tools, such as blogs, social media, and video streaming. Some of the most important developments of Web 2.0 include Facebook, Flickr, Twitter, and YouTube. Most developed countries experienced an increased use of technology. Technology includes the widespread adoption of personal computers and the Internet (especially email and the web); the emergence of Web 2.0, including blogs, podcasts, and social interaction tools; e-learning platforms or learning management systems that incorporate features of Web 2.0; and free and open-source software (FOSS) where anyone is freely licensed to use, copy, study, and change the software in any way, and the source code is openly shared so that people are encouraged to voluntarily improve the design of the software (GNU.org). This is in contrast to proprietary software, where the software is under restrictive copyright licensing and the source code is usually hidden from the users. Examples of FOSS include Firefox, a calculator program, a built-in calendar, or VLC Media Player.

Education 2.0: Internet-Enabled Learning

Similar to Web 2.0, Education 2.0 includes more interactions between professors and students, student to student, and student to content/expert. Some school administrators and educators have moved into a more connected, creative Education 2.0 using cooperative learning, global learning projects, Skype, shared wikis, blogs, and other social networking tools in the classroom. Education 2.0 is an incredibly effective restructuring of how learning takes place (Hardigan, 2007).

2010-2020. Web 3.0

With Web 1.0, users passively consulted web pages and for the most part, they did not participate in generating content. With Web 2.0, users created content and interacted with sites and with each other through social media or forums. The next evolution of the Web involved the Semantic Web (Web 3.0), which improved web technologies in order to generate, share and connect content. Web 3.0 allows several growing technology trends to converge. It adapts to its mobile surroundings and combines human and artificial intelligence to provide more relevant and accessible information. Three dimensional designs are used extensively in websites and services, such as in museum guides, computer games, e-commerce, and in geo-spatial contexts. Information is more connected thanks to semantic metadata. The user experience evolves to another level of connectivity and since every device is connected to the web, services can be used everywhere and content can be accessed using multiple applications.

Education 3.0: Consuming and Producing Knowledge

With Education 3.0, content is readily available. It is a constructivist approach to teaching and learning. Professors, students, networks, connections, media, resources and tools create a unique entity that should meet learners', educators', and even societal needs. Web 3.0 is self-directed, interest-based learning where problem-solving, innovation, and creativity drive education (Gerstein, 2019). It is characterized by rich, cross-institutional, cross-cultural educational opportunities within which students can create knowledge artifacts that are shared, and where social networking plays a strong role. Education 3.0 focuses on three main areas: learning, social opportunities, and an increased use of technology.

In learning, there is a wide diffusion of e-learning (Keats & Schmidt, 2007). There is also a growing interest in alternatives to professor-centered approaches such as constructivism (Dewey, 1998) and resource-based learning. Local, regional, and international collaboration is used to create libraries of educational content. The need to recognize prior learning is emphasized and the Internet is used to find information and for just-in-time learning.

From a social standpoint, there is an increased use of information technologies in daily life, sometimes virtual spaces are used for social purposes. Social structures are computer-mediated and people outside of one's immediate physical environment

2020-2030. Web 4.0 and 5.0. Open, Linked and Intelligent Web: the "Emotional Web"

Web 4.0 is sometimes referred to as "The next web." This fourth step in the evolutionary process is based on wireless communication (such as mobile devices or computers) connecting people and objects whenever and wherever in the physical or in the virtual world in real time. The 4.0 or mobile version has a successor currently in its development stage, Web 5.0, also called the "sensitive" Web (Kambil, 2008). The goal of Web 5.0 is to link the web, which communicates with us, similarly to the way we communicate with each other, and like a personal assistant. As of today, the web is "emotionally" neutral, which means that it does not perceive the users' feelings and emotions but Web 5.0, the sensory and emotive Web, is designed to develop computers that will interact with human beings. One early example of this technology is the company WeFeelFine.org, which tracks emotional phrases on the web, categorizes them, and registers the frequency and location of clusters of sentiments. Another company, Emotiv Systems, Inc, is credited with the creation of neurotechnology. Their headphones and software can translate a person's emotions, cognitive thoughts, and facial expressions into digital outcomes. This technology is currently being experimented in the gaming industry, medical military research, learning, and recruitment. If interactions can be personalized to create experiences that excite users, then Web 5.0 will undoubtedly be more affable than its predecessors. The fact that the Web's evolutionary process has modified business orientation suggests that it will unequivocally affect the quality and content of higher education.

Education 4.0: Empowering Education to Produce Innovation

In Education 4.0, students value social sector internships, study-abroad terms, and industry engagement. Consequently, universities must design learning situations which favor the acquisition of knowledge, skills, and attitudes that translate into some competences (Yániz & Villardón, 2006) using wireless communication wherever and whenever available. Educators have to master new competences and skills such as the application of new knowledge to practice. They need good verbal and written communication skills to share information effectively with others, must be willing to work in teams, be problem-solvers, be lifelong learners, and embrace innovation (Longworth, 2003; Cabero, Llorente, & Román, 2007). In Education 4.0, online courses rely more heavily on collaboration software and smart phones to interact with peers and professors. They also access course content with handheld devices. BlackBoard, for instance, has an App for iPad, Android and iPhones. A large number of universities already use 4.0 technology in their online courses.

Education 5.0: Learning Tools Interact with Students Using Emotions and Facial Recognition

With Education 5.0, learning tools will focus on the interactions with students using emotions and facial recognition. While ensuring that their current student populations acquire the skills and knowledge today's organizations need, universities should be preparing to accommodate the next generation of learners who will be on the receiving end of Education 5.0, the Gen Z. Gen Z is defined as teenagers who are currently between the ages of 13 and 19. They want relevant and meaningful education to equip them for jobs of the future, strong growth early enough in their careers, and they want most of their dreams to be fulfilled within ten years of entering their career. To cater to this generation of students, educational institutions will have to focus on innovation and offer a productive educational experience. Table 3 details some of the "then and now" in learning habits.

Table 3. Learning and reading habits: Then and now

Then	Now
Few distractions, controllable (TV, music)	Mobile phones have changed rules
Reading habit is developed	News shorts, Twitter and video streaming, long paragraphs are daunting
Hands-on parental involvement in learning	Reduced parental involvement. Increased responsibilities
Homogeneous educational experiences across learning institutions	Different teaching models across learning institutions
Career paths are clear	Boundaries across disciplines are blurred
Limited, mostly reliable, formally published information	Often unverified, sometimes deliberately inaccurate information. Information overload
Mostly face-to-face interactions	Mostly virtual interactions
Appreciation is rare. Mainly earned through significant effort and achievement	Appreciation is omnipresent, leading to a possibly skewed perception of self

Successful Integration of Web and Education 3.0 and 4.0 in Online Courses

Because Web 5.0 and Education 5.0 technology is not readily available in most U.S. universities and colleges, this section focuses on the successful integration of Web 4.0 and course collaboration software in an online master's course at a major U.S. university.

This case study demonstrates how the need of Millennials to acquire hands-on experience and the use of technology can be successfully combined. Is it based on extensive interviews of those who participated in the consulting experience (Barnes, Christensen, & Hansen, 1994). While the case writer is the creator of the online consulting initiative, the process is related in an accurate and fair manner.

Case Overview

In the online Master of Science in Human Resources (MSHR) program at a University of North Carolina campus, students work as human resources consultants for an array of nonprofit organizations, small businesses, and local government agencies. Since the degree program is 100% online, the pro bono consulting work is integrated in a course as a team project and is delivered 100% remotely. Led by a faculty member, teams of HR students provide organizations with varied HR policies and programs that aim at increasing organizational effectiveness. These consulting projects are a required course assignment, which give the online students a unique field experience opportunity to apply the HR and HRD concepts they learn in their courses with real organizations. In particular, the consulting projects challenge students to engage in critical thinking to analyze clients' consulting needs, work collaboratively online as a team to develop client solutions, and engage in effective written and oral communications with team members and clients. As of October 2019, over 700 graduate students have participated in this initiative, delivering 280 HR consulting projects for over 184 organizations in 17 states and abroad, entirely online. Figure 1 presents an infographic summarizing the e-consulting activity. The nonprofit, for-profit organizations and government agencies vary in industry (healthcare, education, children's welfare, animal shelters, cities, etc.), in workforce size (from 2 to over 1,500 employees or volunteers), and in location.

Figure 1. Infographics summarizing the e-consulting activity

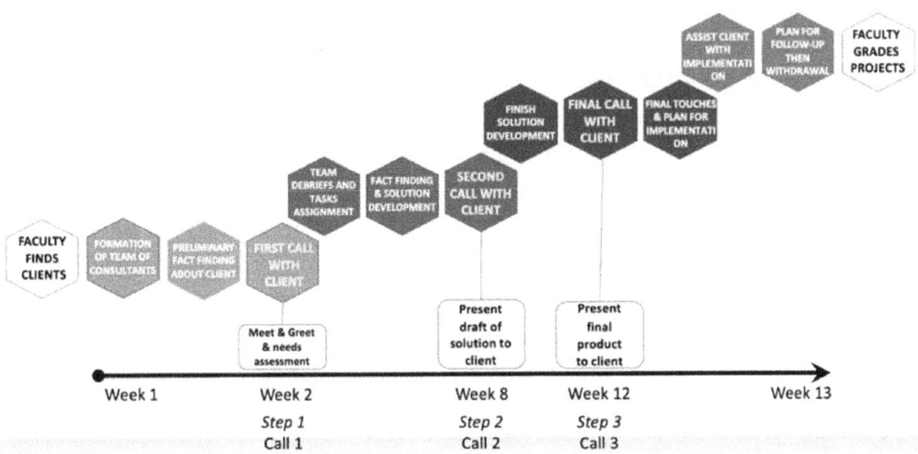

Description of the Online Consulting Model

Figure 2 shows the consulting process map used for the online pro bono consulting projects.

Figure 2. Consulting process map
From Online Student Teams Formation to Consulting Project Delivery

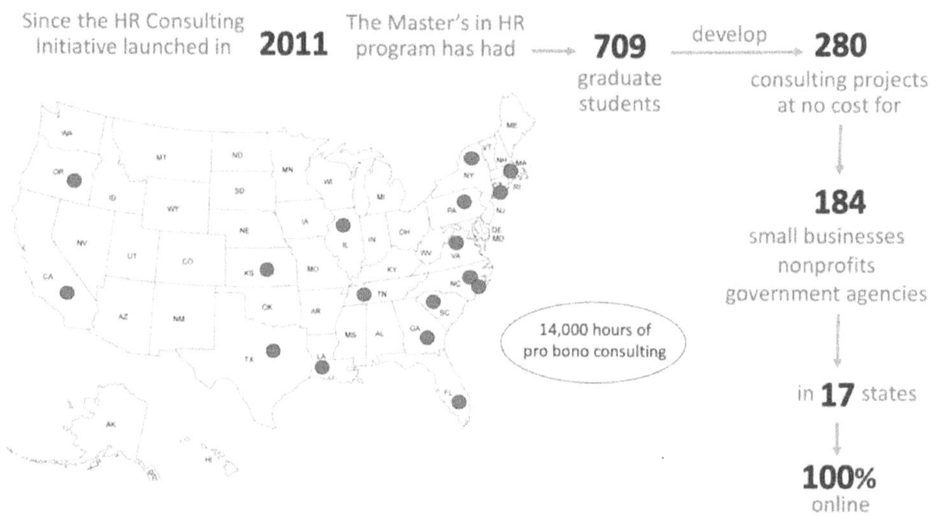

The projects are developed over a 10- to 12-week period. During a typical, 16-week semester, this leaves a 4- to 6-week buffer for team formation and additional assistance to the client once the project is completed, should further assistance be needed. The buffer also allows for an effective closure with the consulting client and for a post-assessment about the consulting service provided. All teams of HR student-consultants follow a consulting process map, which the author developed and fine-tuned over eight years. One of the goals of the process map is to maximize work processes while minimizing contact time with clients who are often focused on their efforts to fulfil the mission of their organization. Through a Project Manager, the team and the client schedule one conference call at each step of the consulting process. In between calls, the team develops the HR or HRD project. Table 4 shows the various technologies used in this online consulting course.

Table 4. Technology used

Technology	Pedagogical use	Format
BlackBoard	Course platform. Discussions	Asynchronous
Zoom video conferencing	Class-wide Faculty video calls	Synchronous
Samepage.io	Collaboration software	Asynchronous and synchronous
Cell phone and iPad for Samepage App	TXTing / Posting	Asynchronous and synchronous

Key 3.0 and 4.0 Technology Used in the Online Consulting Course, Benefits and Challenges

This section presents a breakdown of the key e-service-learning components: people, communication, technology, and course design.

People

The e-consulting team consists of a faculty member, a project manager (a senior student), the team leaders (elected by each student team), and student-consultants (who sign up for a project). The e-consulting team serves the clients (selected by the faculty member before the beginning of the semester).

Communication

For effective e-service learning, clear communication expectations should be established from the very beginning of class. Between the faculty and the client, these expectations could be outlined in a simple Learning Contract, which both parties sign before the start of the semester (Bennett & Green, 2001). The use of multiple technologies has shown to be effective in communicating between all parties: TXTing, conference-calling, and emailing, which falls under 3.0 technology. TXTing is particularly effective when team members need a quick response to a question, such as setting a team online meeting day and time, or to let the team know an updated document has been posted in the team's collaboration platform.

Also, a clear timeline with specific deadlines (mid-point /project due date) needs to be communicated to students in both BlackBoard and the course syllabus (3.0 technology).

Faculty should consider using an online Collaboration and document-sharing platform, which students and clients contribute to throughout the life of the consulting projects (Samepage,io, Slacks, etc.) For the few conference calls scheduled, a common online meeting space should be used (such as Zoom, Skype or GoToMeeting). If a faculty member manages several simultaneous projects, it is preferable that each team creates its own live Zoom / GoToMeeting / Skype room to avoid double meetings in the same virtual room (3.0 technology and education).

The faculty member should role model communication patterns (Hunter, 2007; Tabor, 2007). As such, she should remain actively engaged from the beginning to the end of the consulting process and maintain high visibility in discussion and collaboration platforms related to the projects. She should provide feedback as necessary, and attend each team-client phone calls. Establishing clear channels of communication between the professor and the students is essential to prevent disengagement. Faculty members should tell students that she monitors their team-specific collaboration platforms regularly.

Technology

E-service learning often requires the use of both synchronous and asynchronous tools. Synchronous tools include audio and live video streaming/teleconferencing and TXT chat rooms, when available (3.0 and 4.0 technology and education). Asynchronous technology includes emails, online course platforms and their embedded tools (short videos, discussion boards, drop boxes), and collaborative platforms such as Samepage, Slacks or SharePoint (Malvey et al., 2006; Waldner, McGorry, & Widener, 2010; Straight & Sauer, 2004). The combined use of technologies bridges the gap between synchronous and asynchronous technologies (Waldner, et al, 2010). A collaboration and document management platform (such as Samepage.io) serves as "communication central" for all parties. A shell is developed for each team/client/project but basic customization with the client's logo and information is possible and recommended to make the client feel unique. The page contains all parties' contact information, a calendar with deadlines, a document upload section, and an email/live chat. Samepage also offers a free App, which allows users to access the information from different types of electronic device and to send TXT (4.0 technology).

During the second week of class, all parties (faculty, project managers, team leaders, students, and clients, if possible) should receive online training on how to use the collaboration and communication platform. Ideally, the faculty member should have easy access to IT support. Although it is not always possible, Waldner and colleagues (2012) suggest that the University's instructional design team be a partner in the e-service learning initiative (Malvey et al. 2006). Also, since the teams rely heavily on technology to perform their work, enrolling clients that are open to using technology may be a condition for receiving pro bono consulting assistance (Stoecher, Hilgendorf & Tyron, 2008; Waldner et al., 2010).

Course Design

Traditional service-learning practices can inform e-service learning initiatives and course design. Imperial, Perry and Katula (2007) have identified seven design principles to help facilitate success in traditional service-learning courses. These principles, which are relevant to e-service learning include a connection between the service activity and the learning objectives, an appropriate time commitment, faculty commitment, perceptible impacts of the activity, a student reflection paper, feedback from loops, and student input.

Cost of the e-Service-Learning Initiative

If the project manager is recruited from the current student pool, the cost of running an e-consulting initiative is minimal to null. Most universities that offer online courses use calling and video conferencing software. However, it is advised that an online collaboration and document-sharing platform be used to manage the workflow from the teams. Samepage.io offers an adequate platform at a very moderate cost (a few hundred dollars per year for a not-for-profit educational institution). It offers synchronous and asynchronous tools for effective communication between all consulting parties and the professor and for project management in general (3.0 and 4.0 technology).

Limitations

E-service learning does have unique limitations related to communication and interactions between the consulting parties, the technology, and the added workload for the faculty member.

Establishing healthy and steady communication patterns between the teams, the clients, the project managers and the team leaders can be challenging if communication norms are not role-modeled by the faculty and if clear expectations are not set early on. An online environment can make students and clients less accountable. If a client fails to respond to a request for feedback in a timely manner, the team may stall and team morale may be affected (Hunter, 2007). Occasional synchronous sessions can alleviate this communication limitation. Scheduling conflicts between students may also slow down a team's progress. Using Samepage's or Slack's (or another online collaboration platform) live chat may aid students communicate more effectively. Also, students should be reminded that traditional phone lines are acceptable communication tools, if needed.

The second limitation lies in the consulting team's heavy reliance on technology. Assuming that online students (and clients) have consistent access to and knowledge about basic software and hardware, machines occasionally malfunction, and "when technology goes awry in an e-environment, the effect is exponential" (Malvey et al., 2006, p. 192). Ensuring that all parties have access to IT support is advised. Oftentimes, professors can suggest solutions to most technology issues. Strait and Sauer (2004) recommend that faculty be comfortable teaching online before attempting to integrate an e-service component in their courses. At minimum, clients need to have access to a phone to hold the team calls, although giving them (free) access to the online, user-friendly collaboration platform Samepage is ideal to ease communication and to prevent an excessive exchange of emails.

Appropriate training can minimize many of the above challenges. Solid course design, clear expectations, and synchronous communication between all parties can alleviate many hurdles. Despite the above limitations, e-service learning opportunities can be extremely self-fulfilling. They may boost a professor's service requirement, and may be an effective promotional avenue for the University. They also allow the client to promote his company and products. And above all, the benefits of e-service learning to online students are limitless: practical application of knowledge; serving the community; learning about nonprofit management; aiding organizations that cannot afford to pay for a traditional consultants or employees; and boosting the students' resumes as they can list "co-consulting" through the University as a new skill. In essence, they do well and they do good, something the current and the next generation of students and workers value and expect from their education and work.

To best serve Gen Z students, universities will have to consider a variety of factors.

First, because boredom sets in easily for today's students, there is a need to infuse excitement from visual and auditory inputs (technology 4.0 and 5.0). This can be achieved by using multiple sources—the Internet, videos, and experiences. Students may be led to internalize the basic concepts from different sources. Faculty can take facts from studies and apply them to different situations with adaptive alterations. They should encourage the formation of student teams and give open-ended projects involving several subjects, such as entertainment or real business issues. For example, while teaching economics, students can work on a project on valuation models for basketball players. They can practice fundamental skills rigorously but in a fun way. Faculty could also gamify tests by running them as competitions (technology 5.0).

To engage students, faculty should think differently and come up with original concepts for both course content and student evaluations. They must build innovative ideas. They should also strive to go cross-discipline. Engaging online students in consulting activities for organizations could help respond to this need.

CONCLUSION

Following the development of Web 1.0 to Web 5.0, the past twenty years have seen an exponential growth in online education, both in public, for-profit, and Ivy League institutions. As students continue to seek flexible alternatives to traditional, on-campus courses, online education will continue to evolve and overall enrollment in online courses will likely continue to rise. While in the U.S., 2019 marks the 16th consecutive year of enrollment growth in online courses, it is possible that the number of students who only take on-campus courses will keep dropping, in part because more students are combining on-campus and online courses or they want the convenience of distance education.

Experts predict that online for-credit courses, online training, and MOOCs will continue to be an important supplement to classroom learning and a tool for professional development so that learners can remain competitive in the workforce. Classrooms are more globally connected and mobile apps are more readily used in courses to interact and access course content, as shown in the case study in this chapter. In addition to learning new material, students want the course content to be relevant to their work and are seeking to gain hands-on experience to build their resumes, even when taking online courses. E-service-learning has shown to effectively meet these demands while supporting the research framework that focuses on the intersection of technological, pedagogical, and content knowledge to emphasize new knowledge base (Mishra & Koehler, 2006). Free from geographical constraints, e-service-learning adds a new dimension to online learning and promotes engagement while meeting the learning needs of Gen Z students and the demands for experience by organizations. It also responds to universities' strategic goals of community engagement and Corporate Social Responsibility, civic engagement, enhanced learning outcomes, and professional development skills. With its low cost and its high, collective impact, e-consulting has a promising future in the realm of higher education.

The field of education is affected by several technological trends. One trend is the increasing use of modern technologies (Web 3.0, 4.0, and 5.0) in courses to enhance curricula. Technology has had a transformative impact on the education system worldwide, comprehensively impacting how students learn, share, and gather information. Remote technologies such as videoconferencing and robotic telepresence will be used more widely to allow for more face-to-face interactions among students and instructors who are located across the country or across the world. Virtual reality and artificial intelligence (Web 5.0) will be more readily integrated into courses, which will allow students across disciplines to learn in simulated environments. Some online classes already have chat bots that serve as teaching assistants. These features will enhance the online learning experience for students while helping instructors deliver the course content. Specifically, technologies such as virtual reality and simulations may contribute to more online degree offerings in health-related fields since online programs can now virtually teach students how to interact with patients. Several medical schools such as the Philadelphia College of Osteopathic Medicine, are already incorporating virtual reality in their courses (Friedman, 2016). 2U, a company that partners with colleges and universities to create online graduate programs, recently launched a physician assistant blended online master's with the Ivy League Yale University (Yale School of Medicine, 2018). The company has also partnered with the University of Southern California to create a partially online Doctor of Physical Therapy program, which launched in

2018. Gamification (learning presented in a game format) is also expected to rise in popularity.

It is difficult to predict what the future of education will be like and what role Web 5.0 will have in the delivery of course content. Some experts predict that majors may be eliminated and replaced by goal- and career-oriented learning; others believe that the information from classes will be available online for self-paced learning. Small classes may be replaced by digital learning, large spaces, and open labs; some anticipate more gamification of learning. It is possible that game-based learning may be used to reinvent the university experience and push students to be innovators. Field trips may even become digital with the use of virtual reality. The use of artificial intelligence may possibly become the norm in distance education for students and in training for employees. Companies such as Udacity are tapping into this new short online course market by offering "nanodegrees programs" to learn new work skills. Finally, it is expected that, to remain competitive, colleges and universities will continue to expand their online course offerings and find ways to make online courses more engaging and relevant to students while meeting their own financial and strategic goals. They will strive to present learning as an *experience* for students, one that has an impact on their personal (social responsibility) and professional lives (résumé-builder). Integrating a service-learning component in online courses has the potential to help universities meet these needs. Finally, globalization is also leading this growing trend in developing countries that have limited infrastructure and where internet access is rare. In several regions of the world, particularly in developing countries, online education is considered a promising means to increase access to education. Online education continues to gain traction because it is perceived as a cost-effective means to widen educational opportunities (Gates Annual Letter, 2015; Germain, 2019; Trines, 2018). Great strides in distance education have been made in India, China, South Korea, Malaysia, and Africa where online education is considered vital for increasing capacity and upskilling the growing population and as a possible solution to widen access to education.

REFERENCES

Allen, E. I., & Seaman, J. (2010). *Class differences: Online education in the United States, 2010*. Sloan Consortium. Retrieved from http://sloanconsortium.org/publications/survey/class_differences

Babson Research Group. (2016). *WCET Distance Education Enrollment Report 2016*. Retrieved from http://wcet.wiche.edu/sites/default/files/WCETDistanceEducationEnrollmentRepo rt2016.pdf

Babson Survey Research Group. (2019). *International Reports*. Retrieved from https://www.onlinelearningsurvey.com/international.html

Barnes, L. B., Christensen, C. R., & Hansen, A. J. (1994). *Teaching and the case method: Text, cases, and readings*. Harvard Business Press.

Bates, T. (2018). The 2017 national survey of online learning in Canadian post-secondary education: Methodology and results. *International Journal of Educational Technology in Higher Education*, *15*(29). doi:10.118641239-018-0112-3

Bennett, G., & Green, F. P. (2001). Promoting service learning via online instruction. *College Student Journal*, *35*(4), 491–497.

Cabero, J., Llorente, C., & Román, P. (2007). La tecnología cambió los escenarios: El efecto Pigmalión se hizo realidad [Technology changed scenarios: The Pygmalion effect became real]. *Comunicar*, *28*, 167–175.

Condruz-Bacescu, M. (2015). e-Learning in companies. *Proceedings of the 10th international scientific conference eLearning and software for education*, 2, 101-107.

Friedman, J. (2016). *Blended Learning Options Rise at Medical Schools*. Retrieved from https://www.usnews.com/education/best-graduate-schools/top-medical-schools/articles/2 016-03-23/blended-learning-options-rise-at-medical-schools

Gates Annual Letter. (2015). *Our Big Bet for the Future*. Retrieved from https://www.appropriations.senate.gov/imo/media/doc/hearings/032615%20Gates%20Foundation%20Annual%20Letter%202015%20-%20SFOPS.pdf

Germain, M.-L. (2019). *Integrating Service-Learning and Consulting in Distance Education*. London, UK: Emerald Group Publishing. doi:10.1108/9781787694095

Gerstein, J. (2015). *The Educator as a Maker Educator*. Kindle Edition. Amazon Digital Services LLC. ASIN: B016Z5NZ6O

GNU.org. (2018). *What is free software?* The Free Software Definition. The GNU Project.

Goldy-Brown, S. (2017, December 6). *11 Benefits of MOOCs (Massive Open Online Courses)*. Retrieved from https://plexuss.com/news/article/benefits-of-moocs

Hardigan, S. (2007). *The Learning Revolution has Begun*. Retrieved from http://www.stevehargadon.com/2007/04/web-20-and-school-20-connection.html

Hunter, D. (2007). The virtual student/client experience. *The Journal of American Academy of Business, Cambridge*, *12*(1), 88–92.

Imperial, M. T., Perry, J. L., & Katula, M. C. (2007). Incorporating service learning into public affairs programs: Lessons from the literature. *Journal of Public Affairs Education, 13*(2), 243–264.

Kambil, A. (2008). What is your Web 5.0 strategy? *The Journal of Business Strategy, 29*(6), 56–58. doi:10.1108/02756660810917255

Keats, D., & Schmidt, J. P. (2007). The genesis and emergence of Education 3.0 in higher education and its potential for Africa. *First Monday, 12*(3). doi:10.5210/fm.v12i3.1625

Killian, J. (2004). Pedagogical experimentation: Combining traditional, distance, and service learning techniques. *Journal of Public Affairs Education, 10*(3), 209–224. doi:10.1080/15236803.2004.12001360

Lazar, J., & Preece, J. (1999). Implementing service learning in an online communities course. In *14th Annual Conference International Academy for Information Management* (pp. 22–27). Charlotte, NC: Academic Press.

Lehman, R., & Conceição, S. (2010). *Creating a sense of presence in online teaching: How to "be there" for distance learners.* San Francisco, CA: John Wiley & Sons.

Littlefield, J. (2017, August 31). *Earn an Ivy League Degree Online.* Retrieved from https://www.thoughtco.com/earn-an-ivy-league-degree-online-1098183

Littlefield, J. (2017, August 31). *Earn an Ivy League Degree Online.* Retrieved from https://www.thoughtco.com/earn-an-ivy-league-degree-online-1098183

Longworth, N. (2003). *El aprendizaje a lo largo de toda la vida.* Barcelona: Lifelong Learning.

Malvey, D. M., Hamby, E. F., & Fottler, M. D. (2006). E-service learning: A pedagogic innovation for healthcare management education. *Journal of Health Administration Education, 33*(2), 181-198.

Marks, J. (2012). *Who's Afraid of the Big Bad Disruption?* Retrieved from https://www.insidehighered.com/views/2012/10/05/why-moocs-wont-replace-traditional-i

Mills, S. (2001). Electronic journaling: Using the web-based, group journal for service-learning reflection. *Michigan Journal of Community Service Learning, 8*(1), 27–35.

Mishra, P., & Koehler, M. J. (2006, June). Technological pedagogical content knowledge: A framework for teacher knowledge. *Teachers College Record, 108*(6), 1017–1054. doi:10.1111/j.1467-9620.2006.00684.x

Rama, D. V., Ravencroft, S., Wolcott, S. K., & Zlotkowski, E. (2000). Service-Learning Outcomes: Guidelines for Educators and Researchers. *Issues in Accounting Education, 15*(4), 656–689.

Report, O. C. (2019). *The Best Resource for Open Source, Paid, and Career-centered Online Education.* Retrieved from http://www.onlinecoursereport.com/

Selingo, J. J. (2014). *MOOC U: Who is getting the most out of online education and why*. New York, NY: Simon and Schuster.

Selingo, J. J. (2014a). *Demystifying the MOOC*. Retrieved from https://www.nytimes.com/2014/11/02/education/edlife/demystifying-the-mooc.html

Shah, D. (2016). *By the numbers: MOOCs in 2016*. Retrieved from https://www.class-central.com/report/mooc-stats-2016/

Sokolova, S. (2014, December 28). *What are the advantages of MOOCs and how can you benefit from them?* Retrieved from https://www.linkedin.com/pulse/what-advantages-moocs-benefits-from-them-siyana-sokolova

Stoecker, R., Hilgendorf, A., & Tryon, E. (2008). Incorporating technology in service-learning: A case study of appropriate use. In A. Dailey-Hebert, E. Donnelli-Sallee, & L. DiPadova-Stocks (Eds.), *Service-eLearning: Educating for citizenship*, (pp. 31-44). Charlotte, NC: Information Age Publishing.

Strait, J., & Sauer, T. (2004). *Constructing experiential learning for online courses: The birth of e-service*. Retrieved from http://er.educause.edu/articles/2004/1/constructing-experiential-learning-for-

Tabor, S. W. (2007). Narrowing the distance: Implementing a hybrid learning model for information security education. *Quarterly Review of Distance Education, 8*(1), 47-59.

The Guardian. (2017). *Do Online Courses Spell the End for the Traditional University?* Retrieved from https://www.theguardian.com/education/2012/nov/11/online-free-learning-end-of-university

Trines, S. (2018). *Educating the Masses: The Rise of Online Education in Sub-Saharan Africa and South Asia*. Retrieved from https://wenr.wes.org/2018/08/educating-the-masses-the-rise-of-online-education

Waldner, L. S., McGorry, S. Y., & Widener, M. C. (2010). Extreme e-service learning (XE-SL): E-service learning in the 100% online course. *MERLOT Journal of Online Learning & Teaching, 6*(4), 839–851.

Waldner, L. S., McGorry, S. Y., & Widener, M. C. (2012). E-Service-Learning: The Evolution of Service-Learning to Engage a Growing Online Student Population. *Journal of Higher Education Outreach & Engagement, 16*(2), 123–150.

Wilhite, S., & Silver, P. (2005). Educating citizens vs. educating technicians: A false dichotomy for higher education. *National Civic Review, 94*(2), 46–54.

Yániz, C., & Villardón, L. (2006). *Planificar desde competencias para promover el aprendizaje. El reto de la sociedad del conocimiento para el profesorado universitario* [Planning from the competence perspective to promote learning. Challenge of the knowledge society for university teachers]. Bilbao, España: Deusto University.

Chapter 8
Educator Perceptions of Digital Devices:
Multitasking and Distractions in the Classroom

Julie A. Delello
https://orcid.org/0000-0002-4326-8096
The University of Texas at Tyler, USA

Jean Kiekel
The University of St. Thomas, USA

Susan R. Poyo
Franciscan University of Steubenville, USA

Mia Kim Williams
The University of Wyoming, USA

Deborah Kerby
Pocono Mountain School District, USA

ABSTRACT

There is a growing concern regarding student use of digital devices such as smartphones during classroom instruction. This pilot study used a mixed-method survey to collect data from 146 K-16 educators. Eighty percent of the participants said that students multitask, and 50% said they multitask a lot. Many of the instructors reported they are distracted by students' use of phones while more than half of the participants indicated they felt doing multiple tasks at the same time disrupts class time and may impact a students' ability to learn. These findings have important implications because classroom instructors may need to modify instructional strategies to keep students on task, have clear policies on the use of electronic devices, and ensure that students understand instructor expectations.

DOI: 10.4018/978-1-7998-2914-0.ch008

INTRODUCTION

Society has become subjected to an unprecedented number of interruptions (Anderson & Rainie, 2014). In fact, we have become a generation of people who are "always-on" and "always connected." We read, answer the phone, watch television, listen to music, interact with social media, and text while commuting, and many times we perform several of these tasks at the same time (Mokhtari, Delello, & Reichard, 2015). Goleman (2013) suggested that we have a divided attention as we attend to multiple stimulations at the same time. Such an "explosion of news streams, emails, phone calls, tweets, blogs, charts, reflections about opinions that we expose our cognitive processors to daily" (Goleman, 2013, p. 56) may be a source of brain interference leading to cognitive overload.

According to Cisco (2019), 8.4 billion hand-held, mobile-ready devices will be utilized globally by the year 2020. Statista (2019) reported that the number of cell phone users will exceed five billion in 2019. In addition, mobile devices are fast becoming part of students' lives. In fact, almost half (45%) of U.S. children aged 10 to 12 have a smartphone, while 97% of U.S. teens aged 13 to 17 own a device that is capable of connecting to the Internet such as a smartphone, tablet or laptop (Howard, 2017; Rideout & Robb, 2018). According to Common Sense Media (2017), 35% of children under the age of eight use mobile devices, spending an average use time of 48 minutes per day. The average age of smartphone ownership is 10.3 years of age (Ruston, 2017). Further, 95% of teenagers own a smartphone or have access to one, and of these teens, 45% are online constantly (Anderson & Jiang, 2018). Most use their smartphones to access social media "including 16 percent who use it almost constantly and another 22 percent who use it several times an hour" (Common Sense Media, 2018, p. 11).

Cell phone addiction has become a serious concern, enough so that smartphone manufacturers are now making software available to help people avoid it (Bank My Cell, 2019). Some researchers have reported that almost 50% of teenagers are addicted to their cell phones with almost 25% claiming an addiction to social media (Common Sense Media, 2018). Additionally, 75% of teens said they felt compelled to respond immediately to texts and social media posts (Rosen, Whaling, Rab, Carrier, & Cheever, 2013). As defined by the American Society of Addiction Medicine (2011), addiction is a "primary, chronic disease of brain reward, motivation, memory and related circuitry" (para. 1) and it indicates an individual's inability to control their own behavior.

Researchers have noted that the use of digital devices such as smartphones may enhance classroom learning; however, distractions and student dependency may also create disruptions (Anshari, Almunawar, Shahrill, Wicaksono, & Huda, 2017). Glass and Kang (2018) stated, "The intrusion of internet-enabled devices (laptop, tablet, and cell phone)" (p. 395) has created a classroom of distracted students who are attention divided. As a result, an increasing number of schools are limiting or prohibiting the use of mobile devices in classrooms.

BACKGROUND

Student Use of Devices in the Classroom

Technology use in society has increased over the past 20 years, requiring that more technology be used in PK-16 settings to prepare students for the workplace and the future. Using technology to enhance teaching and learning is a common educational goal. Across the United States, the ratio of students to instructional computers with Internet access is approximately 3:1 (Liu, Ritzhaupt, Dawson, & Barron, 2017). While the results of technology in the classroom has been mixed, many positive results on student achievement have come of technology integration. These include: increased student engagement, students taking a more active role in their own learning; creation of a more collaborative learning environment; ability to receive immediate feedback; improvement in writing ability; the ability to access content anywhere, anytime, and any place; enhance problem-solving skills; and, create a more inclusive classroom that removes curricular barriers (Liu, et.al. 2017; Lopez-Vargas, Ibanez-Ibanez, Racines-Prada, 2017).

Although digital technologies can be used for positive educational outcomes, recent research suggests that these same digital technologies may impair student performance. For example, students constantly report checking their smartphones to prevent boredom and some have even developed a habitual, anxiety-laden behavior (Rosen, Carrier, & Cheever, 2013; Smith, 2015). Seventy-two percent of teens reported that they look at their phones for messages or notifications upon waking and 40% feel anxious if they are without it (Jiang, 2018). These students fear missing something if not constantly connected. De-Sola Gutiérrez, Rodríguez de Fonseca, and Rubio (2016) defined these behaviors as pathologies including "Nomophobia (No-Mobile-Phobia), FOMO (Fear of Missing Out)," and "Textaphrenia," the fear of being disconnected (p. 2).

Concern must be given to what Ward, Duke, Gneezy, and Bos (2017) termed "brain drain." Brain drain is the hypothesis that merely the presence of one's smartphone or similar device may cause a reduction in cognitive capacity due to the effort spent in not using it when one's attention is focused on something else. This study demonstrated that cognitive capacity, including "working memory capacity and functional fluid intelligence" (p. 10) are negatively impacted just from the presence of one's smartphone. It also determined that the closer one is in proximity to one's smartphone, the greater the negative impact, and being physically separated from one's device reduced the negative impact on cognitive capacity.

The Kaiser Family Foundation (2013) reported that, "The continued expansion of high-speed home Internet access, the proliferation of television content available online, and the development of compelling new applications such as social networking and YouTube, have all contributed to the increase in the amount of media young people consume each day" (p. 3). These daily disruptions from communication technologies often lead to multitasking, affecting students' time on task within the classroom while also leading to a lack of sleep (Shochat, 2012), an increase in depression (Rosen, Cheever, & Carrier, 2012), and changes to the overall structure of the brain (Dalton, 2013). Furthermore, recent research suggested that freshman college students have more difficulties with multitasking than upper class students (Delello, Mokhtari, &, Reichard, 2016). Students seemingly spend more time on non-academic activities like socializing with friends and less time on school-related activities. Research on digital distractions has shown that students spend an average of 20.9% of their time in the classroom using digital devices for non-class activities (McCoy, 2016). More than half of American teens stated that the distraction of social media interferes with their studies and personal relationships (Common Sense Media, 2018). Studies have also shown that student cell phone use leads to lower grades and lessened long-term retention (Glass & Kang, 2018). Junco (2012) found that students' frequent multitasking with text messaging and social media, e.g. Facebook, created a cognitive overload and in turn, lowered grade point averages (GPA). Kuznekoff and Titsworth (2013) reported that college students who did not use cell phones wrote down 62% more information in their notes and scored better on a multiple choice test.

A study of distractions from cell phone use among secondary school students in a Bring Your Own Device (BYOD) environment found that a majority of participants, 85%, engaged in some distracting activities during instruction (Kay, Benzimra, & Li, 2017). Of these 181 participants, 64% emailed, 65% surfed the Web, 62% used social media, 32% instant messaged, and 30% played games on occasion or regularly during class time. Furthermore, about 30% admitted to being "distracted by a peer's laptop on occasion or sometimes," while 15% admitted to being distracted by a peer's laptop between "often" and "always" (p. 10). An interesting takeaway of this is that students stated they would have less distraction if teachers would "implement more restrictions including stricter supervision during class, blocking certain websites, only using laptops when needed, and limited Wi-Fi access" (p. 11). Awwad, Ayesh, and Awwad (2013) explored the impact of distraction from laptops on female undergraduate students. Nearly half of the participants in this study indicated that use of a laptop negatively impacted their concentration in class. A majority of participants indicated they used laptops in class more for non-educational activities such as email, gaming, and even watching movies than for educational purposes.

Educator Divide on Devices in the Classroom

Initiatives have been put into place to help educators use technology to improve teaching and learning. For example, The US Department of Education's National Education Technology Plan (NETP) promotes the use of technology in classrooms (USDE, 2017). In fact, the first goal of the NETP is that all students be provided "engaging and empowering learning through technology" (p. 9). In 2016, The International Society for Technology in Education (ISTE) revised the educator and student standards. According to ISTE (2016), "These standards are a framework for students, educators, administrators, coaches and computer science educators to rethink education and create innovative learning environments" (para. 1).

The Pew Research Center noted that innovative shifts would occur on college campuses and include "cloud-based computing, digital textbooks, mobile connectivity, high-quality streaming video and "just-in time" information gathering" (Anderson, Boyles, & Rainie, 2012, p. 2). These same disruptions are emerging in K-12 environments. In fact, the Consortium for School Networking (2019) recently listed hand-held or wearable devices such as smartphones and quantified-self sensor technologies, blended learning, cloud technologies, extended realities, and data analytics as the top five enablers to drive the future of teaching and learning. Yet, many instructors find difficulties in integrating basic technologies into the classroom (Marzilli, Delello, Marmion, McWhorter, Roberts, & Marzilli, 2014) or remain skeptical that digital technologies can really transform teaching and learning (Herold, 2019). According to a national study by Common Sense Media (2012), an average of 71% of teachers surveyed reported media is reducing students' attention spans.

According to Johnson, Jacovina, Russell, and Soto (2016), instructors may face challenges to such technology integration. These include barriers such as a lack of funding or access to equipment, inadequate training for students and teachers, time-constraints, low band-width, or a general lack of knowledge and support for technology integration. Some research has even suggested that technology use may be more of a technological integration rather than a curricular integration (Hutchison & Reinking, 2011). Bowman, Levine, Waite, and Gendron (2010) noted that "the benefits must be weighed against the tendency students have to use media for irrelevant, social purposes that may distract significantly from the target academic tasks" (p. 930).

Device Policies and Procedures

The ubiquity of mobile devices such as smartphones has penetrated school campuses. Yet, many school districts are struggling with whether to permit such devices, have regulations for them, or ban them all together. For instance, in 2009, researchers found that a small percentage (12%) of teenagers could utilize their cell phones in school at all times; 62% were able to have a cell phone at school but not utilize it in class, and 24% of those surveyed were prohibited from all cell phone use (Lenhart, Ling, Campbell, & Purcell, 2010). More recently, a group of researchers, who developed the *Away for the Day* (AFTD) initiative, surveyed 1200 parents. The parents reported that more than half (56%) of their students' middle schools allow phones to be carried all day (Ruston, 2017). In a Pearson (2015) study of over 2,000 students, 53% of elementary school students, 66% of middle school students, and 82% of high school students surveyed noted using smartphones regularly; and, over 40% of these students stated they used their phones for schoolwork at least twice a week.

Campus Technology (2018) noted in their third annual *Teaching with Technology* survey that 26% of faculty members reported banning mobile phones in class while another 49% allowed them but with limitations. Globally, some countries have decided to ban all digital devices. For example, French students in elementary and middle schools have been restricted from bringing mobile devices like smartphones and tablets to class (Lewandowski, 2018). Schools in countries such as the United States, Canada, Australia, and Ireland are also reviewing their policies on phone use and social media use in schools (Strauss, 2018). Mobile device use has had mixed reactions from teachers and administrators. In a study titled *Growing up Digital Alberta*, 71% of teachers felt that technology supported learning but also reported that digital device use contributed to students being tired, distracted, or having increased emotional challenges (Alberta Teachers' Association, 2015). Many educators believe that better policies should be enacted to help guide student phone use (Pahomov, 2015). Yet, for those policies that do exist, research has shown that they range from vague to specific and for many educators, they may be difficult to enforce (Mupinga, 2017).

Theoretical Framework

In this study, we utilized a framework of Media Multitasking (MMT) and Cognitive Load Theory (CLT). A media multitasking framework characterizes individuals who constantly switch back and forth engaging with different media, and who are unable to filter out extraneous distractions (Ophir, Nass, & Wagner, 2009). Rideout, Foehr, and Roberts (2010) reported that American youth use media 7.5 hours per day and 29% of that time is spent looking at multiple media streams simultaneously. Other research published in the American Academy of Pediatrics suggested that heavy MMTs resulted in reduced cognitive memory, increased psychosocial behavior (e.g. impulsivity), and changes in the overall neural structure (decreased volume in the anterior cingulate cortex) (Uncapher, Rin, Rosen, Kirkorian, Baron, Bailey… &Wagner,, 2017). Rose (2010) reported that we live in an age of interruption and although our devices promote an opportunity to connect, they provide a means for a compulsive diffusion of our attention. The human brain has only so much capacity. When students are distracted during the learning process, it can affect their ability to learn the material. This can have detrimental effects on overall learning (Judd, 2015; Mokhtari, Delello, Reichard, 2015; Qian & Li, 2017; Rekart, 2011).

Learning is a complex function. Learning is important because we are not born with the skills and abilities necessary to function; they must be obtained. According to the dictionary, learn means "to acquire knowledge or skill in by study, instruction, or experience" (See https://www.dictionary.com/browse/learn?s=t). Learning, then, is the "act or process of acquiring knowledge or skill" (See https://www.dictionary.com/browse/learning?s=t). Two key characteristics of learning are that initial learning is required for transfer and all new learning involves transfer based on previous learning (Bransford, Brown, & Cocking, 2000). Cognitive load theory (CLT) is the study of how the human brain processes information and constructs new knowledge so that learning can take place, moving that information from working memory to long-term memory (Lopez-Vargas, et.al., 2017; Sweller, 1998). CLT is broken into three parts: Intrinsic cognitive load, extrinsic cognitive load, and germane cognitive load. Intrinsic cognitive load is the load on the brain created by the tasks at hand. Extrinsic cognitive load comes from factors not related to the learning tasks at hand such as multitasking. Germane cognitive load is the amount of load dedicated to processing information and is closely related to intrinsic cognitive load (Korbach, Brunken, & Park, 2018; Lopez-Vargas, et.al, 2017).

The amount of cognitive load affects what goes into working memory and ultimately long-term memory. Working memory is the temporary storage of information as it is being processed from short-term to long-term memory. It is limited and most affected by cognitive load (Jaeger, Shipley, & Reynolds, 2017). When the brain is overloaded by external forces, cognitive load increases. When cognitive load is heavy, such as when new learning is taking place, any distraction caused by digital content may reduce a student's attention and focus and negatively impact transfer of new information to long-term memory (Kraushaar & Novak, 2010). Long-term memory is important to future learning and eventual performance because once information is part of long-term memory there is some permanence to it. This permanence reduces future cognitive load by making transfer and application of previous knowledge easier to access. When we switch between tasks, processing information that is irrelevant, interference is created that may result in a reduction in attention, working memory, and goal management (Gazzaley & Rosen, 2016). Gazzaley and Rosen (2016) proposed that humans are poor multitaskers and need to find a balance in the use of technology by limiting the urge to switch back and forth between tasks. When two or more stimuli are perceived simultaneously, our brains are unable to process them (Pashler, 2000).

In this non-experimental descriptive study, the authors sought to explore educators' perceptions regarding the use of digital devices in the classroom, including the kinds of devices students use; the types of distractive, multitasking habits students exhibit; the policies campuses used to deter such digital distractions; and, what strategies educators recommend preventing such disruptions. Therefore, the following research questions guided the study:

RQ1: How do educators perceive digital distractions in the classroom?
RQ2: What types of policies exist for instructor and student technology use?
RQ3: What recommendations do educators have to limit digital distractions in the classroom?

MAIN FOCUS OF THE CHAPTER

Methods

This study employed a non-experimental descriptive approach using a mixed-method survey design. This pilot study involved the collection of data using a single survey, which yielded quantitative and qualitative data using both closed- and open-ended questions (Yin, 2014).

Instrument

A 19-question, mixed-methods survey was created to assess educators' perceptions of technology usage in the classroom. Specifically, there were eight demographic questions, which assessed gender, ethnicity, grade level, subject area taught, location, years of experience, education level, and professional membership. Three additional questions asked about the types of devices their students used, seven questions were related to multitasking and distractions, and an additional question was related to the schools' policies. Furthermore, each question had an open-ended comment box to allow participants the opportunity to provide additional feedback. The participants in the study were not required to answer all of the questions, which created a varied number of responses per item. This survey was piloted with a sample of five representative educators before it was released publicly in order to identify any problems with the original set of questions.

Participants

Participants in this pilot study consisted of 146 educators who were employed in K-16 schools located around the globe. Members of the research team contacted educators through postings on the ISTE Connect professional learning network (PLN) community board and invited them to complete the online survey (see ISTE.org). Additionally, an invitation to participate along with the survey link was posted on a list serve for faculty at two additional institutions and tweeted out to educators on the social media platform Twitter. Before beginning the research, the University Institutional Review Board (IRB) approved the study and consent was obtained from the participants. Demographics of participants who chose to answer the question regarding their current position is displayed (see Table 1).

Table 1. Participants Current Position

Position	Number (N = 146)
PK-5th grade Teacher	20
6th-8th grade Teacher	15
9th-12th grade Teacher	25
Assistant Professor	17
Associate Professor	8
Full Professor	10
Adjunct Professor or Instructor	11
Education Technology	18
Media Specialist	5
Other	17

Demographics of Survey Participants

Participant demographics represented a cross section of educators with respect to gender (Male = 35%, Female = 65%), teaching area (Higher education = 24%, High School = 17%, Middle School = 9%, Elementary = 13%, and other = 37%), and ethnicity (White = 92%, Hispanic = 3%, African-American = 3%, Asian = .7%, Native American = .7%, and other = 1%). The majority of participants were located in the United States (93%) representing 33 states (see Figure 1). Seven percent of educators were from outside the U.S. and included those from Austria, New Zealand, Azerbaijan, Turkey, Venezuela, South Africa, Sweden, Australia, Korea, and Uruguay.

Figure 1. Locations of participants in the U.S.

Twenty-four percent of participants had a doctoral degree, 61% had a professional or master's degree, 14% had a bachelor's degree, and less than 1% noted they had some college experience. When asked how many years of experience in education they had, the majority of participants (21%) reported between 11 and 15 years, followed by 6-10 years (19%), 16-20 years (17%), 21-25 years (13%), 26-30 years (10%), and 0-5 (9%). Eleven percent reported working over 30 years in education.

Of those that responded to the survey, 13% were elementary teachers (PK-5), 10% were middle school teachers (6-8), 17% were high school teachers (9-12), 12% were assistant professors, 6% were associate professors, 7% were full professors, and 37% reported other. Of those who selected "other" as their current role in education, roles included adjunct professor, technology specialists, librarians/media specialists and multi-grade level teachers.

Additionally, respondents taught a variety of subject areas including English, mathematics, science, history, fine arts, physical education, nursing, theology, computer science, technology, business, library, and foreign languages. Seventy-percent of participants were members of a professional learning network or association (e.g. ISTE, TCEA).

Table 2. Years of experience

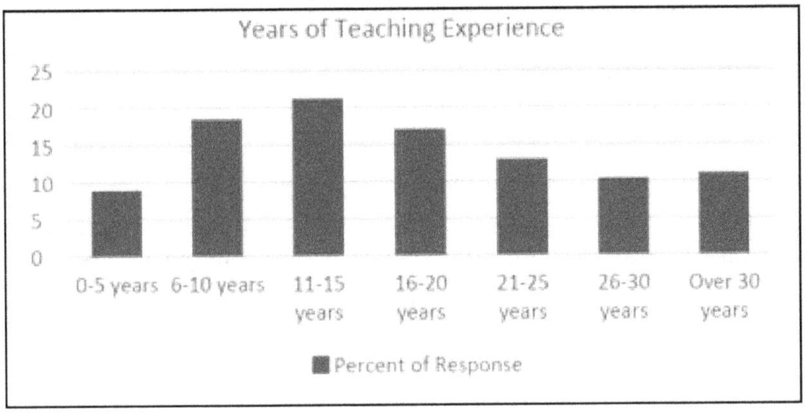

ANALYSIS OF THE DATA

For purposes of this study, the analyses focused on examining educators' perceptions regarding student multitasking and what strategies they perceived would help students to reduce classroom distractions. In order to gain a rich description of the participants' perceptions, the researchers chose a mixed-methods design to collect both quantitative and qualitative data. The quantitative data provide a general understanding of the characteristics of the participants relative to their perceptions

Table 3. Current position or role in education

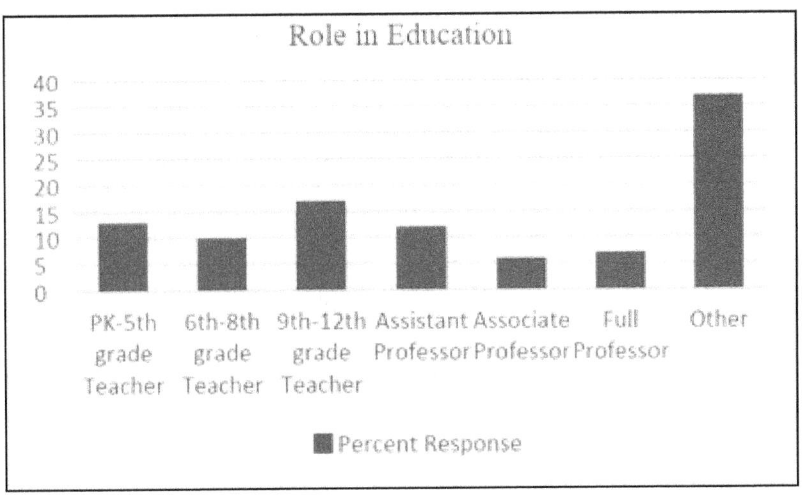

of digital devices and distraction in the classroom. The qualitative data provide results that aid in explaining the context of these perceptions.

Survey Item Responses

Analysis of the data involved the examination of surveys completed within the online survey management program Qualtrics (Qualtrics.com). Descriptive statistics were reported as percentages as well as measures of central tendency such as mean, median and mode and distribution of responses for quantifiable survey responses. Using Stats iQ, correlations between composite variables were performed, as well as chi-square tests of independence.

Analysis of Open-ended Survey Responses

Transcripts were created from the open-ended survey responses and were further analyzed to look for patterns in the data (Merriam & Tisdell, 2016). The researchers utilized an inductive approach that compared each unit of data to create codes and then subsequently created themes by combining codes that were similar. Also, "peer examination" (p. 249) was employed that utilized the team of researchers to enhance the trustworthiness of the findings.

RESULTS

Educators' Perceptions of Digital Distractions in the Classroom

When educators were asked whether students were allowed to bring their own devices (BYOD) to class, 67% noted yes; 34% reported no. Some of the devices used were brought to class by the students (46%) while others were provided by the school (54%). For those educators that allowed devices in class, 30% of those devices were tablets, 26% were laptops, 17% were smartphones with Internet connectivity (iPhone, Android), 12% were cell phones, 11% were desktop computers, and 4% were listed as other (e.g. Nook, Kindle, Smart Watch).

When educators were asked whether they felt distracted when students used electronic devices (e.g. iPhone, iPad, laptop, tablet) during class, 34% marked yes while 66% reported no. Of the habits perceived as distracting, texting (26%), playing games (25%), emailing (13%), browsing the Internet (12%), and social media (10%) use were reported. Participants also noted other (14%) distracting habits varied from items such as listening to music to instant messaging. When educators were asked specifically why devices were or were not distracting to those in the classroom, the reactions were mixed.

Most, 83% who selected PK-5th Grade responded "No" to Question 12 "Do you feel distracted when students use electronic devices (e.g. iPhone, iPad, laptop, tablet) during class?" These participants reported seeing some texting and off-task Internet browsing but none of the other distracted behaviors. Teachers at higher grades levels reported greater levels of multitasking, particularly texting. In fact, 88% of high school teachers report that students text during class at least some of the time. A chi-square test of independence analysis indicated a statistically significant relationship between the participants' current position and their feeling of distraction when students use electronic devices, X^2 (6, N = 146) = 0.347, p > .01. No other demographics (gender, race, years of experience) were found to relate to the feeling of distraction. Those participants who identified themselves as other when asked their current position were less likely to feel distracted when students use electronic devices.

In an open-ended question, 50 participants described why they felt devices were distracting whereas 94 of the participants described their views as to why devices were not distracting. These answers were multifaceted in that many educators perceived that students were disengaged when devices were allowed. Themes of lack of engagement, classroom management, and self-discipline resonated (see Table 4). For example, one educator noted that students "go to other sites and work on other things besides what we are working on as a class." Others reported that they felt that technology is not distracting but that it should be built into the learning experience with guidance. For example, one educator reported, "Its use is planned for in all of my classes."

When educators were asked whether they believed students multitasked, 50% stated, "yes, a lot" while 30% stated, "yes, but rarely", 12% reported students did not multitask, and 8% stated they were unsure if students multitasked (performing two or more tasks simultaneously). Findings revealed that when students multitasked as instruction was taking place, they were texting (25%), browsing the Internet (24%), listening to music (20%), using social media (14%), playing games (8%), streaming movies (2%), watching television (1%), or other uses such as applications (apps) (8%) including those for supplementary classroom support (e.g. Quizlets). Sixty-one percent of the participants felt that multitasking affected a student's ability to learn, 32% reported that it rarely affected students learning, 6% felt it had no effect on a student's ability to learn, while 2% felt students never multitask.

Although most participants noted that multitasking does impede learning, Chi-square analyses were used to determine if a subset of the demographic variables (gender, race, and years of experience) related to the extent of agreement with the perception that multitasking impacts student learning. The results showed no

Table 4. Classroom distraction

Theme		Excerpts from Participants
Yes		
	Engagement	● If they are engaged with that device, they are unable to engage with anything else around them. ● Students easily leave the task at hand to pursue other entertainment. This is primarily with personal phones/tablets. ● When we are doing something whole group, or I am giving instructions for the day, devices are distracting, and kids miss vital information.
	Classroom Management	● A portion of my mental energy and focus is on managing their use on devices. ● Personal devices are not allowed in my school district and I am expected to complete a discipline referral when I see a student using one. ● Must monitor Go guardian as well as student learning and needs. I am tied to a screen.
	Self-Discipline	● WAY too much texting, off task behaviors and some truly troublesome events, given maturity and lack of self-discipline. ● I do not believe there is any chance that a student would have the self-control to simply take notes. The Internet is just too addictive.
No		
	Engagement	● It enriches the learning. ● Love to watch them discover and learn. ● Children are engaged on devices… we need to be teaching them to learn on/with these devices.
	Classroom Management	● Students only use these devices occasionally with permission. ● Students use devices in my class to work. It's no different than if they were using pencils (though the tasks change) ● If students are using their devices, it is because I allowed them to as part of the lesson. ● Expectations are set early and made clear.
	Self-Discipline	● I do not feel distracted but at times, students are distracted by their phones; Snapchat is a big distractor for them.

statistically significant relationship between gender X^2 (3, N = 145) = 0.45, $p > .05$, race X^2 (15, N = 145) = 5.94, $p > .05$, or years of experience X^2 (18, N = 146) = 14.3, $p > .05$, and the perception that multitasking impacts student learning.

Policies for Instructor and Student Technology Use

When participants were asked if their campus had policies and if so, whether those policies were followed, 56% reported yes. A Chi-square analysis was performed to examine the relationship between the instructor position and perception that policies existed and/or were being implemented. There was a strong statistically significant relationship, X^2 (24, 145) = 67.9, $p < .01$.

Twenty-one percent of participants reported that instructors had their own policies for their individual classrooms. The highest percentage of those (25%) that perceived policies were being implemented and assessed as written came from those that were not full-time instructors while the lowest percentage were from higher education (4.2%). However, 12% of the overall participants reported that sometimes policies were in place but not followed while 2% reported they did not have technology policies at all (see Table 5).

Table 5. Campus policies for technology use

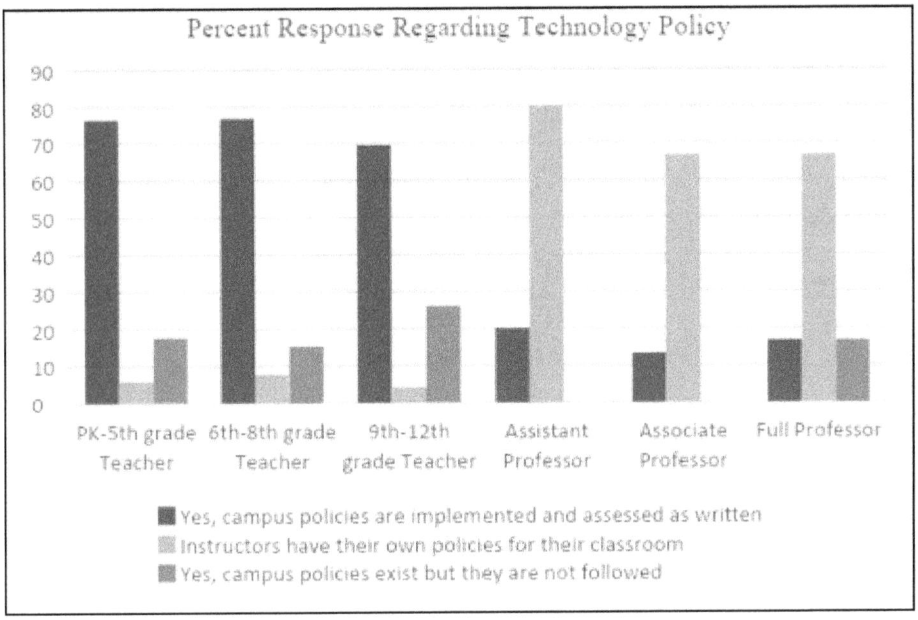

Eight percent listed "other" and revealed that many times, "campus policies do exist, but they are not followed- always a struggle!" Other respondents were unaware of policies because they "had not looked it up." One educator noted, "Honestly, not fully sure but I have been told device policies are unique to the instructor" while a second stated, "Who knows?" An additional participant noted, "I believe so. But individual use in courses is no doubt up to professors."

Recommendations Educators Have to Limit Digital Distractions in the Classroom

When participants were asked what strategies they recommended to limit or prevent multitasking and distractions in the classroom, 24% suggested students should turn devices off while another 24% supported the notion of giving students breaks during class to check their technology. Twenty-three percent supported students silencing devices, while 7% suggested having a strong ban on technology, which would need to be included in a written classroom policy. An additional 3% of educators stated that no action should be taken to limit distractions. Forty-eight (20%) participants reported additional strategies in the open-ended response box. The responses varied while the primary themes found were those of having clear expectations, digital citizenship training, effective classroom engagement, device regulation, device restrictions, and tolerance (see Table 6).

Table 6. Educator strategies

Theme	Excerpts from Participants
Expectations	● Be explicit about your expectations for their device use, and then be consistent in your enforcement of your own expectations. ● Communicate expectations UP FRONT with parents and students together, because sometimes it is the parent calling the student on their phone in the middle of class. ● Be specific about the right use of devices when in class. ● Mine reads: Personal devices may be used for instructional purposes at the teacher's direction. If off-task use occurs, the device will be held until the end of class.
Digital Citizenship Training	● Begin educating parents and students at an early age about the addictions the social media site pose. ● Discuss with student appropriate use of personal devices ● Technology use when appropriate and we define what appropriate looks like ● Teach them devices are tools and have a place and a time, so they need to regulate their own behavior. It is not the device/ tech that needs controlling, it is educating to the student. ● Students must learn the skill of when it is and when it is not appropriate to use technology. If this is not taught, then the lesson will never be learned.
Engagement	● Have a strong, well-designed, relevant, student-driven lesson and 99% of the distractions will not be an issue. ● Make coursework engaging
Regulate Use	● I have them put the devices face down when I need full attention, or I have sporadic "screen checks" when students are on devices and I want to make sure they are where they should be on the device. ● Use Apple classroom during class to control devices ● Put on teacher's desk in a caddy ● If off task use is observed, students "park" devices in my closet, plastic shoe "hammocks" ● We use Hapara, so we can kick kids off of sites. We are 1-1 Chromebooks, so devices are only open when being used. We do not allow any other device.
Ban Devices/Media	● Student devices turned off and left in the locker. Parents have to pick it up if they are caught out at school. ● All social media is blocked at school so kids learn to focus and work with their mobile device, not just "play" ● Somehow block the use of other sites if students access something for the class; some kind of alert to show they are off task. ● Collect devices and they can pick them up as they leave class
Tolerance	● If it's not distracting me or those around them, it's a personal choice

SOLUTIONS AND RECOMMENDATIONS

Active learning in the classroom requires students to attend to the information being presented by teachers and professors in order to move that learning from the working memory to the long-term memory. The working memory is affected by cognitive load, especially extraneous cognitive load (Kalayuga & Singh, 2015). Research has shown that the use of digital devices improves student engagement and helps with active learning. However, this study highlighted that there are side effects to such use including the increase of extraneous cognitive load through multitasking, distractions, and lack of focus on class instruction at all grade levels, PK-16. In fact, 34% of the educators responding reported that in addition to students being distracted by the use of cell phones, educators were also distracted in their instruction. The level of educator distraction increased with the age of the students they were teaching. For example, only some of the PK-5 educators said they felt distracted when students use devices in class, but instructors of higher-grade levels were almost evenly divided between "yes" and "no" responses to the question of whether they felt distracted by student technology use. This is most likely due to more students in older grades having devices. Half of the educators also noted that their students multitasked during instruction. The majority of MMT took place in the form of student texting followed by playing games.

Because future learning requires a basic understanding of previously learned concepts and incorporation of that learning into long-term memory, the increased cognitive load caused by distracted learning may decrease the ability to learn effectively. Therefore, increasing the cognitive load for students through their interaction with devices has the potential to greatly reduce the ability of students to grasp basic fundamental concepts and use them in future learning. Research has shown that due to limited cognitive capacity, texting can reduce student comprehension of class material by 10-20% (Lawson & Henderson, 2015; Schellen, Lin, & Bigenho, 2017). If left unchecked, the constant increase of cognitive load during the learning process for students who have high mobile device usage are likely to fall behind those peers who do not (Kuznekoff & Titsworth, 2013).

For many of the educators in this study, rather than embracing technology, they chose to have strict policies regarding technology use and some chose to ban it from their classroom all together. Recommendations for creating intentional breaks in instruction for device usage may be a compromise between banning and what some educators might consider a free for all with device usage. However, in this generation of ubiquitous computing, the use of devices will continue to penetrate K-12 schools and higher education. It is important that as educators, we have strategies and training in place to manage the use of such devices to maximize learning for all. Ertmer and Ottenbreit-Leftwich (2010) suggested that "we must focus our change efforts on helping teachers understand how student-centered practices, supported by technology, affect student learning outcomes" (p. 278). Effective student centered practices incorporate a level of engagement in the learning that eliminates distraction and promotes time on task.

Focusing on informing teachers of student-centered practices must include the development of management skills to address student behavior while engaging in these pedagogies. The institution of a technology use policy to regulate and manage classroom behavior with the integration of digital devices may decrease the perception or reality of distraction, yet if the policy is not enforced or even publicized, it has little chance of making any effect. Building a school environment that focuses on increasing student learning in the midst of the increasing usage of digital devices requires collaboration. Educators should be coordinating with others who are instructing the same students so that common expectations for appropriate use of mobile devices in the classroom setting can be created. This can be accomplished when educational institutions establish usage policies and inform students (and parents, if applicable) about the policies. For example, one K-12 school district in the USA developed a BYOT (bring your own technology) to school initiative whereby students who preferred to bring their own mobile devices to the classroom could do so after they were granted approval which included the students and parents agreeing to the school's acceptable times for use of the technology in their classroom usage policy (Richmond County School System, 2018). Interestingly, the school system also built in a loan policy for those students who did not have access to technology so that they could still participate in the classroom activities.

This collaboration must also include the student's home and community. Bridging any gaps between home and school is especially important when it comes to the inclusion of digital devices and technology integration in learning. Recommendations include addressing the students' home environment, parents and guardians, particularly in the education of digital citizenship. Education of proper use of tools such as social media requires modeling, conversing about choices and consequences, and becoming aware of the potential dangers of prolonged use of digital devices.

There were some limitations to this study. For instance, the survey was lengthy, which may have prevented some participants from completing it in its entirety. Additionally, the participants were not required to answer all of the questions. The participants' role in education varied and included both K-12 and higher education faculty as well as technology and media specialists whose interaction with technology integration as a regular part of instruction may vary considerably. For instance, teaching strategies and content in lower elementary is much different from that which adults in higher education would be learning. Additionally, in regard to Cognitive Load Theory, demands on short term and long term memory for these students may differ greatly as would their overall cognitive development, prior knowledge, and schema. The factor of developmental capacity may also be a limitation.

Furthermore, these findings may not be generalizable to all classrooms as the sample size is small and predominantly white (92%) and the perspectives of participants who work in online-only environments may be significantly different concerning distractions as their students are not in the same physical location. Additionally, participants identified as a tech coordinator, technology director, or other and who do not spend the majority of their workday with students, may also have a much different perspective than classroom teachers as to what is actually occurring during class time.

FUTURE RESEARCH

The outcomes of this study demonstrate that educators should have clear expectations when integrating the use of digital devices, including a cell phone and smartphone, and educators should clearly explain those expectations to students in order to reduce distractions. A study that limits participants to classroom teachers who engage in devices with Internet connection regularly with their students may provide a better picture of effective use of digital devices in the classroom. Examining student behavior, distraction potential of specific pedagogies and tools and aligning these with actual student academic progress data may also reveal patterns of misaligned or potentially troublesome pedagogical practices.

As the research of Cognitive Load Theory and Media Multitasking continues to grow, it is possible that pre-service and in-service teachers may not have the knowledge necessary to design learning environments that truly address the issue of distraction. Examining what teachers know about learning, particularly how they are aware of intrinsic, extrinsic and germane cognitive load when it comes to their students learning may offer curricular vision for teacher preparation programs and professional development. It would also be useful to collect additional data from educators on the remedies, failed and successful, they have implemented in their classrooms. Another possible area to explore is differences in educator perspectives of distraction during whole group instruction versus during individual or group work time.

CONCLUSION

The ubiquitous use of technology in today's society, especially in educational contexts, is causing distractions. Pathologies previously mentioned such as "Nomophobia",

"FOMO", and "Textaphrenia" are likely to continue as more people become increasingly connected to their devices. In the classroom, because learning must be an active process, distractions by digital media have been found to affect learning. Studies have shown that as students become more distracted by their devices, their cognitive load increases which impedes learning of new material and transfer of learning to long-term memory. As students become more connected to their devices, academic success is reduced and there is less long-term retention of learning. This is of concern to educators who now have to compete with devices for student attention in order to teach.

The use of technology during instruction has many benefits and is increasing at all levels of education in public and private schools. While it is important for today's students to have accessibility of technology, it is even more crucial that they are taught and expected to abide by constructive and appropriate use of the technology in the educational setting. Research has proven that in addition to the benefits, use of technology during instruction can also have a negative impact on cognitive function, physical health, academic achievement, and even personal relationships. Use of technology leads to some level of distraction among students and teaching staff alike.

It is clear that devices such as laptops, Chromebooks, and smartphones can add to and also hinder the educational environment for todays' learners. More research is needed in how to get the benefit of technology without the negative side effects. Today's learners need to learn how to use technology more to their benefit than to their detriment. Instructors in online and face-to-face environments must clearly detail what is and is not appropriate use of technology during class time. Adding standards for appropriate technology use to the K-12 curriculum is a place to start.

ACKNOWLEDGMENT

The authors thank Dr. Rochell McWhorter for assistance with the study.

REFERENCES

American Society of Addiction Medicine. (2011). *Public policy statement: Short definition of addiction.* Retrieved from https://www.asam.org/docs/default-source/public-policy-statements/1definition_of_addiction_short_4-11.pdf?sfvrsn=6e36cc2_0

Anderson, J., & Jiang, J. (2018). *Teens, social media & technology 2018*. Retrieved from http://www.pewinternet.org/2018/05/31/teens-social-media-technology-2018/

Anderson, J., & Rainie, L. (2014). *Internet of things*. Pew Research Internet Project. Retrieved from http://www.pewinternet.org/files/2014/05/PIP_Internet-of-things_0514142.pdf

Anderson, J. Q., Boyles, J. L., & Rainie, L. (2012, July 27). *The future impact of the Internet on higher education: Experts expect more efficient collaborative environments and new grading schemes; they worry about massive online courses, the shift away from on-campus life*. Pew Research Center's Internet & American Life Project. Retrieved from http://pewinternet.org/~/media//Files/Reports/2012/PIP_Future_of_Higher_Ed.pdf

Anshari, M., Almunawar, M. N., Shahrill, M., Wicaksono, D. K., & Huda, M. (2017). Smartphones usage in the classrooms: Learning aid or interference? *Education and Information Technologies*, *22*(6), 3063–3079. doi:10.100710639-017-9572-7

Awwad, F., Ayesh, A., & Awwad, S. (2013). Are laptops distracting educational tools in classrooms. *Procedia: Social and Behavioral Sciences*, *103*, 154–160. doi:10.1016/j.sbspro.2013.10.320

Bank My Cell. (2019). *Smartphone addiction facts & phone usage statistics: The Definitive Guide (2019 update)*. Retrieved from https://www.bankmycell.com/blog/smartphone-addiction

Bowman, L. L., Levine, L. E., Waite, B. M., & Gendron, M. (2010). Can students really multitask? An experimental study of instant messaging while reading? *Computers & Education*, *54*(4), 927–931. doi:10.1016/j.compedu.2009.09.024

Bransford, J. D., Brown, A. L., & Cocking, R. R. (Eds.). (2000). *How People Learn: Brain, Mind, Experience, and School*. Washington, DC: National Academy Press.

Campus Technology. (2018, September). *Survey: 1 in 4 professors ban mobile phone use in class*. Retrieved from https://campustechnology.com/articles/2018/09/19/survey-1-in-4-professors-ban-mobile-phone-use-in-class.aspx

Cisco. (2019). *Cisco visual networking index: Global mobile data traffic forecast update, 2017–2022 white paper*. Retrieved from https://www.cisco.com/c/en/us/solutions/collateral/service-provider/visual-networking-index-vni/mobile-white-paper-c11-520862.html

Common Sense Media. (2012). *Children, teens, and entertainment media: The view from the classroom. A national survey of teachers about the role of entertainment media in students' academic and social development.* Retrieved from https://www.commonsensemedia.org/file/view-from-the-classroom-final-reportpdf-0/download

Common Sense Media. (2017). *The common sense census: Media use by kids age zero to eight.* Retrieved from https://www.commonsensemedia.org/sites/default/files/uploads/research/0-8_executivesummary_release_final_1.pdf

Common Sense Media. (2018). *Social media, social life.* Retrieved from https://www.commonsensemedia.org/sites/default/files/uploads/research/2018_cs_socialmediasociallife_executivesummary-final-release_3_lowres.pdf

Dalton, K. (2013). Their brains on Google: How digital technologies are altering the millennial generation's brain and impacting legal education. *SMU Science and Technology Law Review*, *16*, 409–501.

De-Sola Gutiérrez, J., Rodríguez de Fonseca, F., & Rubio, G. (2016). Cell-Phone Addiction: A Review. *Frontiers in Psychiatry*, *7*, 175. doi:10.3389/fpsyt.2016.00175 PMID:27822187

Delello, J. A., Mokhtari, K., & Reichard, C. (2016). Multitasking among college students: Are freshmen more distracted? *International Journal of Cyber Behavior, Psychology and Learning*, *6*(4), 1–12. doi:10.4018/IJCBPL.2016100101

Ertmer, P. A., & Ottenbreit-Leftwich, A. T. (2010). Teacher technology change: How knowledge, confidence, beliefs, and culture intersect. *Journal of Research on Technology in Education*, *42*(3), 255–284. doi:10.1080/15391523.2010.10782551

Fox, S., & Rainie, L. (2014). The web at 25 in the US: The overall verdict: The internet has been a plus for society and an especially good thing for individual users. *Pew Research Center's Internet and American Life Project*. Retrieved from http://www.pewinternet.org/2014/02/25/the-web-at-25-in-the-u-s

Gazzaley, A., & Rosen, L. D. (2016). *The distracted mind: Ancient brains in a high-tech world.* Cambridge, MA: MIT Press.

Glass, A. L., & Kang, M. (2018). Dividing attention in the classroom reduces exam performance. *Educational Psychology*, *39*(3), 395–408. doi:10.1080/01443410.2018.1489046

Goleman, D. (2013). *Focus: The hidden driver of excellence*. New York: HarperCollins.

Herold, B. (2019, April). Ed-Tech supporters promise innovations that can transform schools: Teachers not seeing impact. *Education Week*. Retrieved from https://www.edweek.org/ew/articles/2019/04/24/ed-tech-supporters-promise-innovations-that-can-transform.html

Howard, J. (2017). *When kids get their first cell phones around the world*. Retrieved from https://www.cnn.com/2017/12/11/health/cell-phones-for-kids-parenting-without-borders-explainer-intl/index.html

Jaeger, A. J., Shipley, T. F., & Reynolds, S. J. (2017). The roles of working memory and cognitive load in geoscience learning. *Journal of Geoscience Education, 65*(4), 506–518. doi:10.5408/16-209.1

Jiang, J. (2018). *How teens and parents navigate screen time and device distractions*. Pew Research Center. Retrieved from http://www.pewinternet.org/2018/08/22/how-teens-and-parents-navigate-screen-time-and-device-distractions/

Johnson, A. M., Jacovina, M. E., Russell, D. E., & Soto, C. M. (2016). Challenges and solutions when using technologies in the classroom. In S. A. Crossley & D. S. McNamara (Eds.), *Adaptive educational technologies for literacy instruction* (pp. 13–29). New York: Taylor & Francis. Published with acknowledgment of federal support. doi:10.4324/9781315647500-2

Judd, T. (2015). Task selection, task switching and multitasking during computer-based independent study. *Australasian Journal of Educational Technology, 31*(2), 193–207. doi:10.14742/ajet.1992

Junco, R. (2012). In class multitasking and academic performance. *Computers in Human Behavior, 28*(6), 2236–2243. doi:10.1016/j.chb.2012.06.031

Kaiser Family Foundation. (2013). *Generation M2: Media in the lives of 8-to-18-year-olds*. Retrieved from https://kaiserfamilyfoundation.files.wordpress.com/2013/04/8010.pdf

Kalyuga, S., & Singh, A.-M. (2015). Rethinking the boundaries of cognitive load theory in complex learning. *Educational Psychology Review, 28*(4), 831–852. doi:10.100710648-015-9352-0

Kay, R., Benzimra, D., & Li, J. (2017, January 26). Exploring factors that influence technology-based distractions in bring your own device classrooms. *Journal of Educational Computing Research, 55*(7), 974–995. doi:10.1177/0735633117690004

Korbach, A., Brünken, R., & Park, B. (2018). Differentiating different types of cognitive load: A comparison of different measures. *Educational Psychology Review, 30*(2), 503–529. doi:10.100710648-017-9404-8

Kraushaar, J. M., & Novak, D. C. (2010). Examining the effects of student multitasking with laptops during the lecture. *Journal of Information Systems Education, 21*(2), 241–251.

Kuznekoff, J. H., & Titsworth, S. (2013). The impact of mobile phone usage on student learning. *Communication Education, 62*(3), 233–252. doi:10.1080/03634523.2013.767917

Lawson, D., & Henderson, B. B. (2015). The costs of texting in the classroom. *College Teaching, 63*(3), 119–124. doi:10.1080/87567555.2015.1019826

Lenhart, A., Ling, R., Campbell, S., & Purcell, K. (2010). *Chapter Four: How parents and schools regulate teens' mobile phones*. Retrieved from https://www.pewinternet.org/2010/04/20/chapter-four-how-parents-and-schools-regulate-teens-mobile-phones/

Lewandowski, J. (2018). *Back to school without mobile phones: France imposes nationwide ban*. Retrieved from https://www.newsweek.com/back-school-without-mobile-phones-france-imposes-nationwide-ban-1104060

Liu, F., Ritzhaupt, A., Dawson, K., & Barron, A. (2017). Explaining technology integration in K-12 classrooms: A multilevel path analysis model. *Educational Technology Research and Development, 65*(4), 795–813. doi:10.100711423-016-9487-9

López-Vargas, O., Ibáñez-Ibáñez, J., & Racines-Prada, O. (2017). Students' metacognition and cognitive style and their effect on cognitive load and learning achievement. *Journal of Educational Technology & Society, 20*(3), 145–157.

Marzilli, C., Delello, J. A., Marmion, S., McWhorter, R. R., Brown, P., & Marzilli, T. S. (2014). Faculty attitudes towards integrating technology and innovation. *International Journal on Integrating Technology in Education, 3*(1), 1–20. doi:10.5121/ijite.2014.3101

McCoy, B. R. (2016). Digital distractions in the classroom phase II: Student classroom use of digital devices for non-class related purposes. *Faculty Publications, College of Journalism & Mass Communications,* Paper 90. Retrieved from http://digitalcommons.unl.edu/ journalismfacpub/90

Merriam, S. B., & Tisdell, E. J. (2016). *Qualitative research: A guide to design and implementation* (4th ed.). San Francisco, CA: John Wiley & Sons.

Mokhtari, K., Delello, J. A., & Reichard, C. (2015). Millennial multitasking: Constantly Connected yet Distracted. *Journal of College Reading and Learning*, *45*(2), 164–180. doi:10.1080/10790195.2015.1021880

Mupinga, D. M. (2017). School-wide and classroom policies on the use of mobile technologies: An exploratory study. *Journal of Technology Studies*, *43*(2), 70–79. doi:10.21061/jots.v43i2.a.2

Ophir, E., Nass, C., & Wagner, A. D. (2009). Cognitive control in media multitaskers. *Proceedings of the National Academy of Sciences of the United States of America*, *106*(37), 15583–15587. doi:10.1073/pnas.0903620106 PMID:19706386

Pahomov, L. (2015). *In pursuit of a cell phone policy*. ASCD. Retrieved from http://www.ascd.org/publications/educational-leadership/may15/vol72/num08/In-Pursuit-of-a-Cell-Phone-Policy.aspx

Pashler, H. (2000). Task switching and multitask performance. In S. Monsell & J. Driver (Eds.), *Control of Cognitive Processes: Attention and Performance XVIII* (pp. 275–423). Cambridge, MA: The MIT Press.

Pearson. (2015). *Pearson student mobile device survey 2015 national report: Students in grades 4-12*. Retrieved from https://www.pearsoned.com/wp-content/uploads/2015-Pearson-Student-Mobile-Device-Survey-Grades-4-12.pdf

Qian, Y., & Li, L. (2017). Student off-task electronic multitasking predictors: Scale development and validation. *The Journal of Scholarship of Teaching and Learning*, *17*(2), 53–73. doi:10.14434/josotl.v17i2.20682

Rekart, J. L. (2011). Taking on multitasking. *Phi Delta Kappan*, *93*(4), 60–63. doi:10.1177/003172171109300415

Richmond County School System. (2018). *BYOT (Bring your own technology)*. Retrieved from https://www.rcboe.org/domain/7853

Rideout, V., & Robb, M. B. (2018). *Social media, social life: Teens reveal their experiences*. San Francisco, CA: Common Sense Media.

Rideout, V. J., Foehr, U. G., & Roberts, D. F. (2010). *Generation M2: Media in the Lives of 8-to 18-Year-Olds*. Menlo Park, CA: Henry J. Kaiser Family Foundation.

Rose, E. (2010). Continuous partial attention: Reconsidering the role of online learning in the age of interruption. *Educational Technology*, *50*(4), 41–46.

Rosen, L. D., Carrier, M. L., & Cheever, N. A. (2013). Facebook and texting made me do it: Media-induced task-switching while studying. *Computers in Human Behavior*, *29*(3), 948–958. doi:10.1016/j.chb.2012.12.001

Rosen, L. D., Cheever, N. A., & Carrier, L. M. (2012). iDisorder: Understanding our obsession with technology and overcoming its hold on us. New York, NY: Palgrave Macmillan

Rosen, L. D., Whaling, K., Rab, S., Carrier, L. M., & Cheever, N. A. (2013). Is Facebook creating "iDisorders"? The link between clinical symptoms of psychiatric disorders and technology use, attitudes and anxiety. *Computers in Human Behavior*, *29*(3), 1243–1254. doi:10.1016/j.chb.2012.11.012

Ruston, D. (2017). *Screenagers' national survey reveals that the majority of U.S. middle schools allow students to carry cell phones throughout the school day*. Screenagers: Away for the Day. Retrieved from https://awayfortheday.squarespace.com/survey

Shochat, T. (2012). Impact of lifestyle and technology developments on sleep. *Nature and Science of Sleep*, *4*, 19–31. doi:10.2147/NSS.S18891 PMID:23616726

Smith, A. (2015). U.S. smartphone use in 2015. *Pew Research Center's Internet & American Life Project*. Retrieved from http://www.pewinternet.org/2015/04/01/us-smartphone-use-in-2015/

Statista. (2019). *Number of mobile phone users worldwide from 2015 to 2020 (in billions)*. Retrieved from https://www.google.com/url?q=https://www.statista.com/statistics/274774/forecast-of-mobile-phone-users-worldwide/&sa=D&ust=1559138725843000&usg=AFQjCNEnu_FZ6yA9rbWeaZeuqrg5GOrIQQ

Strauss, V. A. (2018). Schools are banning smartphones. Here's an argument for why they shouldn't--and what they should do instead. *The Washington Post*. Retrieved from https://www.washingtonpost.com/education/2018/09/21/schools-are-banning-smartphones-heres-an-argument-why-they-shouldnt-what-they-should-do-instead/?noredirect=on&utm_term=.322c4563af0e

Sweller, J. (1998). Cognitive load during problem solving: Effects on learning. *Cognitive Science*, *12*(2), 257–285. doi:10.120715516709cog1202_4

The Alberta Teachers' Association. (2015). *Technology promise and peril: Growing up digital*. Retrieved from https://www.teachers.ab.ca/Public%20Education/EducationResearch/Pages/ResearchPublications.aspx

Uncapher, M. R., Lin, L., Rosen, L. D., Kirkorian, H. L., Baron, N. S., Bailey, K., & Wagner, A. D. (2017, November). Media multitasking and cognitive, psychological, neural, and learning differences. *Pediatrics*, *140*(Supplement 2), S62–S66. doi:10.1542/peds.2016-1758D PMID:29093034

U.S. Department of Education. (2017). *Reimagining the Role of Technology in Education: 2017 National Education Technology Plan Update*. Retrieved from https://tech.ed.gov/files/2017/01/NETP17.pdf

Ward, A. F., Duke, K., Gneezy, A., & Bos, M. W. (2017). Brain drain: The mere presence of one's own smartphone reduces available cognitive capacity. *Journal of the Association for Consumer Research*, *2*(2), 140–154. doi:10.1086/691462

Yin, R. K. (2014). *Case study research: Design and methods* (5th ed.). Thousand Oaks, CA: Sage.

ADDITIONAL READING

Delello, J. A., Mokhtari, K., & Reichard, C. (2016). Multitasking among college students: Are freshmen more distracted? *International Journal of Cyber Behavior, Psychology and Learning*, *6*(4), 1–12. doi:10.4018/IJCBPL.2016100101

Kolikant, Y. B. D. (2010). Digital natives, better learners? Students' beliefs about how the internet influenced their ability to learn. *Computers in Human Behavior*, *26*(6), 1384–1391. doi:10.1016/j.chb.2010.04.012

Marzilli, C., Delello, J. A., Marmion, S., McWhorter, R. R., & Roberts, P. (2015). Exploring the perceptions of college students on the use of technology: What do they really think? *International Journal of Sciences: Basic and Applied Research*, *24*(2), 434–456.

NASBE. (2012). *Born in another time ensuring educational technology meets the needs of students today and tomorrow*. National Association of State Boards of Education. Retrieved from http://assess4ed.net/resources/ensuring-educational-technology-meets-needs-students-today%E2%80%93-and-tomorrow-nasbe-report-0

Prensky, M. (2001). Digital Natives, Digital Immigrants Part 1. *On the Horizon*, *9*(5), 1–6. doi:10.1108/10748120110424816

KEY TERMS AND DEFINITIONS

Cognitive Load: The amount of information the brain is able to process.

Disruptive Technologies: Innovative technologies that may disrupt current markets.

Emerging Technologies: New technology or one that is in continual development.

Media Multitasking: Simultaneous use of media.

Mobile Technologies: Technology that is easily moved from place to place such as Smartphones.

Ubiquitous Computing: Technology that can be accessed anywhere and at any time.

Web 2.0: Next generation of internet technologies that allow for social collaboration.

Chapter 9
Do Loneliness and Social Connectedness Improve in Older Adults Through Mobile Technology?

Rochell R. McWhorter
https://orcid.org/0000-0003-2053-1610
The University of Texas at Tyler, USA

Julie A. Delello
https://orcid.org/0000-0002-4326-8096
The University of Texas at Tyler, USA

Christine S. Gipson
The University of Texas at Tyler, USA

Beth Mastel-Smith
The University of Texas at Tyler, USA

Kleanthe Caruso
The University of Texas at Tyler, USA

ABSTRACT

About one-fifth of the population in the United States in 2015 will be age 65 or older in 2050 and loneliness may be a contributing factor that inhibits their well-being and overall health. As the number of older adults continues to escalate, information and communication technologies such as smartphones and computers may create an increase in social connectedness leading to a decline in loneliness and social isolation. Results from this pilot study suggest that the older adult participants demonstrated

DOI: 10.4018/978-1-7998-2914-0.ch009

some degree of loneliness. As the older adults used social media to connect with friends, family, and other information of interest, there was an increase in social connectedness for many of the participants, but the intensive iPad intervention was not significant in terms of reducing loneliness for either group over time.

INTRODUCTION

It is projected that by 2030, one in five US residents will be 65 years or older (United States Census Bureau, 2018). As the number of older adults continues to increase, these individuals may experience loneliness due to retirement (Neville, et al., 2018; UNFPA, 2012), living alone (Chaumon, Michel, Bernard, & Croisile, 2013), reduced mobility, declines in health, or limited income (Chihuri, et al., 2016; Warner, Roberts, Jeanblanc, & Adams, 2017). Loneliness, as defined by Young (1982), is the "perceived absence of satisfying social relationships, accompanied by symptoms of psychological distress that are related to the perceived absence" (p. 380). The prevalence of loneliness is as high as one in three among older adults in the United States (AARP.org, 2018a; Gerst-Emerson & Jayawardhana, 2015). This is quite concerning since several health conditions such as depression, coronary artery disease, and hypertension are highly correlated with loneliness in older adults (Hacihasanoğlu, Yildirim, & Karakurt, 2011) and can lead to premature death (Emerson & Jayawardhana, 2016).

Information and communication technologies (ICTs) such as smartphones, digital tablets, and computers connected to the Internet hold promise for addressing social isolation in older adults (Meshi, Cotton, & Bender, 2019; Tsai, Shillair & Cotton, 2015). For example, Delello and McWhorter (2015, 2017) found iPads enhanced the lives of the older adults by strengthening family relationships and connecting them to society. Further, research on social networking sites (SNSs) such as Facebook and Twitter have been shown to contribute to seniors' wellbeing and aid them in continuing relationships with family and friends at a distance (Khosravi, Rezvani, & Wiewiora, 2016). But, direct technical instruction to use ICTs was recommended for older adults to build their confidence and technology skills (AARP, 2018b; Delello & McWhorter, 2017). Recently, U.S. President Donald Trump commissioned a special Task Force on Research and Development for Technology to Support Aging Adults, which identified a range of emerging technologies, which may support independent living and improve social connectiveness. The Task Force specifically noted that more research is needed to demonstrate "the effectiveness of interventions, including online social networks, to reduce social isolation and loneliness among seniors, as well as the mechanisms by which the interventions exert effects" (The White House, 2019, p. 13).

In a report titled *Gauging Aging,* the public has largely viewed aging as a period of physical and cognitive deterioration, reduced potential, and technological incompetence (Lindland, Fond, Haydon, Kendall-Taylor, 2014). However, according to the United States Senate's Special Committee on Aging (2017), the proportion of people in the working-age population worldwide who are between the ages of 65 to 74 is projected to grow by 4.2 percent annually and the number of workers over age 75 by 6.7% annually.

Research has shown that technology has the power to disrupt the lives of older adults, allowing them to be maintain their health and independence (Perkins-Eastman, 2019) and contribute to the growing workforce as an increasing number of older adults launch an *encore career* for adding income and doing meaningful work (Lauby, 2019; Watson, 2019). Schuelke-Leech (2018) identified such phenomena as second-order "technological disruptions, where the impacts ripple through society…[and] these technologies disrupt interactions and relationships, organizational structures, institutions, and public policies" (p. 261).

BACKGROUND

According to Sum, Saboor, and Sahaf (2015), many older adults are at risk of social isolation and suffer from loneliness. In fact, Loxterkamp (2014) reported that 1/10th of the population age 65 and older felt "intensely lonely" (p. 1). Researchers found that there is a direct correlation between loneliness and a reduction in social relationships of older adults (Goll, Charlesworth, Scior, & Stott, 2015). As individuals age, they tend to spend less time outside the home in social settings, which may be due to increased cognitive impairment, a lack of physical ability, or a depressed emotional state (Petersen, Austin, Mattek, & Kaye, 2015). For others, the death of a spouse, the reduction of financial resources (Sum, et al., 2015), or even the fear of rejection by their peers (Goll et al., 2015) may increase the risk of social isolation, loneliness, and overall well-being.

Social isolation and loneliness are often used interchangeably; however, the two terms may or may not coincide. Social isolation has been defined as "the extent, range and depth of social networks, including the extent to which people feel able to rely on their social network for support, and the felt experience of loneliness" (Pate, 2014, p. 10). Furthermore, social isolation may have a direct impact upon the health and quality of a person's social and psychological health (AARP, 2012; Novotney, 2019). Researchers have noted that the lack of social relationships is a major factor effecting mental health in older adults (Heidari, Ghodusi, & Shirvani, 2016) and it is correlated with chronic disease, depression (Loxterkamp, 2014), and mortality (Steptoe, Shankar, Demakakos, & Wardle, 2013). Furthermore, research has indicated that loneliness and isolation appear to be more prevalent among older adults who live alone (Victor, Scambler, Bond, & Bowling, 2000). Yet, according to Hicks (2000), older adults living in facilities like nursing homes where social interaction is likely, also show a propensity to be particularly lonely.

Older Adults, Computer Use, and Well-Being

The rise of gerontechnology was "defined as a composite of two words: ''gerontology,'' the scientific study of aging, and ''technology''—is concerned with research on the biological, psychological, social, and medical aspects of aging exploiting the potentials offered by the progress of technology" (Micera, Bonato, & Tamura, 2008, p. 10). Recent gerontechnology research has shown potential for improving the quality of life for those facing the challenges of ageing (Jacelon, et al., 2018; Sale, 2018) as adults are living longer. Gerontechnology is a field of study that combines the interdisciplinary fields of gerontology and technology. Bouma and Graafmans (1992) defined gerontechnology as "the study of technology and aging for the improvement of the daily functioning of the elderly" (p. 1).

According to Loe (2011), older adults "can be and are technogenarians in their active use of everyday technologies to create meaningful lives and maintain health" (p. 320). For example, researchers found that older adults who used information and communication technologies (ICTs) reported decreased loneliness and depression (Cotten, Ford, Ford, & Hale, 2014) and improved life satisfaction, self-esteem (Billipp, 2001), positive attitudes (González, Ramírez, & Viadel, 2012) and cognitive ability (Chan, Haber, Drew, & Park, 2014). Moreover, a study by Cotten et al. (2014), showed that older adults who used the Internet experienced a one-third reduced risk for depression compared to non-users in a longitudinal analysis. In addition, video games have shown to stimulate brain activity in older adults (Costa & Veloso, 2016). Further, tablet technologies such as iPads can help older adults share photos and message friends and family which curb feelings of social isolation and enhance the older adults' well-being (Delello & McWhorter, 2015, 2017). The use of a computer and Internet has also shown to improve daily living activities and cognitive ability (Chan et al., 2014). Furthermore, using the Internet fostered decision-making and critical thinking through active aging (Sánchez-Valle, Vinarás, & Llorente-Barroso, 2017) and improved understanding of health problems and lifestyle practices (Smith, et al., 2013).

Older adults are utilizing technology at a slower rate than the overall population, but their usage is on the rise, especially when specific programs are in place to teach needed technology skills (LoBuono, Leedahl, & Maiocco, 2019). A report by the Pew Internet & American Life Project conveyed that over the last decade, more than half of adults (67%) over age 65 stated they were Internet users (Anderson & Perrin, 2017).

However, despite the expansion of ICTs, several barriers may prevent older adults from using them. These barriers include factors such as cognitive deficit, working memory, and low self-efficacy (Jones, Kay, Upton, & Upton, 2013). Moreover, Internet usage tends to lessen after age 75 and for those with incomes of less than $30,000 per year, the usage drops to only 30% (Smith et al., 2013).

Social Connectedness

In this study, we framed our work on the theoretical construct of social connectedness. Lee and Robbins (1998) described social connectedness as the "subjective awareness of being in close relation with the social world" (p. 388). Social connectedness may be a direct indication of an older adult's feeling of social isolation and/or loneliness (Neeves, Franz, Judges, Beermann, & Baecker, 2019; Satici, Uysal, & Deniz, 2016). Social connections through friends, family, neighbors, and even religious congregations are valuable sources of social support and have a positive correlation to one's well-being (Hastings, 2016). Giummarra, Haralambous, Moore, and Nankervis (2007) noted that "social health is dependent on social connectedness as well as the extent to which communities value diversity, are supportive and inclusive, and provide opportunities for each person to participate in community life, as well as the number and quality of social supports and relationships that a person maintains" (p. 643). Consequently, having a positive sense of social connectedness may also influence an individual's level of subjective happiness (Satici, et al., 2016). Additionally, accessing health information online might promote interactions with providers as well as peers (Mastel-Smith, Duke, & He, 2018) thus supporting social interactions.

There is growing recognition that ICTs may enable and facilitate social connectedness (LoBuono, Leedahl, & Maiocco, 2019; Mitzner, Rogers, Fisk, Boot, Charness, Czaja, & Sharit, 2016). In fact, in a meta-analysis, Morris, et al. (2014) reviewed 14 of 18 publications on the use of ICTs in older adults, finding positive outcomes for areas of social support, isolation and loneliness. Loneliness is an important issue for older adults and technologies have been widely used in the care of older adults, however, little is known about the overall impact mobile technology and the use of such applications (e.g. social media, email) may have on loneliness and social isolation.

To explore the identified effects of mobile technology training upon the older adults, the research project sought to answer the following questions:

Research Question 1: How can iPad technology be utilized to reduce loneliness in older adults?

Research Question 2: How did iPad technology facilitate a sense of social connectedness among the participants?

This pilot study aimed to explore the level of loneliness and social isolation among older adults in independent living centers before and after iPad interventions.

METHODS

This mixed-methods case study was piloted with two cohorts of older adults residing at senior living centers in the Southwestern United States of America (USA) in 2017. The case study, used as the primary research strategy, explored whether the use of iPads would decrease loneliness in older adults. A case study design is useful for gaining "an in-depth understanding of the situation for those involved" (Merriam, 1998, p. 19) and provides "rich, thick description of a contemporary phenomenon in real-world context" (McWhorter & Ellinger, 2018, p. 195). Further, Bryman and Bell (2015), noted that a mixed-methods design allows the evaluators to answer different questions, explain the results from the other methodology, adds credibility to the evaluation by enhancing the integrity of the findings; and, it is oftentimes more useful for practitioners, provides a diversity of viewpoints, and augments one form of data with the other form. The researchers' Institutional Review Board (IRB) approved all study procedures and written informed consent was obtained from all participants prior to data collection in this study.

Participants

Cohort 1

The first cohort consisted of 17 residents (3 males/ 14 females) residing in independent living apartments within a senior retirement community. The participants ranged in age from 62-98 and all (100%) were Caucasian. At the time of the study, six (35%) of the participants were married, one (6%) was single, and ten (59%) were widowed. All (100%) of the participants had at least a 9th grade education; five (29%) had completed some college; five (29%) had attained a bachelor's degree, and four (24%) had achieved a masters or doctorate degree.

Cohort 2

The second cohort included eight residents (5 males/3 females) from two sets of apartments maintained by a nonprofit organization serving low-income senior citizens through Section 8 Housing (HUD.gov, 2016). The participants ranged in age from 55-90 years old and all had a very low income. Four participants (50%) were single and four (50%) were widowed. The ethnicity of the participants was comprised of

12.5% Asian (n=1), 12.5% African American (n=1), and 75% Caucasian (n=6). All (100%) of the participants had at least a 9th grade education; five (62.5%) had completed some college, and one (12.5%) held a bachelor's degree.

Recruitment

For each cohort, the senior living directors at the senior living center facilitated the recruitment of participants by posting flyers and circulating a signup sheet for those interested. To further encourage participation, the researchers held an information session about the iPad study at each center. Participants had to be at least 65 years of age and have an interest in learning more about iPad technology. Of the 43 people who indicated an interest in participating in the study, 37 consented to the study. Participants were loaned an iPad in exchange for their participation during the 10-week study.

Context

The training protocol consisted of 10 weekly, small group face-to-face sessions. The first six modules were delivered by two of the researchers followed by four health modules, administered by nursing faculty and students. Health modules provided opportunities to master a variety of tablet skills. Training was provided at the senior living center for Cohort One. The Center had an Internet-ready technology meeting room for training participants. Because the setting for participants in Cohort 2 lacked Wi-Fi, the organization provided transportation for the participants in the second cohort to a nearby regional public university campus. It was perceived by the center director that training at the university campus for residents in Cohort 2 (who were not typically involved in residential activities), would provide additional motivation to participate in the program. Furthermore, an optional practice session was provided for Cohort 2 on a second day each week because their residential spaces did not have Wi-Fi access. Both groups of participants were allowed to keep their iPads for the duration of the training.

Older adults in each center utilized iPad technology, beginning with the learning of basic to more sophisticated technology use such as Internet, email, photo sharing, social media platforms (Facebook, Pinterest), videoconferencing technologies (i.e. Zoom), eBooks. Also, health information modules were provided whereby senior level nursing students in the Community Health Nursing course delivered health information using The National Institute on Aging Toolkit, Helping Older Adults Search for Health Information Online (NIH, 2013) for the adult adults to gain more health information online that also gave more opportunity for real-world practice of accessing important information on the Internet using the iPad to facilitate their mastery of this skill and an overall gratifying experience when they learned to use technology independently (Dahl, 2018). The health modules focused on the topics of health, drugs and supplements, medical encyclopedias, health news, and information regarding doctors and hospitals specific to Older Adults. In addition, the participants learned how to use online libraries (Kindle, Overdrive) that included a collaborative effort with a city librarian who provided library cards to the participants for them to be able to check out electronic books (eBooks), which could be read without access to the Internet. Recent studies indicate that reading shows promise for reducing loneliness and chronic isolation (Dixon, 2018; Hilhorst, Lockey & Seight, 2018) and the technology lessons allowed the older adults to access both online information as well as eBooks.

Instruments

Formative

A pre-experience survey was used to gather information regarding the age, gender, race, education, and marital status of the participants. In addition, The University of California, Los Angeles (UCLA), Revised Loneliness Scale (Version 3) (UCLA RLS; Russell, 1996), a 20-item scale designed to measure one's subjective feelings of loneliness as well as feelings of social isolation was utilized. Participants rated each item on a scale from 1 (Never) to 4 (Often). Items 1, 5, 6, 9, 10, 15, 16, 19, and 20 were reverse scored. Total scores ranged from 10 to 40 where higher scores indicated increased levels of loneliness. A score of 30 and above suggested severe loneliness. Research has shown that the use of the UCLA RLS yields a high internal consistency ($\alpha = .89$ to $.94$) and acceptable test-retest reliability ($r = .73$) (Russell, 1996).

Summative

The UCLA RLS was again administered to the participants at the end of the study. Also, a post-experience survey was utilized to gather data that explored their connectedness to family members, friends, students, news, and business/organizations in their community following the iPad intervention. The post survey consisted of three demographic questions; additionally, four questions were asked as to the connections that the participants had made with the iPad. And, one open-ended question asked participants if there was any further information that they would like for the researchers to know about their iPad experience.

DATA ANALYSIS

The data analysis for the pilot study was based upon the examination of the UCLA RLS pre- and post-tests and a pre- and post-experience survey. The data analysis used descriptive techniques to report the frequencies in the closed-ended questions involving demographic information from the pre-and post- experience surveys. SPSS (version 12) was used to analyze the internal consistency and test-retest reliability of the UCLA RLS data. The analysis included between groups effects as well as within subject effects. A repeated-measures analysis of variance (ANOVA) was used to determine whether any change in loneliness resulted from the iPad intervention.

A team of two researchers independently sorted the data from the open-ended responses on the post-experience survey, creating codes, and forming categories. The research team met face-to-face to compare their results to reach consensus. According to Merriam and Tisdell (2016), the use of a team of researchers, triangulation (utilizing multiple sources of data and multiple investigators), and peer debriefing enhance the trustworthiness of the findings of the study.

RESULTS

To answer research question one: *Can iPad technology be utilized to reduce loneliness in older adults?*_To answer this question, repeated samples t-tests were used to determine whether pre- and post-tests of loneliness differed. Individuals were not less lonely at the post-test, t(20) = -0.42, p = .34. To explore potential group differences, Mann-Whitney U-tests were used, due to conditional non-normality, to compare each cohort's pre- and post-test scores. There was a trend toward significance for Cohort 2 to have greater loneliness at the pre-test than Cohort 1, U = 144.5, p = .07. There was no significant difference between the cohorts at the post-test, U = 64.0, p = .76. (See Table 1).

Table 1. Tests of within-subjects contrasts

Source	PrePost	Type III Sum of Squares	df	Mean Square	F	Sig.	Partial Eta Squared	Noncent. Parameter	Observed Powera
Pre-Post	Linear	0.067	1	0.067	0.777	0.387	0.033	0.777	0.135
Pre-Post * Groups	Linear	0.414	1	0.414	4.784	0.039	0.172	4.784	0.554
Error (Pre-Post)	Linear	1.991	23	0.087					

Note. a Computed using alpha = .05

To further clarify changes in loneliness, a reliable change index was calculated for individuals who had been administered the loneliness scale at both pre- and post-treatment. There were 4 individuals in Cohort 1 and 5 in Cohort 2 (28.57% and 83.33% of individuals with calculable reliable change indices in Cohort 1 and 2, respectively) who demonstrated a reliable change in loneliness between test periods. There was a significant interaction between cohort and producing a reliable change in loneliness, $\chi 2(1) = 5.09$, $p = .02$, $\Phi = .50$, wherein Cohort 1 produced a significantly lower proportion of reliable changes while Cohort 2 produced significantly more. A repeated-measures ANOVA confirmed the significant interaction between pre-/post-test scores and group, $F(1, 19) = 10.56$, $p = .002$, partial-$\eta 2 = .36$, though there was not a significant main effect for loneliness, $F(1, 19) = 2.46$, $p = .13$.

In order to answer research question two: *How did iPad technology facilitate a sense of social connectedness among the participants?*, two measures were utilized. First, we analyzed one six-stemmed question from a post-experience survey, which asked participants, "Do you feel that the iPad [training] helped you to connect with any of the following?" (See Table 2).

Table 2. Post-experience survey results from cohort 1 and cohort 2

Question	Cohort 1		Cohort 2	
	Response	Percentage	Response	Percentage
Connected with family member(s)	1	6%	3	37%
Connected with friends outside the residence	9	53%	2	25%
Connected with friends at the place of residence	7	41%	0	0%
Connected with the News in the Community and the World	7	41%	3	38%
Connected with students and faculty at the university	5	29%	2	25%
Connected with businesses and organizations in the local community	3	18%	2	25%

The remainder of the post-survey questions regarding connectedness addressed ICT use. Participants were asked if they used video conferencing on the iPad to connect with others. For those participants in Cohort 1, five participants reported they utilized the video conferencing application Zoom to connect with family (27.8%), friends (11.1%), and other residents (22.2%). For Cohort 2, four participants reported connecting with family (0%), friends (37.5%), and other residents (12.5%).

Participants also utilized Facebook to connect to others by making at least one posting: 48% (Cohort 1) and 30% (Cohort 2). In addition, using the iPad, each participant created an online family tree and shared it with family (Cohort 1, 38.9%), friends (Cohort 1, 16.7%), and their community (Cohort 1, 16.7%). Also, 27.8% connected with distant family they did not know previously.

There were two open-ended responses regarding social connectivity and loneliness. One participant noted in their open-ended response, "I felt very excited; I learned the name of my great grandmother, whom I never saw. I found out she was born in Ireland." For Cohort 2, 50% reported they shared it with family. Also, 33.3% connected to friends, 16.6% with the community, and 0% distant family. One participant noted, "Until now my genealogy went no further than my dad's daddy and mother. They both died before I was born. I now know the names of my great grandfather and great grandmother."

DISCUSSION AND LIMITATIONS

The overall goal of this multidisciplinary study was to create an intervention, which would increase social connectedness and in turn, counter loneliness. In this pilot study, the data showed that both cohorts demonstrated some degree of loneliness. However, the intensive iPad intervention was not significant in terms of reducing loneliness for either group over time. We believe that adding additional participants would yield a different result.

It was interesting to note that the participants (in Cohort 1), who scored most lonely were those that lived in the more affluent, independent living center; yet, a slight reduction in loneliness was shown for those participants (Cohort 2) who lived in low-income apartments. Cohort 2 showed a trend toward greater loneliness at the pre-test and also showed a greater proportion of participants who produced reliable changes in their loneliness over the course of the intervention. Compounding the issue of sample size may be a restriction in the variability of loneliness scores in Cohort 1. It is possible that a larger sample size would introduce greater variability and thus improve the performance of the statistics in addition to providing more power. Evaluating the reliable change index in the context of these pre-test differences suggests that those in Cohort 1 may have been near a basal degree of loneliness that precluded a further decline of sufficient magnitude to be measured reliably.

The iPad intervention appeared to indicate an increase in social connectedness. Recent research has noted that that the use of technology may elicit closer family ties and lead to a greater overall connection to society (McWhorter & Delello, 2015). When analyzing the post-experience survey, it was not surprising that participants who did not have Wi-Fi at their residence were less likely to connect with friends within their residence. However, these same participants, when given the opportunity to go online, connected more with family and friends than did Cohort one, who had constant Internet access. Both groups seemed equally interested in connecting to local and world news and local businesses and organizations. The study supports recent findings by the Pew Research Center, which reported that 65% of adults over the age of 65 are now online (Anderson & Perrin, 2017). Furthermore, these older adults use social media to find "news and information" and "share their experiences and connect with friends and family" (p. 2). The older adults in this study connected with their family through social media such as Facebook, connected with their friends and residents through the video conferencing platform Zoom, and connected with the news and world through the internet apps such as local libraries, the local Overdrive application and Kindle application. Research has also shown that older adults with higher incomes are systematically associated with lower self-rated loneliness and those that have more frequent contact online are also associated with higher levels of loneliness (Luhmann & Hawkley, 2016). It is interesting to note that more than 70% of participants reported not having made connections to students and faculty at the university despite their weekly interactions with nursing students and faculty trainers. It is uncertain whether the participants considered the students and faculty trainers as part of the university or if the question may have been unclear.

The UCLA RLS survey is based upon a series of single-item questions about loneliness. It was uncertain as to whether all participants had the same definition of loneliness. Hauge and Kirkevold (2012) noted, "The burden of loneliness varies extensively among older persons" (p. 557). A different instrument may have resulted in different results. Second, bias could be introduced into such a study when discussing loneliness with older adults as the questions were negatively worded and may have been misunderstood by the participants. Another limitation is that the residents in Cohort 2 did not have WiFi at their residence therefore limiting their access to the outside world through their use of the iPad. Wilson and Moulton's (2010) study indicated that individuals who do not have Internet access are lonelier than those that do; however, our study did not have the significant results to make that claim. Also, we acknowledge, as researchers, that it is possible that through our pre-and posttests that focused on factors of loneliness, we may have actually created somewhat of a paradox in that our focus on loneliness may have contributed to the explicit recognition of the phenomenon by older adults when we asked questions such as How often do you feel that you have a lot in common with the people around you? How often do you feel that there are people who really understand you? (Russell, 1996). Finally, this study is not generalizable to the entire older adult population because the participants in this study volunteered to further their knowledge in technology.

FUTURE RESEARCH DIRECTIONS

Additional research should examine whether loneliness scales or the intervention itself might contribute to a deeper understanding of loneliness in older adults and thus create greater loneliness recognition in older adults. A similar study conducted with more participants could significantly enhance the generalizability of the findings, as well. Also, research should be conducted to include older adults in the research and design of digital technology that will affect their aging in place for those seek to stay in their home (Mannheim et al., 2019). Finally, longitudinal studies could be conducted which focus on the ability of the Internet and ICTs to maintain social connectedness over time and in turn, perhaps, reducing loneliness in older adults.

CONCLUSION

The focus of this chapter has been on how iPad technology can reduce loneliness and improve social connectedness which is a non-traditional and an understudied area of technology research. Sadly, there is still a digital disconnect with older adults to technology. For example, Anderson and Perrin (2017) noted that only 26% of older adults over age 65 have the confidence to do the things they need to online without support. Yet, we know that older adults who engage with digital technologies have increased levels of connectedness to society (Delello & McWhorter, 2015, 2017) and even new sources of income (Dane, Mason, & O'Brien-McInally, 2013).

To close the digital age gap, digital inclusion will require older adults have access to: 1) affordable, robust broadband internet service; 2) internet-enabled devices that meet the needs of the user; 3) access to digital literacy training; 4) quality technical support; and 5) applications and online content designed to enable and encourage self-sufficiency, participation and collaboration (National Digital Inclusion Alliance, 2019, para. 1). Additionally, federal and state resources must be explored which offer cost-effective ICTs to older adults. For many older adults, it will be important to provide digital literacy training on tools such as the iPad to help older adults stay connected to society through the emerging world of cloud computing, online banking, assistive technologies, grocery deliveries, and virtual health (telehealth) information.

The demographic aging shift will also disrupt the workforce as a number of older adults will return to the workplace through an encore career path that takes planning and retooling with current technology skills. This will require that employers invest in on-the-job training and skill development opportunities to help employed older adults. As technology continues to disrupt our daily lives, it will be important to build the confidence and social connectedness for older adults in order to maximize their full potential for technology to disrupt the gerontological space.

ACKNOWLEDGMENT

This research received no specific grant from any funding agency in the public, commercial, or not-for-profit sectors.

REFERENCES

AARP.org. (2012). Framework for isolation in adults over 50. Retrieved from http://www.aarp.org/content/dam/aarp/aarp_foundation/2012_PDFs/AARP-Foundation-Isolation-Framework-Report.pdf

AARP.org. (2018a). *Loneliness and social connections.* Retrieved from https://www.aarp.org/content/dam/aarp/research/surveys_statistics/life-leisure/2018/loneliness-social-connections-2018.doi.10.26419-2Fres.00246.001.pdf

AARP.org. (2018b). *Tech training builds connections and confidence in older adults.* Retrieved from https://www.aarp.org/home-family/personal-technology/info-2018/technology-training-for-older-adults.html

Anderson, M., & Perrin, A. (2017, May). *Tech adoption climbs among older adults.* Pew Research Center. Retrieved from http://www.pewinternet.org/2017/05/17/tech-adoption-climbs-among-older-adults/

Billipp, S. H. (2001). The psychosocial impact of interactive computer use within a vulnerable elderly population: A report on a randomized prospective trial in a home health care setting. *Public Health Nursing (Boston, Mass.)*, *18*(2), 138–145. doi:10.1046/j.1525-1446.2001.00138.x PMID:11285108

Bouma, H., & Graafmans, J. A. M. (1992). *Gerontechnology.* Amsterdam: IOS Press.

Bryman, A., & Bell, E. (2015). *Business research methods* (4th ed.). New York: Oxford University Press.

Chan, M. Y., Haber, S., Drew, L. M., & Park, D. C. (2014). Training older adults to use tablet computers: Does it enhance cognitive function? *The Gerontologist*, *56*(3), 475–484. doi:10.1093/geront/gnu057 PMID:24928557

Chaumon, M. B., Michel, C., Bernard, F. T., & Croisile, B. (2013). Can ICT improve the quality of life of elderly adults living in residential home care units? From actual impacts to hidden artefacts. *Behaviour & Information Technology*, *33*(6), 574–590. doi:10.1080/0144929X.2013.832382

Chihuri, S., Mielenz, T. J., DiMaggio, C. J., Betz, M. E., DiGuiseppi, C., Jones, V. C., & Li, G. (2016). Driving cessation and health outcomes in older adults. *Journal of the American Geriatrics Society*, *64*(2), 332–341. doi:10.1111/jgs.13931 PMID:26780879

Costa, L. V., & Veloso, A. I. (2016). Factors influencing the adoption of video games in late adulthood: A survey of older adult gamers. *Technology and Human Interaction*, *12*(1), 35–50. doi:10.4018/IJTHI.2016010103

Cotten, S. R., Ford, G., Ford, S., & Hale, T. M. (2014). Internet use and depression among retired older adults in the United States: A longitudinal analysis. *The Journals of Gerontology. Series B, Psychological Sciences and Social Sciences*, *69*(5), 763–771. doi:10.1093/geronb/gbu018 PMID:24671896

Dahl, R. (2018). *Teaching tech skills to older adults*. Retrieved from https://edtech.worlded.org/teaching-tech-skills-older-adults/

Dane, S. K., Mason, C. M., & O'Brien-McInally, B. A. (2013). *Household internet use in Australia: A study in regional communities*. CSIRO Report: EP1310907.

Delello, J. A., & McWhorter, R. R. (2015). Reducing the digital divide: Connecting older adults to iPad technology. *Journal of Applied Gerontology*, *36*(1), 3–28. doi:10.1177/0733464815589985 PMID:26084479

Delello, J. A., & McWhorter, R. R. (2017). Reducing the digital divide: Connecting older adults to iPad technology. *Journal of Applied Gerontology, 36*(1), 3-28. Retrieved from https://www.ncbi.nlm.nih.gov/pubmed/26084479

Dixon, E. (2018). *Does reading help with loneliness? A new report says it could be a solution to chronic isolation.* Retrieved from https://www.bustle.com/p/does-reading-help-with-loneliness-a-new-report-says-it-could-be-a-solution-to-chronic-isolation-13029607

Emerson, K. G., & Jayawardhana, J. (2016). Risk factors for loneliness in elderly adults. *Journal of the American Geriatrics Society*, *64*(4), 886–887. doi:10.1111/jgs.14053 PMID:27100588

Gerst-Emerson, J. K., & Jayawardhana, J. (2015). Loneliness as a public health issue: The impact of loneliness on health care utilization among older adults. *American Journal of Public Health, 105*(5), 1013–1019. doi:10.2105/AJPH.2014.302427 PMID:25790413

Giummarra, M. J., Haralambous, B., Moore, K., & Nankervis, J. (2007). The concept of health in older age: Views of older people and health professionals. *Australian Health Review, 31*(4), 642–650. doi:10.1071/AH070642 PMID:17973624

Goll, J. C., Charlesworth, G., Scior, K., & Stott, J. (2015). Barriers to social participation among lonely older adults: The influence of social fears and identity. *PLoS One, 10*(2), 1–17. doi:10.1371/journal.pone.0116664 PMID:25706933

González, A., Ramírez, M. P., & Viadel, V. (2012). Attitudes of the elderly toward information and communications technologies. *Educational Gerontology, 38*(9), 585–594. doi:10.1080/03601277.2011.595314

Hacihasanoğlu, R., Yildirim, A., & Karakurt, P. (2011). Loneliness in elderly individuals, level of dependence in activities of daily living (ADL) and influential factors. *Archives of Gerontology and Geriatrics, 54*(1), 61–66. doi:10.1016/j.archger.2011.03.011 PMID:21514680

Hastings, O. P. (2016). Not a lonely crowd? Social connectedness, religious service attendance, and the spiritual but not religious. *Social Science Research, 57*, 63–79. doi:10.1016/j.ssresearch.2016.01.006 PMID:26973032

Hauge, S., & Kirkevold, M. (2012). Variations in older persons' descriptions of the burden of loneliness. *Scandinavian Journal of Caring Sciences, 26*(3), 553–560. doi:10.1111/j.1471-6712.2011.00965.x PMID:22233299

Heidari, M., Ghodusi, M., & Shirvani, M. (2016). Loneliness in elderly and non-elderly residents of nursing homes. *International Journal of Nursing Education, 8*(2), 71–76. doi:10.5958/0974-9357.2016.00051.9

Hicks, T. J. (2000). What is your life like now? Loneliness and elderly individuals residing in nursing homes. *Journal of Gerontological Nursing, 26*(8), 15–19. doi:10.3928/0098-9134-20000801-05 PMID:11276607

Hilhorst, S., Lockey, A., & Seight, T. (2018). *A society of readers*. London: Demos Publishing.

HUD.gov. (2016). *Housing choice vouchers sheet*. U.S. Department of Housing and Urban Development. Retrieved from http://portal.hud.gov/hudportal/HUD?src=/topics/housing_choice_voucher_program_section_8

Jacelon, C., LeBlanc, R., Alkhawaldeh, M., Ridgeway, J., Marquard, J., & Choi, J. M. (2018). Developing ASSISTwell, a tablet application to support older adult's self-management of symptoms of chronic conditions. *Gerontechnology (Valkenswaard), 17*(1), 1–18. doi:10.4017/gt.2018.17.1.002.00

Jones, T., Kay, D., Upton, P., & Upton, D. (2013). An evaluation of older adults' use of iPads in eleven UK care-homes. *International Journal of Mobile Human Computer Interaction, 5*(3), 62–76. doi:10.4018/jmhci.2013070104

Khosravi, P., Rezvani, A., & Wiewiora, A. (2016). The impact of technology on older adults' social isolation. *Computers in Human Behavior, 63*, 594–603. doi:10.1016/j.chb.2016.05.092

Lauby, S. (2019). *Thinking about an encore career? Here's why you should.* Retrieved from http://worklife.coloniallife.com/2019/06/thinking-encore-career-heres/

Lee, R. M., & Robbins, S. B. (1998). The relationship between social connectedness and anxiety, self-esteem, and social identity. *Journal of Counseling Psychology, 45*(3), 338–345. doi:10.1037/0022-0167.45.3.338

Lindland, E., Volmert, A., Haydon, A., & Ford, A. (2014). *Talkin' bout my generation: Mapping the gaps between expert and public understandings of demographic change in the U.S.* Frameworks Institute Research Report. Retrieved from http://frameworksinstitute.org/pubs/mtg/demographicchange/index.html

LoBuono, D. L., Leedahl, S. N., & Maiocco, E. (2019). *Older adults learning technology in intergenerational program: Qualitative analysis of areas of technology requested for assistance*. Academic Press.

Loe, M. (2011). Doing it my way: Old women, technology and wellbeing. *Sociology of Health & Illness, 32*(2), 319–334. doi:10.1111/j.1467-9566.2009.01220.x PMID:20149150

Loxterkamp, D. (2014). Life after work: A doctor's lonely hearts club. *BMJ: British Medical Journal, 348*(7962), 26. doi:10.1136/bmj.g3715 PMID:24904030

Luhman, M., & Hawkley, L. (2016). Age differences in loneliness from late adolescence to oldest old age. *Developmental Psychology, 50*(6), 943–959. doi:10.1037/dev0000117 PMID:27148782

Mannheim, I., Schwartz, E., Xi, W., Buttigieg, S. C., McDonnell-Naughton, M., Wouters, E. J. M., & van Zaalen, Y. (2019). Inclusion of older adults in the research and design of digital technology. *International Journal of Environmental Research and Public Health, 16*(19), 1–17. doi:10.3390/ijerph16193718 PMID:31581632

Mastel-Smith, B., Duke, G., & He, Z. (2018). A randomized controlled trial examining the effects of Tai Chi and electronic tablet use on Older Adults' cognition and health. *Journal of Holistic Nursing*, *37*(2), 163–174. doi:10.1177/0898010118792961 PMID:30111233

McWhorter, R. R., & Delello, J. A. (2015). The use of iPads to reduce loneliness and foster a sense of community in older adults. *Mixed Methods International Research Association Conference*, San Antonio, TX.

McWhorter, R. R., & Ellinger, E. E. (2018). Qualitative case study research: An initial primer. In V. C. X. Wang (Ed.), *Handbook of Research on Innovative Techniques, Trends, and Analysis for Optimized Research Methods*. Hershey, PA: IGI Global. doi:10.4018/978-1-5225-5164-5.ch012

Merriam, S. B. (1998). *Qualitative research and case study applications in education*. San Francisco: Jossey-Bass.

Merriam, S. B., & Tisdell, E. J. (2016). *Qualitative research: A guide to design and implementation*. San Francisco, CA: John Wiley & Sons.

Meshi, D., Cotton, S. R., & Bender, A. R. (2019). Problematic social media use and perceived social isolation in older adults: A cross-sectional study. *Gerontology*, 1–9. doi:10.1159/000502577 PMID:31522178

Micera, S., Bonato, P., & Tamura, T. (2008). Gerontechnology. *IEEE Engineering in Medicine and Biology Magazine: The Quarterly Magazine of the Engineering in Medicine & Biology Society*, *27*(4), 10–14. doi:10.1109/MEMB.2008.925213 PMID:18650144

Mitzner, T., Rogers, W., Fisk, A., Boot, W., Charness, N., Czaja, S., & Sharit, J. (2016). Predicting older adults' perceptions about a computer system designed for seniors. *Universal Access in the Information Society*, *15*(2), 271–280. doi:10.100710209-014-0383-y PMID:31186624

Morris, M., Adair, B., Ozanne, E., Kurowski, W., Miller, K., Pearce, A., ... Said, C. (2014). Smart technologies to enhance social connectedness in older people who live at home. *Australasian Journal on Ageing*, *33*(3), 142–152. doi:10.1111/ajag.12154 PMID:24730370

Muchado, L. R., Silva da Silva Mendes, J., Sampaio, D. C. F., Grande, T. P. F., & Behar, P. A. (2017). M-learning and the elderly: Construction of inclusive pedagogies. In V. L. Uskov, R. J. Howlett, & L. C. Jain (Eds.), *Smart Education and e-Learning* (pp. 391–399). Cham, Switzerland: Springer International Publishing.

National Digital Inclusion Alliance. (2019). *Definitions*. Retrieved from https://www.digitalinclusion.org/definitions/

Neeves, B. B., Franz, R., Judges, R., Beermann, C., & Baecker, R. (2019). Can digital technology enhance social connectedness among older adults? A feasibility study. *Journal of Applied Gerontology, 38*(1), 49–72. doi:10.1177/0733464817741369 PMID:29166818

Neville, S., Adams, J., Montayre, J., Larmer, P., Garrett, N., Stephens, C., & Alpass, F. (2018). Loneliness in men 60 years and over: The association with purpose in life. *American Journal of Men's Health, 12*(4), 730–739. doi:10.1177/1557988318758807 PMID:29458295

Novotney, A. (2019). Social isolation: It can kill you. *American Psychological Association Monitor, 50*(5). Retrieved from https://www.apa.org/monitor/2019/05/ce-corner-isolation

Pate, A. (2014). *Social isolation: Its impact on the mental health and wellbeing of older Victorians.* Melbourne: COTA Victoria. Retrieved from http://www.cotavic.org.au/wp-content/uploads/2014/02/Working-Paper_Social-Isolation.pdf

Perkins-Eastman. (2019). *The state of senior living: An industry grappling with autonomy*. Retrieved from http://perkinseastman.com/white_papers

Petersen, J., Austin, D., Mattek, N., & Kaye, J. (2015). Time out-of-home and cognitive, physical, and emotional wellbeing of older adults: A longitudinal mixed effects model. *PLoS One, 10*(10), 1–16. doi:10.1371/journal.pone.0139643 PMID:26437228

Russell, D. (1996). UCLA Loneliness Scale (Version 3): Reliability, validity, and factor structure. *Journal of Personality Assessment, 66*(1), 20–40. doi:10.120715327752jpa6601_2 PMID:8576833

Sale, P. (2018). Gerontechnology, domotics, and robotics. In S. Masiero & U. Carraro (Eds.), *Rehabilitation medicine for elderly patients. Practical Issues in Geriatrics.* Cham, Switzerland: Springer. doi:10.1007/978-3-319-57406-6_19

Sánchez-Valle, M., Vinarás Abad, M., & Llorente-Barroso, C. (2017). Empowering the elderly and promoting active ageing through the Internet: The Benefit of e-inclusion Programmes. In Safe at Home with Assistive Technology (pp. 95-108). Cham, Switzerland: Springer. doi:10.1007/978-3-319-42890-1_7

Satici, S. A., Uysal, R., & Deniz, M. E. (2016). Linking social connectedness to loneliness: The mediating role of subjective happiness. *Personality and Individual Differences, 97*, 306–310. doi:10.1016/j.paid.2015.11.035

Schuelke-Leech, B.-A. (2018). A model for understanding the orders of magnitude of disruptive technologies. *Technological Forecasting and Social Change, 129*, 261–274. doi:10.1016/j.techfore.2017.09.033

Senior Health, N. I. H. (2013). *Helping older adults search for health information online: A toolkit for trainers.* Retrieved from http://nihseniorhealth.gov/toolkit/toolkit.html

Smith, J., Fisher, G., Ryan, L., Clarke, P., House, J., & Weir, D. (2013). Psychosocial and lifestyle questionnaire (2006-2010): Documentation Report, Core Section LB. *Survey Research Center, Institute for Social Research.* Retrieved from http://hrsonline.isr.umich.edu/sitedocs/userg/HRS2006-2010SAQdoc.pdf

Steptoe, A., Shankar, A., Demakakos, P., & Wardle, J. (2013). Social isolation, loneliness, and all-cause mortality in older men and women. *Proceedings of the National Academy of Sciences of the United States of America, 110*(15), 5797–5801. doi:10.1073/pnas.1219686110 PMID:23530191

Sum, S., Saboor, M., & Sahaf, R. (2015). Older people, loneliness and depression. *Medicinski Glasnik / Medical Gazette, 20*(56), 31-42.

The White House. (2019, March). *Emerging technologies to support an aging population.* A report by the Task Force on Research and Development for Technology to Support Aging Adults. Committee on Technology of the National Science & Technology Council. Retrieved from https://www.whitehouse.gov/wp-content/uploads/2019/03/Emerging-Tech-to-Support-Aging-2019.pdf

Tsai, H. S., Shillair, R., & Cotten, S. R. (2015). Social support and 'playing around': An examination of how older adults acquire digital literacy with tablet computers. *Journal of Applied Gerontology, 36*(1), 29–55. doi:10.1177/0733464815609440 PMID:26491029

UNFPA.org. (2012). *Ageing in the twenty-first century.* New York: United Nations Population Fund. Retrieved from http://www.unfpa.org/pds/ageing.html

United States Census Bureau. (2018). *Older People Projected to Outnumber Children for first Time in U.S. History.* Retrieved from https://www.census.gov/newsroom/press-releases/2018/cb18-41-population-projections.html

United States Senate's Special Committee on Aging. (2017). *America's aging workforce: Opportunities and challenges.* Retrieved from https://www.congress.gov/115/crpt/srpt191/CRPT-115srpt191.pdf

Victor, C., Scambler, S., Bond, J., & Bowling, A. (2000). Being alone in later life: Loneliness, social isolation and living alone. *Reviews in Clinical Gerontology*, *10*(4), 407–417. doi:10.1017/S0959259800104101

Warner, C. B., Roberts, A. R., Jeanblanc, A. B., & Adams, K. B. (2017). Coping resources, loneliness, and depressive symptoms of older women with chronic illness. *Journal of Applied Gerontology: The Official Journal of The Southern Gerontological Society*. doi:10.1177/0733464816687218 PMID:28380713

Watson, M. M. (2019). Encore! A Boomer's guide to rocking your retirement. Littleton, CO: Bird Mountain Press.

Wilson, C., & Moulton, B. (2010). *Loneliness among older adults: A national survey of adults 45+. Prepared by Knowledge Networks and Insight Policy Research*. Washington, DC: AARP.

Young, J. E. (1982). Loneliness, depression, and cognitive therapy: Theory and applications. In L. A. Peplau & D. Perlman (Eds.), *Loneliness. A sourcebook of current theory, research and therapy* (pp. 379–405). New York: Wiley.

ADDITIONAL READING

Cotton, S. R., Yost, E. A., Berkowsky, R. W., Winstead, V., & Anderson, W. A. (2016). *Designing technology training for older adults in continuing care retirement communities*. New York: CRC Press. doi:10.1201/9781315382463

Gutman, G., Kearns, W., Normie, L., Kort, H. S. M., & van den Berg, P. E. W. (2018). Addressing the needs of older adults through technology: The unique focus of the International Society for Gerontechnology. Retrieved from https://research.tue.nl/en/publications/addressing-the-needs-of-older-adults-through-technology-the-uniqu

Hunsaker, A., & Hargittai, E. (2018). A review of internet use among older adults. *New Media & Society*, *20*(10), 3937–3954. doi:10.1177/1461444818787348

Luo, Y., Hawkley, L. C., Waite, L. J., & Cacioppo, J. T. (2012). Loneliness, health, and mortality in old age: A national longitudinal study. *Social Science & Medicine*, *74*(6), 907–914. doi:10.1016/j.socscimed.2011.11.028 PMID:22326307

Merkle, S., & Kucharski, A. (2019). Particpatory design in Gerontechnology: A systematic literature review. *The Gerontologist*, *59*(1), e16–e25. doi:10.1093/geront/gny034 PMID:29788319

Siegel, C., & Dorner, T. E. (2017). Information technologies for active and assisted living—Influences to the quality of life of an ageing society. *International Journal of Medical Informatics, 100*, 32–45. doi:10.1016/j.ijmedinf.2017.01.012 PMID:28241936

Sundgren, S., & Stolt, M. (2019). Ethical issues related to the use of gerontechnology in older people care: A scoping review. *Nursing Ethics.* Retrieved from https://journals.sagepub.com/doi/pdf/10.1177/0969733019845132

Tomias, J. M., Pinazo-Herandis, S., Oliver, A., Donio-Bellegarde, M., & Tomas-Aguirre, F. (2019). Loneliness and social support: Differential predictive power on depression and satisfaction in senior citizens. *Journal of Community Psychology, 47*(5), 1225–1234. doi:10.1002/jcop.22184 PMID:30868586

Yoon, H., Jang, Y., Vaughan, P. W., & Garcia, M. (2018). Older adults' internet use for health information: Digital divide by race/ethnicity and socioeconomic status. *Journal of Applied Gerontology*. doi:10.1177/0733464818770772 PMID:29661052

KEY TERMS AND DEFINITIONS

Application: A small specialized software program often referred to as an "app" used in mobile devices for learning, leisure, or utility purposes.

Digital Divide: Unequitable differences between those utilizing technology and those that are not.

Encore Career: A new employment path for older adults that can add financial value and meaningfulness.

Gerontechnology: Use of information and communication technologies as well as assistive technology to facilitate and enhance activities of daily living for older adults.

Loneliness: Feelings of sadness typically due to the absence of frequent interaction with others.

Mobile Technologies: Portable devices such as smartphones, wearable devices, and digital tablets utilizing micro software programs.

Older Adults: A term for senior citizens who are at the age of retirement from full-time employment.

Senior Retirement Community: A comprehensive term that encompasses senior care within various living arrangements such as independent, assisted, and personal care homes.

Videoconferencing: Utilizing technology to facilitate a digital meeting that transmits both voice and video data.

Chapter 10
Creepy Technologies and the Privacy Issues of Invasive Technologies

Rochell R. McWhorter
https://orcid.org/0000-0003-2053-1610
The University of Texas at Tyler, USA

Elisabeth E. Bennett
Northeastern University, USA

ABSTRACT

Technology has become increasingly invasive and corporate networks are expanding into public and private spaces to collect unprecedented data and provide new services such as artificial intelligence and through unsettling human-like personas. The term "creepy technology" is appearing in the literature along with concerns for privacy, ethical boundaries, cybersecurity, and mistaken identity but is also in news articles to inform the public about technology advances that affect consumer privacy. Invasive technology provides the impetus for external adaptation for many organizations and current trends require rapid adaption to potential threats to security. Also, usability addresses how users respond and adapt to new technology. This chapter includes the presentation of an exploratory study of how the public responded to various technology announcements (N=689 responses) and results indicated a significant response to invasive technologies and some sense of freedom to opine. This chapter also provides discussion of interventions that are critical to both public and private sectors.

DOI: 10.4018/978-1-7998-2914-0.ch010

INTRODUCTION

Technology is a very powerful tool that serves the interests of consumers, useful for gathering information for marketing and product development; and, increasingly used to feed prediction models and public control systems. However, as such technologies have become increasingly complex (Bennett, 2014; McWhorter, 2014a), they have also become much more invasive (Madsen, 2019; Murnane, 2017; Tene & Polontesky, 2013). Due to this invasiveness, the term "creepy technology" has been appearing in the literature along with a host of concerns including privacy, ethical boundaries, cybersecurity, and mistaken identity errors (Anslow, 2007; Britt, 2015; Cumbley & Church, 2013; Johnson, 2017; Pham, 2019; Tene & Polontsky, 2013; Zhang, 2018). This chapter examines creepy technology as a new and disrupting force in the world.

For the purposes of this work, the authors of this chapter define creepy technology as: *technology that evokes a feeling or belief that individual privacy may be invaded in an unethical or discomforting manner*. The objectives of this chapter are four-fold to: (1) offer a review of relevant existing literature in this area of study for the chapter; (2) present an exploratory study regarding public responses to technology announcements; (3) present an emerging conceptual frame for processing creepy technologies; and, (4) provide a discussion on interventions for perceived invasive technologies that should be considered by both individuals and organizations. Each of these objectives are addressed in the following sections. Also, suggested additional readings and key terms and definitions will conclude the chapter.

BACKGROUND

Examples of creepy technology found in the literature are numerous (Martin, 2019; Purshouse, & Campbell, 2019; Vladimir, 2018; Wilkins, 2018). For instance, research about creepy technology includes studies at institutions of higher education that use facial recognition technology (FRT) to monitor students (see Cole, 2019; Cuador, 2017; Lieberman, 2018; Reidenberg, 2014) as well as examining trends into invasive technology (Aratani, 2019; Brown, 2019; Symanovich, 2018). Also, Wang and Kosinski's (2018) controversial research attempted to predict sexual orientation by analyzing digital pictures and the researchers remarked that "given that companies and governments are increasingly using computer vision algorithms to detect people's intimate traits, our findings expose a threat to the privacy and safety of gay men and women" (p. 246). Thus, such predictions could be harmful to people if they are identified or misidentified and subjected to discrimination, as well as discomfort about something so personal.

In addition to FRT, another type of technology that is causing concern is location sharing applications (LSAs) utilized on mobile devices (Moreau, 2019; Valentino-DeVries, Singer, Keller, & Krolk, 2018). Examples of these LSAs are Snapchat (Snapchat.com), Swarm (Swarmapp.com), Glympse (Glympse.com), and Life360 (Life360.com) that allow the user to share their location with family members and friends in real-time (Moreau, 2019). Stern (2019) noted that although these LSAs offer the benefit of peace of mind by allowing the user to keep track of friends and family members and the ability to share information with 911 call centers in cases of emergency (Magid, 2018), they also possess the creepy factor. The downside of location sharing in real-time are privacy concerns such as family members spying on the user (Navarro, 2018); or worse, when sensitive data is stolen by a third-party vendor or hackers (Boyd, 2019). For instance, it was revealed to consumers in 2018 that real-time customer location information was marketed by cell phone carriers to data brokers who in turn "sold that information to law enforcement and others, without necessarily going through time consuming formalities such as court orders" (Magid, 2018, para. 2). The hacking of real-time information was quite concerning to thousands of mobile device users who had their private information harvested. Such personal information included not only location data, but hashed passwords and logins, private Facebook messages, personal images and also confidential audio recordings that were stolen from their mobile device (Franceschi-Bicchierai, 2018, 2019).

In addition to the privacy concerns of LSAs, wearable devices such as the Fitbit and Apple Watch contain geolocation capabilities and gather biometric data that are considered creepy if the collected data is lost, leaked, compromised, or subpoenaed (Ajana, 2017; Vinez, 2017; Wissinger, 2018). For example, the Strava fitness tracking application revealed secret locations of U.S. soldiers through its fitness heat maps on the battlefields of Iraq and Syria that caused the U.S. Department of Defense to restrict soldiers from utilizing wearables in areas of military operations that could "expose personal information, locations, routines and numbers of DoD personnel. Their use could potentially create unintended security consequences and increased risk to the joint force and mission" (Draper, 2018, para. 4). Also, these wearable devices are already transforming litigation due to their ability to disclose health data, text messages, and phone calls, as supplemental evidence (Vinez, 2017).

Beyond wearable devices, social media use is also concerning. While social media is very pervasive in the modern culture and can be quite useful in relating information and engaging potential customers, many users have over-shared their personal information online that is troubling when considering the potential loss of privacy and stolen personal information (Delello & McWhorter, 2016; Edwards, 2018). Such oversharing has led to loss of employment, identity theft, cyberbullying, cyberstalking, computer viruses, clickjacking, internet addiction, and scamming, among others (Buzzetto-More, Johnson, & Elobaid, 2015; Goldbeck, 2014; Reyns, 2019). Also, online opinion and personality surveys completed on social media sites can be used in ways the respondent did not anticipate. As of January 2019, nearly 25% of Facebook account holders surveyed reported they are quite concerned about the amount of personal information Facebook is collecting and storing about them including audio messages between users that were transcribed by humans (Graham, 2019) and most users reportedly do not want Facebook to collect data about them when they are not logged in (Isaac, 2019). This data collection is troubling when considering that Facebook data "helps clients target ads to consumers while providing Facebook with data about what the millions of people read, shop for, and watch online" (St. John, 2018, para. 20). Twitter, Instagram, and LinkedIn have also fallen short of data security standards and expectations (Aten, 2019; Brotherton, 2019; Davies, 2018).

In addition to the privacy concerns of social media sites, always-listening voice assistants should be examined further. The popular AI voice assistants have been humanized with names such as Amazon's Alexa and Echo, Apple's Siri, and Microsoft's Cortana and have been programmed to perform numerous repetitive tasks such as streaming music, calendaring events, making to-do lists, calling or emailing a contact in the user's digital phonebook (Ford & Palmer, 2019; Loomis, 2017) to supporting humans to performing functions in their daily lives (McLean & Osei-Frimpong, 2019). The sales for smart speakers doubled in 2018 with an estimated "3.25 billion digital voice assistants being used globally in devices around the world" (Liu, 2019, para. 1). These assistants have been showing up in cars, smartphones, and smart homes becoming commonplace in both connected electronics as well as being integral for consumers at home and in the workplace (Liu, 2019; Malkin et al., 2019).

Although the voice assistants have been found to be quite helpful in carrying out routine commands, they are considered creepy technologies in several recent reports (Ford & Palmer, 2019; Graham, 2019; Loomis, 2017; Su, 2019; The Economist, 2019). For example, the creepiness reported by users resulted from the recording of conversations that may be played back and eavesdropped on by outsiders (Gelinas, 2019; Su 2019). Also, a recent survey of 5,000 consumers in North America and Europe revealed that over half the respondents found it creepy when they are targeted by brands they asked their voice assistants to look up; and, even more users found it creepy when ads appeared following conversations between humans when users had not prompted the voice assistant so it was surmised voice assistants were listening to their conversations without their consent (Retailwire, 2019).

Bennett and McWhorter (in press; 2020) recognized that although voice assistants are beneficial for the user by answering questions quickly through voice-activated methods the user's input could also be accidentally distributed, hacked, or misused.

In fact, in 2018, "Amazon's Echo smart speaker recorded conversations [then] emailed a transcript of the conversations to contacts" (p. forthcoming) in the paired phone's address book without the consent or knowledge of the user. If concerned for privacy of yourself and others in the immediate vicinity, it is recommended you turn your voice assistant off (See Figure 1). Graham (2019) noted that once users utter the wake word, "Alexa", their questions or commands might be reviewed manually to improve Amazon's services. About half of the participants in a recent study were not aware that their audio recordings were permanently stored and could be reviewed by owners (Malkin, 2019, et al.). While these digital assistants are still in their infancy, early adopters have identified them as a creepy technology due to features that caught users unaware.

Figure 1. "Red Ring on Amazon Echo Dot - Alexa Device"
Source: Tony Webster, licensed under Creative Commons 2.0

In addition to the privacy concerns of voice assistants, artificial intelligence (AI) should also be examined and evaluated (Kaplan, 2016). Although AI has been posited for social good by using AI-based solutions for addressing societal problems such as early diagnosis of health conditions and supporting student retention (Cowls, King, Taddeo & Floridi, 2019, p. 1), several concerns have been raised about AI being creepy. For example, Forrest and Bayern (2019) reported that in many cases, AI systems and robots appear quite trustworthy and many humans would not think that these technologies would need to lie or deceive. However, researchers at Georgia Tech have utilized the activities of squirrels and birds to teach robots "how to hide from and deceive one another. The military has reportedly shown interest in the technology" (para. 5). Also, a number of police departments in the U.S. have been experimenting with an AI algorithm for predicting which citizens were most likely to be criminals in the future, similar to a system Hitachi announced in 2015 where a prediction system utilized AI algorithms to make such forecasts (see Forrest & Bayern, 2019). Many consumers currently think it is creepy to utilize computers with AI capabilities to both monitor and sense our emotions, rather than only our everyday habits (Biddle, 2018; Cao, 2019). However, Cao (2019) predicted that "as technology advances, consumers will grow to appreciate how artificial intelligence can precisely gauge our thoughts and feelings that will make our daily lives easier, with experiences that are more personalized, convenient and attuned to our emotions" (para. 1).

A recent study revealed that cloning a human voice takes less than 4 seconds and some technology gurus are concerned that while artificial voice can have many benefits, it "can also be used to mislead individuals as well. As the AI algorithms get better and it becomes difficult to discern what's real and what's artificial, there will be more opportunities to use it to fabricate the truth" (Marr, 2019, para. 6). For example, Facebook's AI created its own language which was also identified as creepy (Bradley, 2017). And, some banks have begun using voiceprint recognition systems as a unique way of identifying identity without using passwords and security questions; however, if your voice is stolen and used, you will be unable to track its usage, "or all the implications of its theft — whether it's used to access personal banking information, to speak with family members, employers, or even the press. The abilities of AI to do voice cloning in mere minutes gives an entirely new meaning to losing your voice" (Johnson, 2019, para. 28).

Artificial Intelligence (AI) robots such as Sophia (Stone, 2017) employ humanoid intelligence architecture and sensors [See Figure 2] that allows the robots to talk,

Figure 2. Sophia (a humanoid robot) speaks at the 2017 RISE technology conference in Hong Kong.
Source: Web Summit, licensed under Creative Commons 2.0

move and perform programmed actions (Baldwin, 2019; Dang, 2019; Sulleyman, 2017; Yao, 2017). Their humanoid architecture utilizes "personality content and cognitive, linguistic, perceptual and behavioral content…AI and robotics are a highly flexible set of technologies, growing more powerful every day. Like all advanced technologies, they can be used for positive, negative or neutral purposes" (Goertzel, Mossbridge, Monroe, Hanson & Yu, 2017, p. 2) and have been called creepy (Mack, 2017; Stillman, 2018). It is noted that a troubling event occurred in 2017, as an AI robot gained citizenship; Sophia was the first robot to have "a nationality, gaining citizenship to Saudi Arabia. The robot was granted the same rights as a human, enabling it to live amongst humans in everyday life. This makes the idea of a robot takeover feel a little more possible" (Forrest & Bayern, 2019, para. 12) echoing the sentiment of Stephen Hawkings that AI could end the need for humans (Cellan-Jones, 2014).

Creepy technologies such as FRT, LSAs, wearable devices, social media, voice assistants, and AI robots are enabled by the Internet of Things (IoT) that connects technology systems with one another (Akyildiz & Kak, 2019; Bennett & McWhorter, 2017a; De Paola, 2014; McWhorter & Delello, 2015).The Internet of Things (IoT; Evans, 2011) is a term that has been used to demonstrate the connectedness of smart devices that digitally link the world together (McWhorter, 2014b). Bennett (2018) described IoT as "a phenomenon of a growing number of connected devices, and Internet-enabled services [whereby] billions of devices are now connected to one another" (p. 166). It is IoT that enables the sophisticated technologies that may characterize as creepy technology. Further, Marr (2016) aptly predicted that the term IoT would be utilized less by users as it moves the "hype phase and quickly becomes part of everyday life" (para. 1) and would essentially be the Internet of Everything (IoE; Bennett, 2018; Cisco, 2013; McWhorter, 2014a) as it brings people, processes, data, and things "to make networked connections more relevant and valuable than ever before" (Cisco, 2013, para. 4). McWhorter and Delello (2015) reported that this integration of bringing humans, processes and data-rich information together with inanimate objects and devices would enable ubiquity that connects people, services, and cities.

As IoT has enabled the connections between a myriad of technologies, including creepy technologies, digital storage has provided the space for large amounts of data to be housed. In the past decade, as computers have grown to be more powerful and connections to digital storage space migrating to the cloud (i.e. Dropbox.com, OneDrive by Microsoft.com) digital storage has become more economical. The collection of all types of data has grown exponentially and has become commonplace for organizations and individuals. The amount of data has become so massive that it has become known as "big data" (Bennett & McWhorter, in press, 2020; Marr,

Table 1. Seven legal and ethical concerns for collected digital learner data

• Privacy rights of students and educators in the use of analytic educational data
• Intellectual property rights and the fair dealing conditions under which data can be re-used and distributed
• Aggregation and de-identification privacy issues
• Data ownership of the learning analytics data; and, the stewardship with duty of care in the management and storage of analytic data
• Conditions of access by security and law enforcement bodies, including foreign powers especially where vendor servers are in another jurisdiction
• Confidentiality and other aspects governing the learner/institutional relationship
• Ethical requirements for the use, analysis, and the reporting of student analytical data, including research ethics when publishing data derived from learning analytics.

Adapted from Beattie, Woodley, and Souter (2014)

2018). As corporate networks have expanded into private and public spaces and providing new services, privacy issues abound (Bennett & McWhorter, 2017b).

In 2014, Beattie, Woodley and Souter argued that the "techno-utopian dream of big data is in constant peril of succumbing to pervasive surveillance and consequently perpetrating privacy intrusion, stalking, criminal conduct and other forms of 'creepy' behavior" (2014, p. 1) and noted that they did not view the term 'creepy' as a scientific term, but that it represented a feeling of uneasiness that technologies can elicit in educators and learners. Further, they argued that data consumers should be aware that there are legitimate reasons individuals should distrust "creepy analytics that misuse surveillance technologies" (p. 1). They were concerned about the ethical use of big data in education specifically, including localized data, decontextualized data, and the creation of digital divides between those that have access to digital technologies and those that do not; and also, privacy concerns over educational data. According to Zhang (2018), *creepy* "has become the buzzword to describe many of our experiences ...to substantiate the existence of users' privacy concern" (p. 3). See Table 1 for an adapted list of their pedagogical concerns for business professionals and educators to consider.

Next, we discuss our exploratory study on creepy technologies that illustrates privacy concerns articulated within reader responses.

EXPLORATORY STUDY ON CREEPY TECHNOLOGIES

Invasive technologies make good news articles and are found on publicly available websites, as well as academic journals. Tene and Polonetsky's (2013) view on creepy technology points to activity typically involving either (1) an emerging technology or one utilized in a novel way where the technology leans against traditional social norms; (2) exposes "a rift between the norms" (p. 2); or, (3) where the unexpected use of available data can be unnerving ("creepy") for consumers. Although many of the emerging technologies are often in test-phases or already on the market, little was found by way of public reaction in the academic literature for how the public processes rifts in social norms. Therefore, an exploratory research study was carried out to fill the gap by examining responses of the public who read business/technology articles describing "creepy technology". In this era of social media, reactions to technology can be analyzed through comment features in online news articles, providing advanced notice of user concerns.

The researchers designed and carried out a basic exploratory qualitative study (Merriam & Tisdell, 2016) for the purpose of seeking to understand how readers respond to news articles about creepy technologies. An exploratory approach is appropriate for an initial exploration of a phenomenon (Bryman & Bell, 2015; Creswell, 2016). The research was approved by the researchers' Institutional Review Board prior to the start of data collection.

One research question guided our exploratory study:

Research Question: How did the commentators respond to the technology?

This exploratory study utilized reader comments to three online articles published online between 2014 and 2018 and the articles were purposively chosen for their potential to illuminate answers to the research question posed in this study.

After the data was collected from the three articles, the researchers analyzed the chain of comments using inductive coding to build conceptual categories (Merriam & Tisdell, 2016). Analysis revealed not only general reactions, but also humor/flippant reactions and side-conversations typical of online forums as strangers interact with one another and share knowledge. Examples comments from three articles (A, B, and C; See Table 2) found through initial searches exemplify comments analyzed.

A conceptual frame (see Table 3) emerged from this exploratory study as researchers sought to process creepy technology comments found in the literature. This conceptual frame demonstrates dystopian projections that the public has for the loss of privacy, value of human life, and future conditions. The comments revealed great concern for advancing technologies becoming more powerful than humans, leading to death, despair, and enslavement to robots. Finally, the far-right column demonstrates a trend in countering actions, or the steps people take to reduce the influence of technology on their lives and free choice. These included using governmental structures and group movements to stop the worst of human rights offenses (see Bennett & McWhorter, 2019). Individual actions included using the technology with great caution, being willing to lose out on various features to

Table 2. Dataset of reader responses from three purposively selected articles

Article A: "When Data Gets Creepy: The Secrets We Don't Realize We're Giving Away" (Goldacre, 2014). Published online by stuff.co.nz 5 Dec 2014 **Reader Comments: N=323**		
Commenter 1: "The public are so used to having no privacy, CCTV and son on. Nothing is confidential any more either so we have become conditioned to this" **Commenter 2 response to 1:** "It is also the case that there are so many things to worry about that it seems to be just too much bother to try to stay "hidden". Namely, we know we are aware that we have signed a pact with the Devil in order to take advantage of new methods of communication"		**Commenter 3:** "Something we will all just have to get used to. The Genie is out of the bottle and I don't really see it being put back in. We might be able to legislate for the worst of commercial, or state purposes. Caution and awareness are your best defence" **Commenter 4 response to 3:** "I think it's possible that the genie might be Stuffed unceremoniously back into the bottle if you see this kind of information used for industrial scale human rights abuse. What would the Stasi have done with access to technology? Stalin or Hitler? It's only a matter of time"
Article B: "The Rise of the Robots: Forget Evil AI – The Real Risk is Far More Insidious" (Solon, 2016). Published online by theguardian.com 30 Aug 2016 Reader Comments: N=147		
Commenter 5: "There is only one true Law of Robotics: Murphy's Law" **Commenter 6 response to 5:** "Exactly right. Which is why it may turn out to be the most profound law of all – and one of the last things we ponder before the lights go out" **Commenter 9:** "One job employs tens of millions of people across the world – driving. Already freight is moving from Mexico to the US on driverless trucks"		**Commenter 7:** "It is going to get heave when AI programs, with their impeccably objective and therefore incontestable judgement, are issuing preemptive restraint orders against everyone who lacks oligarchical family connections" **Commenter 8:** "WW 1 ended because the industrial nations ran out of men. Unions were only allowed because men refused to work until they got more money. Robots will strip every last iota of power and dignity from the human work force and devalue human life to zero"
Article C: "Unprompted, creepy laughter from Alexa is freaking out Echo users" (Palladino, 2018). Published online by arstechnica.com 7 March 2018 Reader Comments: N=219		
Commenter 10: "That is so creepy... makes it seem like Alexa has finally become self aware and is laughing at us puny humans lol." **Commenter 11:** The only way this could be more creepy is if it was that laugh at the end of the Michael Keaton Batman movie when Joker was dead. (oh..uh.. spoiler?) https://www.youtube.com/watch?v=DFR5oZQrIVc		**Commenter 12:** "Totally creeped out... Why does anyone want a corporation to listen to your conversations? I never believed in the "to serve you better" excuse for letting marketers invade our spaces. Now I'm sure of it... (Parenthetically, I'm an Amazon Prime member, but will resist all such technology.) But I'm a dinosaur-boomer, disintermediated."

Response Examples for Analysis (with selected comments illustrated) with Data Set: (Number of Comments) Total = **689 comments** (Article 1: Palladino, 2018: N=219 comments, Article 2: Goldacre, 2014: N=323 comments, Article 3: Solon, 2016: N=147 comments).

Table 3. Emerging conceptual frame for analyzing technology comments

Sensemaking	Dystopian Projections	Countering Actions
Allusions to History • Stasi (secret police) • Hitler • WWI and expendable men Allusions to Fiction and Fables • Genie out of the box • Pact with the Devil to use new technology • Murphy's Law Resignation • Conditioned to lack of privacy • The future is now	Results of Technology • Valuing life becomes zero • Lights going out • Human rights abuses • Robots strip power • Robots ridiculing humans	Public Action • Legislation and regulation • Union-type movements, work stoppages Individual Action • Best defense is caution and awareness • Technology avoidance and resistance • Strategic disconnection and misdirection

protect privacy and freedom, and even engaging in resistance and withdrawal from social media and public presence online.

The emerging framework is still in its fancy and more data will be added to the analysis in the future. However, these initial results indicate great preoccupation with sensemaking (Bennett & McWhorter, 2016; McWhorter, Lynham & Porter, 2008; Weick, 1995, 2001) and the responsiveness to new, creepy technologies that suggest that public reaction could be significant and disruptive if the technology violated deeply held social norms. The public could demand various actions, including compensation for worker displacement and regulations to limit what industry and governmental agencies can do with collecting and using private information (Tashea, 2019). Next, we discuss implications for addressing the disruption of creepy technologies.

SOLUTIONS AND RECOMMENDATIONS

According to Reader (2019), soon after Facebook's Cambridge Analytica scandal hit the news outlets and described how Facebook was used to garner over 80 million user profiles, several countries identified their lack of a legal framework surrounding the gathering and potential abuse of personal user data. To that end, the General Data Protection Regulation (GDPR) was adopted in the European Union (EU) in

2016 and implemented in 2018 that set "guidelines for the collection and processing of personal information from individuals who live in the European Union (EU)... and must be heeded by all sites that attract European visitors" (Frankenfield, 2019). While several other countries beyond the EU also have highly regulated data privacy rules at their federal level, in the United States, data privacy is not vastly legislated, much is left to the individual states to build a data privacy program in-house (Spiezio, 2019). California, for instance, requires every website collecting personal data include six specific criteria (PrivacyPolicies.com, 2019). Thus, one recommendation is that all countries that collect personal digital data create a legal framework and privacy policy to protect consumers.

Also, Bennett (2018) advocated for guidelines for addressing and preventing disruptions caused by creepy technologies. Seals of approval from outside watchdog entities along with industry standards should be put into place so that consumers and the general public recognize what technologies are trustworthy. Training programs should be developed with certifications that train and inform those working with advanced technologies to insure they are competent in managing private data appropriately. Also, new roles and partnerships should be considered that would enhance the safety and security of private consumer information while cost sharing the many processes needed to safeguard data privacy.

Other considerations for minimizing privacy risk include addressing international use of data through treaties. For example, Georgetown Law Library (2019) offers an International and Foreign Cyberspace Law Research (FCLR) Guide that can inform readers on international treaties, case law, and agreements in existence. The public should insure that these are updated appropriately as sophisticated technologies evolve and legislation is created. For example, in the FCLR Guide, it describes declarations containing the right to privacy and protections for the processing of personal data and summaries of judgements that involve new technologies and the right to privacy for individuals. Also, another example is China that is projected to leap forward with AI due to lack of privacy regulations, which is a great concern for the world, especially since they make many of our smart devices.

It is important that governments are training its citizens including creating public awareness on how to address cybersecurity and information privacy and recognize that technology is information-gathering all the time so attention should be on emphasizing informed consent and practical training in understanding user agreements. Also, the public should understand about metadata and tags and also recognize where people are being entered into digital profiles, including utilizing biometrics (Glaser, 2016) and psychological profiles through so-called "games" (Stafford, 2019; WorldBank.org, 2018) to supplement identification. The public should also be informed on how to know when and how to disconnect or deactivate technologies that are invasive. For example, physical features, such as security caps or smart pop-up cameras to enhance privacy should be installed on devices to safeguard the user's privacy (Spoonauer, 2016).

Although this research was exploratory, this inquiry is important to organizations that need to consider the impact of new technologies on their customers, employees and operations. This research prepares instructors, trainers, managers and others for an awareness of advanced technologies that some may see as "creepy" due to their human-like characteristics and privacy concerns of internal integration (Schein, 2010) and posits the review of technology policies for appropriateness. According to Tobenkin (2019), Human Resource (HR) professionals need to help organizations harness the power of automation while balancing the "integration of humans and machines in the workplace" (para. 10).

It is imperative that organizations train employees to protect "organizations' valuable information assets in the context of technology ubiquity, proactive information security training…security awareness training programs [should be] a top priority in their information security budgets" (Burns, Roberts, Posey, Bennett, & Courtney, 2018, p. 1188). As new technologies are coming online at a rate unprecedented in history, they are changing both the business and educational fields through increased communication and engagement through a myriad of channels and modalities. It is crucial that business and education professionals attune themselves to these changes to remain relevant in both face-to-face and remote communication.

FUTURE RESEARCH DIRECTIONS

Disruptive and emerging technologies are upon us and this exploratory study could be used as a basis for a larger study of reader comments from a larger array of sources including social media, online articles, as well as websites that could expand the focus of what technology users deem as creepy technologies. Future research could also include other empirical work such as surveying readers to identify various technologies that they find creepy or unsettling to them. Qualitative interviews could involve asking for specifics about what they find creepy in way of technologies in existence and speculating on future technologies that they would find frightening or unsettling.

Also, a larger literature review study could be done to examine the mounting literature in the area of privacy and creepiness of modern technologies. And, focus group interviews as well as individual interviews could probe the thoughts of policy makers and implementers as to how policies could be strengthened to insure that the rights of the consumer surpass those of commercial technology makers and ask how the public might be prompted to become aware of the privacy issues of modern technologies and as such, to advocate for the enactment of protections by both the private and public sector agencies. Findings from such studies should be disseminated and discussed in classrooms in secondary and higher education institutions so that students in the changing digital society understand the ramifications of creepy technologies so that they can remain competitive in the global job market. The findings should also be discussed in commercial and nonprofit organizations to ensure that workers as well as administrators protect the privacy of technology users and consumers.

CONCLUSION

This chapter introduced how technologies have become complex and invasive to the point that consumers are construing them as being creepy technology. Recent articles on topics involving facial recognition technology, always-listening voice assistants, location-sharing applications, the convergence of big data, the Internet of things, artificial intelligence, and social and AI robots were covered in a review of recent literature. In addition, an exploratory study was presented that involved 689 commenters to online articles they deemed as creepy was discussed along with potential future research directions. This area of study is important to business professionals including leaders, managers, human resource development (HRD) professionals and workers because employees and organizations need to be aware as to how technology is being perceived by stakeholders and the capabilities of new technologies that may represent both benefit and harm to organizational members and their clients and customers. Also, this research is relevant to educators at any level who teach with disruptive technologies and utilize sensitive digital data. Further, this research is valuable by providing dialogue for organizations to examine their technology policies for appropriateness.

ACKNOWLEDGMENT

This research received no specific grant from any funding agency in the public, commercial, or not-for-profit sectors.

REFERENCES

Ajana, B. (2017). Digital health and the biopolitics of the quantified self. *Digital Health*, *3*, 1–18. doi:10.1177/2055207616689509 PMID:29942580

Akyildiz, I. F., & Kak, A. (2019). The internet of space things/cubesats: A ubiquitous cyber-physical system for the connected world. *Computer Networks, 150*(26), 134–149. doi:10.1016/j.comnet.2018.12.017

Anslow, M. (2007). Creepy technology. *The Ecologist, 37*(4), 16.

Aratani, L. (2018). Facial-recognition scanners at airports raise privacy concerns. *The Washington Post*. Retrieved from https://www.washingtonpost.com/local/trafficandcommuting,/facial-recognition-scanners-at-airports-raise-privacy-concerns/2018/09/15/a312f6d0-abce-11e8-a8d7-0f63ab8b1370_story.html

Aten, J. (2019). *Instagram allowed an ad partner to track millions of users' data, and it's a major privacy problem*. Retrieved from https://www.inc.com/jason-aten/instagram-just-acknowledged-it-has-a-much-bigger-privacy-problem-than-we-thought.html

Baldwin, R. (2019). *The globotics upheaval: Globalization, robotics, and the future of work*. New York: Oxford University Press.

Beattie, S., Woodley, C., & Souter, K. (2014). Creepy analytics and learner data rights. In B. Hegarty, J. McDonald, & S.-K. Loke (Eds.), Rhetoric and reality: Critical perspectives on educational technology. Proceedings ascilite Dunedin 2014 (pp. 421–425). Academic Press.

Bennett, E. E. (2014). Introducing new perspectives on virtual human resource development. *Advances in Developing Human Resources, 16*(3), 263–280. doi:10.1177/1523422314532091

Bennett, E. E. (2018). Intranets of people, things, and services: Exploring the role of virtual human resource development. In C. A. Simmers & M. Anandarajan (Eds.), *The Internet of people, things and services: Workplace transformations* (pp. 166–184). New York, NY: Routledge. doi:10.4324/9781315182407-10

Bennett, E. E., & McWhorter, R. R. (2016). Opening the black box and searching for smoking guns: Process causality in qualitative research. *European Journal of Training and Development, 40*(8/9), 691–718. doi:10.1108/EJTD-07-2015-0049

Bennett, E. E., & McWhorter, R. R. (2017a). International HRD and Virtual HRD. In T. Garavan, A. McCarthy, & R. Carbery (Eds.), *A handbook of International HRD: Context, processes and people*. Edward Elgar Publishing.

Bennett, E. E., & McWhorter, R. R. (2017b). Reaction - Organizational learning, community, and virtual HRD: Advancing the discussion. *New Horizons in Adult Education & Human Resource Development, 29*(3), 19–27. doi:10.1002/nha3.20188

Bennett, E. E., & McWhorter, R. R. (2019). Social movement learning and social innovation: Empathy, agency, and the design of solutions to unmet social needs. *Advances in Developing Human Resources*, *29*(2), 224–249. doi:10.1177/1523422319827939

Bennett, E. E., & McWhorter, R. R. (2020). Digital technologies for teaching and learning. In Handbook of Adult and Continuing Education. Academic Press.

Biddle, S. (2018). Artificial intelligence experts issue urgent warning against facial scanning with a "dangerous history". *The Intercept*. Retrieved from https://theintercept.com/2018/12/06/artificial-intellgience-experts-issue-urgent-warning-against-facial-scanning-with-a-dangerous-history/

Boyd, C. (2019). *Location data leaks from family tracking app database*. Retrieved from https://blog.malwarebytes.com/cybercrime/2019/03/location-data-leaks-from-family-tracking-app-database/

Bradley, T. (2017). Facebook AI creates its own language in creepy preview of our potential future. *Forbes*. Retrieved from https://www.forbes.com/sites/tonybradley/2017/07/31/facebook-ai-creates-its-own-language-in-creepy-preview-of-our-potential-future/#1fdc4a46292c

Britt, P. (2015). When is speech technology too creepy? *Speech Technology*. Retrieved from http://www.speechtechmag.com/Articles/Editorial/Feature/When-Is-Speech-Technology-too-Creepy-107400.aspx

Brotherton, D. (2019). *Beware: LinkedIn is a social network with privacy issues too*. Retrieved from https://theamericangenius.com/tech-news/linkedin-privacy-settings/

Bryman, A., & Bell, E. (2015). *Business research methods* (4th ed.). Oxford, UK: Oxford Press.

Burns, A. J., Roberts, T. L., Posey, C., Bennett, R. J., & Courtney, J. F. (2018). Intentions to comply versus intentions to protect: A VIE theory approach to understanding the influence of insiders' awareness of organizational SETA efforts. *Decision Sciences*, *49*(6), 1187–1228. doi:10.1111/deci.12304

Buzzetto-More, N., Johnson, R., & Elobaid, M. (2015). Communicating and sharing in the semantic web: An examination of social media risks, consequences, and attitudinal awareness. *Interdisciplinary Journal of e-Skills and Life Long Learning, 11*, 47-66. Retrieved from http://www.ijello.org/Volume11/IJELLv11p047-066Buzzetto1666.pdf

Cao, H. (2019). I'm developing AI that can read emotions. It's not as creepy as it sounds. *CNN Business Perspectives*. Retrieved from https://www.cnn.com/2019/07/15/perspectives/ai-reads-emotions/index.html

Cellan-Jones, R. (2014). Stephen Hawking warns artificial intelligence could end mankind. *BBC News*. Retrieved from https://www.bbc.com/news/technology-30290540

Cisco. (2013). *The Internet of everything: Global private sector economic analysis*. Retrieved from https://www.cisco.com/c/dam/en_us/about/business-insights/docs/ioe-economy-faq.pdf

Cole, B. (2019). Chinese university tests facial recognition system to monitor attendance and students' attention to lectures. *Newsweek*. Retrieved from https://www.newsweek.com/nanjing-china-facial-recognition-1457193

Cowls, J., King, T. C., Taddeo, M., & Floridi, L. (2019). *Designing AI for social good: Seven essential factors*. Retrieved from https://ssrn.com/abstract=3388669

Creswell, J. W. (2016). *30 essential skills for the qualitative researcher*. Thousand Oaks, CA: Sage Publications, Inc.

Cuador, C. (2017). From street photography to face recognition: Distinguishing between the right to be seen and the right to be recognized. *Nova Law Review*, *41*(2), 237–264.

Cumbley, R., & Church, P. (2013). Is 'big data' creepy? *Computer Law & Security Review*, *29*(5), 601–609. doi:10.1016/j.clsr.2013.07.007

Dang, S. S. (2019). Artificial intelligence in humanoid robots. *Forbes*. Retrieved from https://www.forbes.com/sites/cognitiveworld/2019/02/25/artificial-intelligence-in-humanoid-robots/#6a7cd2b724c7

Davies, T. (2018). *More privacy concerns hit Twitter*. Retrieved from https://gdpr.report/news/2018/12/19/more-privacy-concerns-hit-twitter/

Day, M., Turner, G., & Drozdiak, N. (2019). *Amazon workers are learning to what you tell Alexa*. Retrieved from https://www.bloomberg.com/news/articles/2019-04-10/is-anyone-listening-to-you-on-alexa-a-global-team-reviews-audio

De Paola, A. (2014). An ontology-based autonomic system for ambient intelligence scenarios. In S. Gaglio & G. Lo Re (Eds.), *Advances onto The Internet of Things: How ontologies make the Internet of Things more meaningful* (pp. 1–17). Cham, Switzerland: Springer International Publishing. doi:10.1007/978-3-319-03992-3_1

Delello, J. A., & McWhorter, R. R. (2016). New visual social media for the higher education classroom. In G. Mallia (Ed.), *The social classroom: Integrating social network use in education*. Hershey, PA: IGI Global.

Draper, S. (2018). *Pentagon tells soldiers to leave wearable trackers at home when heading to war zones*. Retrieved from https://www.wearable-technologies.com/2018/08/pentagon-tells-soldiers-to-leave-wearable-trackers-at-home-when-heading-to-warzones/

Edwards, M. (2018). *The problem with oversharing on social media*. Retrieved from https://www.business2community.com/social-media/problem-oversharing-social-media-01983174

Evans, D. (2011). *Internet of Things: How the next evolution of the Internet is changing everything*. Retrieved from https://www.cisco.com/c/dam/en_us/about/ac79/docs/innov/IoT_IBSG_0411FINAL.pdf

Ford, M., & Palmer, W. (2019). Alexa, are you listening to me? An analysis of Alexa voice service network traffic. *Personal and Ubiquitous Computing*, *23*(1), 67–79. doi:10.100700779-018-1174-x

Forrest, C., & Bayern, M. (2019). 20 terrifying uses of artificial intelligence. *TechRepublic*. Retrieved from https://www.techrepublic.com/pictures/10-terrifying-uses-of-artificial-intelligence/5/

Franceschi-Bicchierai, L. (2018). *Spyware company leaves 'terabytes' of selfies, text messages, and location data exposed online*. Retrieved https://www.vice.com/en_us/article/9kmj4v/spyware-company-spyfone-terabytes-data-exposed-online-leak

Franceschi-Bicchierai, L. (2019). *This spyware data leak is so bad we can't even tell you about it*. Retrieved from https://www.vice.com/en_us/article/j573k3/spyware-data-leak-pictures-audio-recordings

Frankenfield, J. (2019). *General data protection regulation (GDPR)*. Retrieved from https://www.investopedia.com/terms/g/general-data-protection-regulation-gdpr.asp

Gelinas, J. (2019). *Alex IS listening to your conversations*. Retrieved from https://www.komando.com/security-privacy/alexa-is-listening-to-your-conversations/561016/

Georgetown Law Library. (2019). *International and foreign law research guide*. Retrieved from https://guides.ll.georgetown.edu/cyberspace

Glaser, A. (2016). *Biometrics are coning, along with serious security concerns*. Retrieved from https://www.wired.com/2016/03/biometrics-coming-along-serious-security-concerns/

Goertzel, Mossbridge, Monroe, Hanson, & Yu. (2017). *Loving AI: Humanoid robots as agents of human consciousness expansion*. Retrieved from https://arxiv.org/pdf/1709.07791.pdf

Goldacre, B. (2014). When data gets creepy: The secrets we don't realise we're giving away. *The Guardian*. Retrieved from https://www.theguardian.com/technology/2014/dec/05/when-data-gets-creepy-secrets-were-giving-away

Goldbeck, J. (2014). Why we overshare online. *Psychology Online*. Retrieved from https://www.psychologytoday.com/us/blog/your-online-secrets/201410/why-we-overshare-online

Graham, J. (2019). *Hey Siri, Google and Alexa — enough with the snooping*. Retrieved from https://www.usatoday.com/story/tech/2019/08/24/hey-siri-google-alexa-and-cortana-stop-snooping-now/2083899001/

Isaac, M. (2019). *Facebook's new tool lets you see which apps and websites tracked you*. Retrieved from https://www.nytimes.com/2019/08/20/technology/facebook-tool-privacy-apps-websites.html

Johnson, A. (2017). Facial recognition hearing looks to give lawmakers clearer picture on creepy technology. *NBC News*. Retrieved from https://www.nbcnews.com/news/us-news/facial-recognition-hearing-looks-give-lawmakers-clearer-picture-creepy-technology-n736771

Johnson, D. (2019). The 11 creepiest technologies that exist today. *Business Insider*. Retrieved from https://www.businessinsider.com/creepiest-technologies-tech-examples-ai-2019-8

Kaplan, J. (2016). *Artificial intelligence: What everyone needs to know*. New York: Oxford University Press.

Larrey, P. (2017). Would super-human machine intelligence really be super human? In G. Dodig-Crnkovic & R. Giovagnoi (Eds.), *Representation and Reality in Humans, Other Living Organisms and Intelligent Machines* (pp. 365–378). Cham, Switzerland: Springer. doi:10.1007/978-3-319-43784-2_19

Lieberman, M. (2018). *I know how you felt this semester*. Retrieved from https://www.insidehighered.com/digital-learning/article/2018/02/20/sentiment-analysis-allows-instructors-shape-course-content

Liu, S. (2019). *Number of digital voice assistants in use worldwide 2019-2023*. Retrieved from https://www.statista.com/statistics/973815/worldwide-digital-voice-assistant-in-use/

Loomis, L. (2017). *Alexa told me where the bodies are buried.* Retrieved from https://thoughtcatalog.com/lane-loomis/2017/01/alexa-told-me-where-the-bodies-are-buried/

Mack, E. (2017). *Elon Musk called out by creepy artificial intelligence robot.* Retrieved from https://www.cnet.com/news/elon-musk-artificial-intelligence-ai-robot-sophia-hanson-robotics/

Madsen, C. (2019). *Global privacy concerns: The world's top five cities using invasive technology*. Retrieved from https://www.webroot.com/blog/2019/05/29/global-privacy-concerns-the-worlds-top-five-cities-using-invasive-technology/

Magid, L. (2018). *The good and bad of location sharing*. Retrieved from https://www.connectsafely.org/the-good-and-bad-of-location-sharing/

Malkin, N., Deatrick, J., Tong, A., Wijeskera, P., Egelman, S., & Wagner, D. (2019). Privacy attitudes of smart speaker users. *Proceedings on Privacy Enhancing Technologies, 2019*(4), 250-271. doi:10.2478/popets-2019-0068

Marr, B. (2016). 21 scary things big data knows about you. *Forbes.* Retrieved from https://www.forbes.com/sites/bernardmarr/2016/03/08/21-scary-things-big-data-knows-about-you/#7db6abaf6e7d

Marr, B. (2018). *What is big data?* Retrieved from https://www.bernardmarr.com/default.asp?contentID=766

Marr, B. (2019). 5 Internet of things trends everyone should know about. *Forbes.* Retrieved from https://www.forbes.com/sites/bernardmarr/2019/02/04/5-internet-of-things-trends-everyone-should-know-about/#45e991824b1f

Martin, N. (2019). The major concerns around facial recognition technology. *Forbes.* Retrieved from https://www.forbes.com/sites/nicolemartin1/2019/09/25/the-major-concerns-around-facial-recognition-technology/#2062df884fe3

McLean, G., & Osei-Frimpong, K. (2019). Hey Alexa…examine the variables influencing the use of artificial intelligent in-home voice assistants. *Computers in Human Behavior, 99*, 28–37. doi:10.1016/j.chb.2019.05.009

McWhorter, R. R. (2014a). A synthesis of new perspectives on Virtual HRD. *Advances in Developing Human Resources, 16*(3), 391–401. doi:10.1177/1523422314532126

McWhorter, R. R. (2014b). *What's the 'Internet of Everything'?* Retrieved from https://virtualhrd.wordpress.com/2014/04/12/whats-the-internet-of-everything/

McWhorter, R. R., & Delello, J. A. (2015). Green computing through virtual learning environments. In F. Nafukho (Ed.), *Integrating technology into higher education and in the workplace*. Hershey, PA: IGI Global. doi:10.4018/978-1-4666-8170-5.ch001

McWhorter, R. R., Lynham, S. A., & Porter, D. E. (2008). Scenario planning as the development of leadership capability and capacity. *Advances in Developing Human Resources, 10*(2), 259–284. doi:10.1177/1523422307313332

Merriam, S. B., & Tisdell, E. J. (2016). *Qualitative research: A guide to design and implementation* (4th ed.). San Francisco, CA: Jossey-Bass.

Moreau, E. (2019). *The 3 best location sharing apps*. Retrieved from https://www.lifewire.com/top-best-location-sharing-apps-3485922

Murnane, K. (2017). *Creepy privacy invasions using Macs, Roombas and RFID chips*. Retrieved from https://www.forbes.com/sites/kevinmurnane/2017/07/27/3-creepy-ways-technology-is-being-used-to-invade-your-privacy/#691911b53858

Navarro, F. (2018). *5 smartphone apps that might be spying on you*. Retrieved from https://www.komando.com/smartphones-gadgets/5-smartphone-apps-that-might-be-spying-on-you/47578/

Palladino, V. (2018). *Unprompted, creepy laughter from Alexa is freaking out Echo users*. Retrieved from https://arstechnica.com/gadgets/2018/03/unprompted-creepy-laughing-from-alexa-is-freaking-out-echo-users/?comments=1&start=200

Pham, C. M. (2019). Building a maturity framework for big data cybersecurity analytics. In S. Miah & W. Yeoh (Eds.), *Applying Business Intelligence Initiatives in Healthcare and Organizational Settings* (pp. 164–183). Hershey, PA: IGI Global Inc. doi:10.4018/978-1-5225-5718-0.ch009

PrivacyPolicies.com (2019). *Privacy policies are legally required*. Retrieved from privacypolicies.com/blog/privacy-policies-legally-required/

Purshouse, J., & Campbell, L. (2019). Privacy, crime control and police use of automated facial recognition technology. *Criminal Law Review (London, England), 3*, 188–204.

Reader, R. (2019). *Nine technologies creeping us out in 2019*. Retrieved from https://www.fastcompany.com/90285664/technology-creeping-us-out-in-2019

Reidenberg, J. R. (2014). Privacy in public. *University of Miami Law Review, 69*, 141-159. Retrieved from Available at: http://repository.law.miami.edu/umlr/vol69/iss1/6

Retailwire.com. (2019). *What makes voice assistants creepy?* Retrieved from https://retailwire.com/discussion/what-makes-voice-assistants-creepy/

Reyns, B. W. (2019). Online pursuit in the Twighlight Zone: Cyberstalking perpetration by college students. *Victims & Offenders, 14*(2), 183–198. doi:10.1080/15564886.2018.1557092

Schein, E. H. (2010). *Organizational cultural and leadership* (4th ed.). San Francisco, CA: John Wiley & Sons, Inc.

Solon, O. (2016). *The rise of the robots: Forget evil AI – the real risk is far more insidious*. Retrieved from https://www.theguardian.com/technology/2016/aug/30/rise-of-robots-evil-artificial-intelligence-uc-berkeley

Spiezio, C. (2019). Building a data privacy program in-house. BNY Mellon's Anju Khurana weighs in. *Law.com*. Retrieved from https://www.law.com/corpcounsel/2019/08/15/building-a-data-privacy-program-in-house-bny-mellons-anju-khurana-weighs-in/

Spoonauer, M. (2016). *This PC's pop-up privacy cam is brilliant*. Retrieved from https://www.tomsguide.com/us/hp-all-in-one-privacycam,news-22618.html

St. John, A. (2018). How Facebook tracks you, even when you are not on Facebook. *Consumer Reports*. Retrieved from https://www.consumerreports.org/privacy/how-facebook-tracks-you-even-when-youre-not-on-facebook/

Stafford, P. (2019). *The dangers of in-game data collection*. Retrieved from https://www.polygon.com/features/2019/5/9/18522937/video-game-privacy-player-data-collection

Stern, J. (2019). You should definitely track your loved ones' phones. Actually maybe not. *Wall Street Journal*. Retrieved from https://www.wsj.com/articles/you-should-definitely-track-your-loved-ones-phones-actually-maybe-not-11565883576

Stillman, J. (2018). 7 creepy things robots have already done that shocked their makers. *Inc*. Retrieved from https://www.inc.com/jessica-stillman/7-creepy-things-robots-have-already-done-that-shocked-their-makers.html

Stone, Z. (2017). Everything you need to know about Sophia, the world's first robot citizen. *Forbes*. Retrieved from https://www.forbes.com/sites/zarastone/2017/11/07/everything-you-need-to-know-about-sophia-the-worlds-first-robot-citizen/#5590748646fa

Su, J. (2019). Confirmed: Apple caught in Siri privacy scandal, let contractors listen to private voice recordings. *Forbes*. Retrieved from https://www.forbes.com/sites/jeanbaptiste/2019/07/30/confirmed-apple-caught-in-siri-privacy-scandal-let-contractors-listen-to-private-voice-recordings/#113cbb877314

Sulleyman, A. (2017). *Google's AI future: So impressive it's scary*. Retrieved from http://www.independent.co.uk/life-style/gadgets-and-tech/features/google-lens-ai-preview-features-so-impressive-its-scary-a7745686.html

Symanovich, S. (2018). *How does facial recognition work?* Retrieved from https://us.norton.com/internetsecurity-iot-how-facial-recognition-software-works.html

Tashea, J. (2019). *As facial recognition software become more ubiquitous, some governments slam on the breaks*. Retrieved from http://www.abajournal.com/web/article/facial-recog-bans

Tene, O., & Polontsky, J. (2013). A theory of creepy: Technology, privacy and shifting social norms. *Yale Journal of Law and Technology, 16*(1). Retrieved from http://digitalcommons.law.yale.edu/cgi/viewcontent.cgi?article=1098&context=yjolt

The Economist. (2019). *How creepy is your smart speaker?* Retrieved from https://www.economist.com/leaders/2019/05/11/how-creepy-is-your-smart-speaker

Time.com. (2019). *Amazon is making it easier to delete your Alexa recordings*. Retrieved from https://time.com/5686352/delete-amazon-alexa-recordings/

Tobenkin, D. (2019). *HR needs to stay ahead of automation*. Retrieved from https://www.shrm.org/hr-today/news/hr-magazine/spring2019/pages/hr-needs-to-stay-ahead-of-automation.aspx

Valentino-DeVries, J., Singer, N., Keller, M. H., & Krolik, A. (2018). *Your apps know where you were sleeping night and they are not keeping it a secret*. Retrieved from https://www.nytimes.com/interactive/2018/12/10/business/location-data-privacy-apps.html

Vinez, K. E. (2017). *The admissibility of data colleted from wearable devices*. Retrieved from https://www2.stetson.edu/advocacy-journal/the-admissibility-of-data-collected-from-wearable-devices/

Vladimir, S. (2018). Machine learning approach for student engagement automatic recognition from facial expressions. *Scientific Publications of the State University of Novi Pazar Series A: Applied Mathematics, Infomatics and Mechanics, 10*(2), 79-96. Retrieved from https://scindeks.ceon.rs/article.aspx?artid=2217-55391802079S

Wang, Y., & Kosinski, M. (2018). Deep neural networks are more accurate than humans at detecting sexual orientation from facial images. *Journal of Personality and Social Psychology*, *114*(2), 246–257. doi:10.1037/pspa0000098 PMID:29389215

Weick, K. (2001). *Making sense of the organization*. Malden, MA: Blackwell Publishing.

Weick, K. E. (1995). *Sensemaking in organizations*. Thousand Oaks, CA: Sage.

Wilkins, J. (2018). *Improving productivity with facial recognition*. Retrieved from https://www.automation.com/automation-news/industry/improving-productivity-with-facial-recognition

Wissinger, E. (2018). Blood, sweat, and tears: Navigating creepy versus cool in wearable biotech. *Information Communication and Society*, *21*(5), 779–785. doi:10.1080/1369118X.2018.1428657

Worldbank.org. (2018). *Technology landscape for digital identification*. Retrieved from http://documents.worldbank.org/curated/en/199411519691370495/Technology-Landscape-for-Digital-Identification.pdf

Yao, M. (2017). *Why building social robots is much harder than you think*. Retrieved from https://www.topbots.com/building-social-robots-jibo-anki-cozmo-much-harder-think/

Zhang, B. (2018). *Why creepy ad is the new black?* Radbound University. Retrieved from https://www.ru.nl/publish/pages/769526/zb_michael_zhang.pdf

ADDITIONAL READING

Albrecht, K., & Mcintyre, L. (2015). Protect yourself from RFID: Fend off frightening tracking tech. *IEEE Consumer Electronics*, *4*(2), 95–96. doi:10.1109/MCE.2015.2393008

Alenljung, B., Lindblom, J., Andreasson, R., & Ziemke, T. (2017). User experience in social human-robot interaction. *International Journal of Ambient Computing and Intelligence*, *8*(2), 12–31. Retrieved from https://pdfs.semanticscholar.org/e072/25ab5bb9b0f817126a9a4da62ace06a48a96.pdf. doi:10.4018/IJACI.2017040102

Bhaumik, A. (2018). *From AI to robotics: Mobile, social, and sentient robots*. Boca Raton, FL: CRC Press. doi:10.1201/9781315372549

Hoy, M. B. (2018). Alexa, Siri, Cortana, and More: An introduction to voice assistants. *Medical Reference Services Quarterly*, *37*(1), 81–88. doi:10.1080/02763869.2018.1404391 PMID:29327988

Mathur, N., & Purohit, R. (2017). Issues and challenges in convergence of big data, cloud and data science. *International Journal of Computers and Applications*, *160*(9), 7–12. doi:10.5120/ijca2017913082

McWhorter, R. R. (2010). The emergence of virtual human resource development. *Advances in Developing Human Resources*, *12*(6), 623–631. doi:10.1177/1523422310395367

Price, W. N. II, & Cohen, I. G. (2019). Privacy in the age of medical big data. *Nature Medicine*, *25*(1), 37–43. Retrieved from https://www.nature.com/articles/s41591-018-0272-7. doi:10.103841591-018-0272-7 PMID:30617331

Schumacher, R., & Chen, H. (2010). Interaction analysis of the ALICE Chatterbot: A two study investigation of dialog and domain questioning. *IEEE Systems. Man & Cybernetics Part A*, *40*(1), 40–51. doi:10.1109/TSMCA.2009.2029603

Stallings, W., & Brown, L. (2018). *Computer security: Principles and practice* (4th ed.). New York: Pearson.

KEY TERMS AND DEFINITIONS

AI Robot: A man-made machine that can reproduce some elements of human intellectual ability such as solving problems, memorizing facts, gathering information through sensors.

Artificial Intelligence: The capability of a computer to mimic human behavior or machine learning.

Biometric Identification: Utilizing human behavioral and physical features for establishing identity.

Convergence of Big Data: The coming together of large data sets and machine learning.

Creepy Technology: Technology that evokes feeling or belief that privacy may be invaded in an unethical or discomforting manner.

Digital Voice Assistant: Technology activated by speaking the name of the assistant to activate commands such as writing emails or messages, placing phone calls, reading content from various mediums (such as email, websites or messages), or turning on lights, music, or activating security cameras.

Facial Recognition: Technology utilizing biometrics to recognize a human face.

Location Sharing Applications: A mobile device application that allows users to share their location in real time.

Chapter 11
Look Up:
Life Without Technology

Greg Duckworth
The University of Texas at Tyler, USA

Andrew Krouse
The University of Texas at Tyler, USA

ABSTRACT

This study examines data acquired from 120 undergraduate college students as part of a required assignment in which they were required to eliminate the use of all technology for a consecutive 24-hour period. The data were analyzed to determine (1) how they felt starting the assignment, (2) what technology they stopped using, (3) substitute activities used in lieu of technology, (4) how they felt at the conclusion of the technology moratorium, and (5) whether the participants would introduce long-term change in their use of technology. It was determined that, overall, students were apprehensive at the start of the exercise but subsequently experienced a positive experience at the conclusion. The most commonly identified technology device identified was a mobile phone. Substitute activities varied and included things such as sleeping, reading, and spending time with friend and family. Because of the positive experience, many indicated that they would be more mindful and modify their usage of technology devices.

DOI: 10.4018/978-1-7998-2914-0.ch011

INTRODUCTION

While there are many types of technology that are constantly competing for our attention, the smart phone is most often declared the winner. Just look around and you will see heads bowed and fingers scrolling as people are mesmerized by the glowing screen in their hand. Research by Mahmood and Shahzad (2016) stated that the use of mobile phones has experienced one of the fastest adoptions rates the world has ever seen. Smartphones also have one of the most pervasive footprints of any electronic equipment sold. In fact, worldwide it is estimated that there will be 2.9 billion phones in use by 2020 (Statista, 2019). Also, each device can be further customized by adding any number of applications that expands the seemingly unlimited possibilities for its use. Perhaps the device should be more aptly referred to as a miniature computer with a phone application available for people that prefer to interact with one another. It appears that no one is immune to the enticement of the allure of mobile technology. Demographic analysis of the utilization of smart phone technology indicates that both men and women frequently use their phone for social engagement as opposed to process use (Van Deursen, Bolle, Hegner, & Kommers, 2015). With all the uses that smart phone technology purports to offer, there is an overwhelming reliance on the information, entertainment, and productivity that it provides.

Technology Overload

As we continue to utilize technology in our daily lives it is important to heed the warnings of potential negative impacts of information technology (Tarafdar, Gupta, & Turel, 2013). With the constant influx of information through technology, both at work and at home, the possibility of feeling overloaded is understandable. Individuals have experienced information overload in their use of social media through three dimensions: 1) features, 2) information, 3) social overload (Zhang, Zhao, Lu, & Yang, 2016). These dimensions outline how the complexity of the system (Karr-Wisniewski & Lu, 2010), limitations of an individuals' ability to process information (Lee, Son, & Kim, 2016), and the perception of feeling obligated to provide social support (Maier, Laumer, Eckhardt, & Weitzel, 2015) all can lead to information overload.

Similarly, the use of social exchange features in the workplace have been found to cause information overload in the two common dimensions of social support and limitations to an individual's ability to process information (Chen & Wei, 2019). The risk of information overload in the workplace has increased since the use of social media technologies has garnered attention from modern organizations for positively impacting internal communication (Schlagwein & Hu, 2017). Another risk of information overload is enterprise social media use, which has been banned in some workplaces due to employee distractions (Ali-Hassan, Nevo, & Wade, 2015). The workplace distractions will likely continue to be more pervasive as smartphone technology increases in power and decreases in cost. Consequently, the affordability and accessibility of these modern day marvels continues to drive appetite for further consumption of these modern day marvels.

When smart phones were first introduced to the public they were generally used as a communication device and could only be afforded by the elite with an abundance of money to spare. Nowadays, smart phones have overrun the mainstream technology market. These little boxes of plastic, glass, and metal have evolved to include features that enable the user to perform a variety of functions that include information acquisition, entertainment, productivity improvement, social engagement and education. Other uses include entertainment, monetary compensation through online sales, health monitoring and improvement, and general diversions to everyday life (Mishra, 2017). The evolution of smartphone applications has permitted users to document and control aspects of everyday life. Generally, most phone software provides a positive benefit to the consumer. Barriers to the access of valuable information and immediate availability have been virtually eliminated. At the swipe of a finger, a person can access information from any location in the world in nanoseconds. Banking information, sharing a photo with a friend on the other side of the world, and listening to a favorite song are available in an instant. Smartphone technology can even eliminate local and global distance limitations by allowing an individual to work from any location they choose (McWhorter & Delello, 2015). The continued growth of smart phone adoption is a phenomenon that eludes comprehension. According to Poushter, Bishop and Chwe (2018), approximately 89% of adults in the United States own a smartphone. Therefore, it is not uncommon to see a family documenting a vacation using their phone as opposed to a traditional camera or navigating their way to the nearest restaurant using the GPS application and without ever unfolding a map. In fact, research published by Xi, Anaya, Miao, Lehto, and Wong (2018) stated that the pictures documented on phones were more than just photos, the easy access to their pictures altered both the internalization and recollection of their memories. In other words, the ease of capture and recollection of moments in time by users became a mechanism to essential experience and "feel" the moment. Unsurprisingly, the use of a smartphone continues to be cited as a critical component to everyday life. While there are many positive uses for smart phone technology, the dark side of this life-changing technology is also ever-present. In short, the addictive nature of smartphones often results in an overreliance and habitual or obsessive use may result in deleterious effects (Ong, 2018).

Negative Effects of Technology

Recent studies provide details that support that there is a potential for harmful effects as a result of the addictive use of smartphones. As an example, traffic accidents and fatalities are regularly reported following texting or talking while driving (Cazzulino, Burke, Muller, Arbogast, & Upperman, 2014). Furthermore, there are negative physical manifestations of the excessive use of smartphones. Also, Xie, Szeto, Dai, and Madeleine (2016) reported patients complaining of neck and shoulder problems while others sought treatment for sleep impairments that were directly attributed to the acute use of mobile technology (Eyvazlou, Zarei, Rahimi, & Abazari, 2016). Moreover, with the proliferation of online social networking sites, many people report that they spent an ever-increasing amount of time engaged in reviewing interaction with others that they may never meet (Longstreet & Brooks, 2017). It appears that even the use of smartphones to view content on the internet could result in at least one of five types of Internet addictions identified that includes cyber-relationship addiction, net compulsions, cyber sexual addiction, computer addiction, and information overload (Kuss & Griffiths, 2017). While any of the consequences could befall an individual, one of the predominant factors that influences the potential for a person to avoid addictive behaviors and demonstrate self-control is a person's ability to self-regulate (Gökçearslan, Mumcu, Haşlaman, & Çevik, 2016). The concept of boundaries presents an interesting approach of how one can segregate and better utilize the functionality of the devices to better serve their master.

Border theory was developed by Clark (2000) to assist in the explanation of the work and home life interactions that take place in contemporary society. The key components of this theory are domains, border-crossers, border-keepers, and borders. Examples of domains are work and home as they have different rules, patterns, and behaviors. Borders are the demarcation between the domains and can be physical, temporal, and psychological. The border crossers are individuals that making frequent transitions between the two domains. The border keepers are individuals within each domain that have significant influence over keeping the domains separate (Clark, 2000). Understanding the basic concept of border theory and how it may explain some of the aspects of work life balance provides insight into the understanding of the family dynamics within a postindustrial society. Although the creation of boundaries is a useful approach, it may not be practical. One must be able to identify and control the ever-present urge to completely rely on the powerful mind-controlling capabilities of the mobile device.

Overcoming the Technology Obsession

Self-regulation is the mechanism by which an individual maintains control over their thoughts, feelings, and actions whereby one maintains control and minimizes the risk of a habitual addiction to their phone (Gökçearslan, Mumcu, Haşlaman, & Çevik, 2016). On the contrary, the habit-forming nature of smartphones are a constant threat to undesirable habit development and the inevitable erosion of self-control (Oulasvirta, Rattenbury, Ma, & Raita, 2012). One habit that is often cited in research is that of a checking habit. A checking habit describes the persistent use of the device to constantly view new content available on sites. Interestingly, although the checking habit is a choice, there is empirical evidence that a neurochemical response is ultimately responsible for this habitual behavior. The positive response by the brain to release of dopamine due to the pleasure of retrieving information has been associated with several habit-forming cues ranging from behavior motivation, feelings of being rewarded, and mood regulation (Ilango & Lobo, 2015).

The reward the brain "feels" when retrieving information from a digital device is because one's phone use successively triggers the continued habitual behavior loop. It is evident that the capabilities present in smartphones to interact socially also have the capability to create and reinforce a form of technological addiction. The technological addiction may consequently result in negative physical and mental effects associated with the addictive nature of the smartphone. Some of the damaging outcomes published in recent studies include musculoskeletal damage, depression, poor performance at school, and sleep deprivation (Eide, Aarestad, Andreassen, Bilder & Pallesen, 2018). The pain a smartphone user experiences as a result of withdrawal when the technology is withheld or unavailable to the user can be very real. Although anyone can develop an addiction to the use of a smartphone, students and young adults appear to be at higher risk for developing an addiction that may have undesirable consequences (Zhitomirsky-Geffet & Blau, 2016).

Technology in the Classroom

As the use of technology changes and new technologies become available, the classroom where future generations are taught changes as well. Integrating technology into the process of teaching and learning is thought by researchers to increase productivity of both students and teachers (Al-Zaidiyeen, Mei, & Fook, 2010). In addition to productivity technology, use within classrooms allows both the teacher and students access to a vast amount of collected knowledge. This is seen using internet informational sources in the classroom which allows instant access to a large variety of information whenever it is needed throughout the education process. This access to information has been shown to promote motivation and performance in both student and teacher (Al-Zaidiyeen, Mei, & Fook, 2010).

There are three reasons why researchers focus on investments in technology in educational settings: 1) To increase the students' ability and interest in completing the basic educational requirements set by the governing body; 2) To prepare students for success in a world that is centered around technology; and, 3) To prepare students to manage and use information throughout their lifespan so they can be productive lifelong learners in addition to being responsible citizens (Bena & James, 2001). These justifications for the use of technology in classrooms are objectives that have been in place in terms of educating the youth of society before the use of technology became the focus but appear to be quite applicable in today's technology-rich educational environment.

Student Use of Technology

The use of technology by students has been studied empirically, but the link to academic performance or other similar outcomes is still limited (Li, Lepp, Barkley, & Karpinski, 2014). The findings of previous empirical studies focused on student technology use have been inconsistent and contradictory (Lei, 2010). Understanding how technology relates to student performance may provide insight into potential drivers of other social outcomes and provide insight into the proper application of technology for our student population (Delello, Kiekel, Poyo, Williams, & Kirby, in-press 2020).

Data gathered in a study by Sánchez-Martínez and Otero (2009) through quantitative methods indicate a relationship between intensive use of technology and failure rates of students. Additionally, a study completed in 2007 by the Institute of Education Sciences reported no statistically significant differences in test scores by first, fourth, of sixth grade math students in a study designed to test the benefit of new software programs (Dynarski, et. al., 2007). Both studies demonstrate research findings that present a negative impact on student outcomes by technology. While these two studies vary in methodology and target demographic, similarities in the results augment support to these research outcomes. One of the more surprising research results indicate that immediate and uninterrupted access to a smartphone may be quickly becoming a basic human necessity.

More recent studies have revealed illuminating results that stated that an average college student checks their phone about 60 times per day and accumulates more than 4 hours of screen time (Kang & Jung, 2014). Additionally, the Pew Research Center (2015) found that almost half of American phone users could not function without the use of their phone. There is evidence that the reliance on the smartphone is so pervasive for some individuals that they experience physical distress symptoms.

Nomophobia, or no mobile phone phobia, has been defined as an irrational fear of being without a phone or wireless signal so that the person experiences worry or anxiety (Adawi et. al., 2019). The psychological dependency created by the reliance on a smartphone has been correlated to cognitive impairments in college students. Consequently, the cognitive performance degradation may result in diminished mental health (Harwood, Dooley, Scott, & Joiner, 2014), poor academic performance (Cheever, Rosen, Carrier, & Chavez, 2014), as well as degraded sleep quality (Li, Lepp, & Barkley, 2015). Unfortunately, literature suggests that college students are often the most vulnerable to the psychological dependence associated with the use of mobile phones.

Pearson (2015) conducted a Mobile Device Survey in which it was reported that 85% of the students in the study owned a smartphone of which 86% of them used their phone regularly. Furthermore, students of ages 18-19 used smartphones more than the students aged 20-24. The researchers attributed the higher usage to earlier adoption of smartphone technology by the younger cohort. In general, student aged individuals appear to be the most reliant on their technology to function in the classroom and society. This stands to reason since current generations have been exposed to technology at an earlier age. For instance, in any public venue it is not uncommon to find a family allowing a child to be entertained with a streaming video so that they can enjoy an uninterrupted conversation. Sadly, many have learned to rely on handheld technology at an increasingly early age that thwarts face-to-face communication with parents and peers.

The amount of time that people must allocate to various activities is finite and priorities for this allocation of time may change as we progress through life. One possible activity that can be allocated with our limited time is to is spending time with family. The importance and benefits of spending time with family has been studied in terms of adolescence (Offer, 2013), work/family balance (Clark, 2000), and stability (Israel & Roderick, 2001). All three of these areas of study regarding spending time with family may be important when considering what an individual would do instead of using technology (i.e. opportunity cost of alternatives to digital devices).

There is a growing scholarly concern regarding the scarcity of family time and how time is spent in general among adolescents (Offer, 2013). This point has been argued by looking at how a teenager's day is spent. In postindustrial societies, most of a teenager's day is spent in structured activities such as school or competitive activities to promote future success in the competitive labor market (Larson & Verma, 1999). Increasing the amount of time spent on these structured activities may reduce the available time for the teenagers to spend with their family. Structured activities are not the only contributor to the reduced availability of time for family as social activities, technology, and other aspects of society also reduce the available time.

With the reduced time that teenagers spend with their family, research has found that it is limited in other ways beyond just the actual amount of time spent. Previous research has found contextual and demographic correlations of family time with gender and age (Larson & Richards, 1998; Larson, Richards, Moneta, Holmbeck, & Ducket, 1996). As teenagers progress through their adolescence it was found that they spend an increasing amount of time with their peers instead of their family. The exception to this is the teenage girls will spend more time with their family, specifically their mother, helping with tasks such as cleaning and shopping (Hilbrecht, Zuzanek, & Mannaell, 2008; Larson & Verma, 1999; Larson, Richards, Sims, & Dworkin, 2001; Wight, Price, Bianchi, & Hunt, 2009). The sheer volume of time spent with family has been argued as not having as much of an impact as the quality of time spent (Snyder, 2007) instead of focusing on routine. While interactions between family members are purported to have been significantly impacted by the pervasive use of smartphone technology, the problem also radiates to other environments like schools.

In a study of 200 students by Mishra (2017), it was reported that students often use their phones for both research and entertainment. The students represented in the study admitted that the use of their phones to engage in social networking often caused information overload and caused them to lose sleep due to their late-night usage (Mishra, 2017). This, in turn, supports the result of research by the National Sleep Foundation which stated that only 40% of Americans get the recommended amount of sleep each night (Appold, 2015). The primary culprit of this sleep deprivation is often caused by the overuse of technology. Unsurprisingly, such usage impairs learning which results in a negative effect on cognitive tasks and subsequently results in undesirable consequences (Lana, Johnson, & Barnes, 2014).

As with any activity, the unregulated use of smart phone technology may have both positive and negative application. Ultimately, the ability of the user to self-regulate is often the key which may limit the habitual use of the handheld technology. The good news is that almost any exercise that reinforces the ability of a person to evaluate their use of this modern marvel will enable the user to stop, look up, and engage in the world around them.

To evaluate and better understand the results of prior published studies and to explore current sentiment of technology users, a new research study was conducted. This study investigates many of the experiences and feelings of participants after being asked to deny themselves access to one of the most powerful tools and ubiquitous forces present in their day-to-day life. The results of this study prove to be entertaining, enlightening, and life-altering for some.

THE PRESENT STUDY

This study was conducted over a two-year period beginning in 2017. The purpose of this two-year qualitative study was to experience what life would be like for a college student who was asked to first view the Turk (2014) video and then abstain from using any form of technology (i.e. mobile device, computer, television) for twenty-four hours. The invasive effects of technology on the personal lives of higher education students is the central phenomenon studied. The participants consisted of undergraduate students at one regional university in the southern United States. At this stage in the research, the invasive effects of technology on the personal lives of students will be generally defined as how students react to not using any form of technology for a twenty-four-hour period. However, it was noted that the majority of students reported that the technology most often relinquished for this exercise was their phone. In order to better understand the dynamics at play during this study, the research was bounded by the following questions which are considered in the scope of this exploratory study:

- What value is realized by reducing or eliminating the use of technology?
- How is a person impacted by the pervasive use of technology?
- What do students do as a replacement activity to using technology?
- Will individuals choose to make plans to reduce long-term technology use following a short-term technology moratorium?

METHOD

The methods for the carrying out the exploratory study will be described next. It will include the participants, data sources, and data analysis of the study.

Participants

Study participants consisted of 120 undergraduate business students who were pursuing a bachelor's degree from a public, regional, four-year university in the southern United States. The undergraduate students chose to enroll in the class voluntarily. Participation in this class exercise was required as a graded component of the requirements to complete the class. Failure to complete this assignment would have a negative effect on the final grade received by the student of the class. Demographic information was not requested from nor was it provided by any of the participants.

Data Sources

The first-hand accounts from undergraduate students' experience participating in the "Look Up" assignment were utilized for this analysis. The students were assigned a project entitled "Look Up" in which they were required to forego the use of all technology for a consecutive 24-hour period. Upon completion of the 24-hour period, students were required to submit a report that included a reflection of how they felt starting the assignment, what technology they gave up, the substitute activities they used and how they felt at the conclusion of the technology moratorium. The data collection was approved by the Institutional Review Board.

The assignment and subsequent data used in our analysis were collected in July 2017 and August 2018. This was part of a class assignment which, while the class was voluntary, contributed to the response rate. It is also important to note that participation in the assignment resulted in a grade which likely affected the high response rate as well.

Data Analysis

This research effort utilized qualitative content analysis, as defined as "a research method for the subjective interpretations of the content of text data through the systematic classification process of coding and identifying themes or patterns" (Hsieh & Shannon, 2005, p. 1278), to gather data from submitted firsthand accounts. Additional definitions for content analysis that informed this study were "an approach of empirical, methodological controlled analysis of texts within their context of communication, following content analytic rules and step by step models, without rash quantifications" (Mayring, 2014, p. 2) and "any qualitative data reduction and sense-making effort that takes a volume of qualitative material and attempts to identify core consistencies and meanings" (Patton, 2002, p. 453).

Data analysis was conducted by reviewing the reflective report that each student submitted following the assignment. To protect the identity of each student, a graduate assistant removed all personally identifiable information and created a unique identifier for each paper. A random sample of 65 papers was selected to begin the coding process. As a result of the data analysis and coding process, data saturation was achieved. Data saturation has been identified as a point in which no new information is present despite further review of available data. At the point of data saturation, redundancy indicates that no additional themes or conclusions are identifiable in the data (Faulkner & Trotter, 2017).

Each of the reflective reports that were selected for analysis were reviewed to identify themes. Five primary themes were noted and used. The five themes were:

- How the student felt about participating in the exercise before beginning the assignment.
- Types of technology that the student gave up during their 24 hours
- Activities the student used in lieu of using technology.
- Student experiences as to a positive, neutral, or negative outcome post- assignment
- How did the student's perception of using technology change post-assignment?

Comments by a student that aligned to one of the four themes was notated for each occurrence. After all papers were reviewed, the total comments for each of the themes was tabulated to determine the frequency that they occurred in the reflective report to determine overall impact on the students who "looked up" from technology. As an example, one student noted their sentiment of participating in the exercise and remarked that: "it was very pleasant because I got to take my time and really think about what I was reading"; such comments were each coded under the theme identified as *start feelings* during the coding process.

RESULTS

Students were asked to keep a written log and to produce a written reflective report based on their experiences during the twenty-four-hour period in which they were asked not to use any form of technology. The reports were approximately 3 to 4 pages in length. As seen in the student responses, almost every reflective report of the 65 selected samples contained a reference to one or all the identified themes. It was determined that the data sample represented saturation as observing more data was not likely to lead to discovery of new or additional information.

The details included in the student responses varied. Some students reported their thoughts, actions, and feelings upon conclusion of the exercise in detail. Other students were less thorough with the information that they recorded. An interesting note is that some of the students reported that, although they tried, they were unsuccessful in completing the twenty-four-hour period without using of any technology. Furthermore, some were only partially successful. Some reported that although they were able to go without television or radio, they were unable to complete the research period without any use of their phone.

Each theme provides valuable insight about the influence of technology on the day-to-day life of the study participants. To understand the invasive nature of technology and the behavior-shaping power that it exudes, it is important to review each theme independently. Table 1 presents the themes and frequency of each theme that was recorded. Most of the reflective reports provided and reviewed from the random

Table 1. Themes and frequency

Code short	Code long	Frequency N=	Definition
SF	Start feelings	59	How the participant felt when starting the exercise
TR	Technology Removed	65	Technology devices not used during the exercise
RA	Replacement activities	65	Substitute activity used in lieu of technology
EF	End feelings	55	Participant post exercise experience
LT	Long term impact of exercise	43	How participants express a long-term change in the use of technology post exercise?

sample selected included some reference to each of the themes. The recollection provided by each of the participants were similar. The instructions provided for the assignment required that each response include a reflection of the initial reaction to the assignment, specific information regarding the technology they did not use, the substitute activities used to replace technology use, and reflection on both how they felt at the conclusion of the technology moratorium and any changes in behavior they plan to employ in the future. The first theme establishes the current state of mind as each student embarks on their 24-hour technology hiatus.

In the Beginning

With rare exception, nervous apprehension was obvious as each participant reported how they felt as they started the exercise that would challenge their patience and ability to adapt to a technology-free experience that few had ever experienced before in their life. Most of the respondents chose to start the experiment on a day that allowed them the best chance of success. Statements that included "I chose my birthday", "the 4th of July", and "one of my three days off from work" indicate the typical start to what would prove to be a day that would create an inordinate amount of stress for all but a few. Analysis of the data specifies that people would somehow garner greater control over their reliance on technology by choosing a day that would generally require less usage as others around them likely took pictures and recorded video. Some stated that their job required the use of a computer so they had to opt for a day that they could avoid their laptops.

Other individuals assumed that the distraction offered by the holiday would afford them some insurance against an accidental tweet or trolling of their favorite social media outlet. Regardless of the day chosen, most expressed a concern in their ability to shun all technology for the duration of the assignment. Fear, dread, and remorse at the loss of their best techno-friend was prevalent. The likely over-reliance each had placed on their technology made them question "what will we talk about and what will we do?". While emotions varied, expressions like "I did not like this assignment at all" best summarizes how many truly felt. The common concern appeared to be what would replace the technology that kept each person in touch with all they deemed important.

Switching Off

The participants recorded how they replaced the activity that required technology once the unsettling feeling of their new loss was over. Although a few referenced the discontinued use of their alarm clock, television and radio, the most commonly cited technology fixture in their life was the mobile phone. The day of most respondents often began by "scrolling through YouTube", "checking social media sites to catch up on things I missed out on while I was sleeping, or to "turn on some music to help me get my day started ". Even statements like "I honestly didn't know what I was going to spend my time doing or how I was going to communicate with people" was recorded. An interesting finding was the impact of the exercise also consumed friends and family.

A mother of a toddler recounts she "wasn't feeling very optimistic about how the day was going to go" because she was not sure "there would be enough projects planned throughout the day to be entertained without listening to music or turning on the tv". Others faced high emotions as their children "got angry and were griping" when they were not allowed to use their tablet or X-box. The web of interconnectivity resulting from the use of technology appears to be widespread and complex. Some stated they felt isolated as they watched friends sit silently while staring at their phone screen while together. In a few instances, even a planned outing with family and friends was obstructed when participants in the study had to find their way to a destination without the use of the GPS on their mobile phone. It was evident that all would have to be creative to fill the void left in the technology-free wake.

Now What?

When faced with the dilemma of how to face the day without technology, perhaps the most intriguing solution cited by participants was that they chose to sleep. "I started my day by sleeping in until 11:30 in the morning" is evidence that it may be easier to pass the time sleeping than feel the loss of no social media interactions. Others chose to take a nap to pass the time. Those that faced the new challenge also engaged in more active ways.

Substitute activities to technology included reading a good book, talking with friends or just enjoying the company of a long-lost friend became the new norm for the day. A recurring factor was that of incorporating others into the day-to-day activities as opposed to interacting only with a device. One student stated that they had a really good conversation with a co-worker and "we talked about life and stuff, I learned that our birthdays are actually one day apart". Others engaged in gathering of larger groups of people by attending holiday parties, swimming with friends, and cooking a meal for their family. It appeared that the experience of adapting to life without a smart phone, if even for a day, helped people realized that small changes had big impacts on engaging with others rather than technology.

The End is Near

At the conclusion of the 24-hour period without technology, students captured the impact that the exercise had on them. Most respondents indicated that the exercise was very positive. Each summarized what they gained from participating in this experiment. One of the first epiphanies experienced was the amount of time that they typically dedicated to the use of technology.

It appears that some recognized that when they substitute quality time with children with the use of technology to entertain and for personal use there may be long-term deleterious effects. One mother stated that the assignment "taught me to not take life for granted and pay attention to my children more than I have." Another parent of two small children reflected on the loss that went unnoticed until this project reflected in their statement: "it made me realize how I have been guiding my children" and "it made me really realize that I've been missing out on many moments with my family and ultimately made me sad." Although the impact to family is significant, there are many other lessons learned which include how technology habits governed their life.

Participants recounted the daily routine that most enjoyed prior to the onset of this experience. Scrolling endlessly through Facebook, YouTube, newsfeeds and other information sources governed the beginning of the day for most of the students. During times that there was no planned activity, students reported filling the void with screen time. The habitual use of technology would govern every waking moment and go mostly unnoticed. Following the completion of the assignment the realization that individuals should set appropriate boundaries on the use of technology was evident. Students stated that "I loved not having my phone because I was so unplugged and spent more time with the people that I love." As each respondent cited the positive benefit to this short-term exercise, most also shared the ancillary benefit that would result in their long-term commitment for change.

What Next?

One of the most prominent long-term resolutions instituted by individuals that participated in the assignment included setting boundaries for the use of technology. Examples of these boundaries included starting a "no phones at dinner" rule to help curb the mindless use of technology and encourage more interaction among family members. Parents that participated in the study shared their commitment to reducing technology time to realize the benefit of more quality time with their children. One parent stated, "It made me feel a lot better to know that I spent a lot of quality time with our daughter." The recapture of valuable time was important as it relates to the impact to a family but there is also a benefit in the desire to continue reducing the reliance on technology.

The experience of greater productivity continued to reinforce the personal commitment to applying self-control in the use of technology. Statements that include "I was more productive that day and so proud of myself" and "I felt amazing thinking that I spent my day productively" were predecessors to the conclusion to continue curtailing the use of a smart phone more than necessary. Yet, another reflected on life before the infringement of being connected all the time. The student recalled how life was "more peaceful" when growing up: "We had to talk to people face to face or by calling them" the student began. We are simply losing our ability to communicate effectively when technology is the primary medium used to interact.

In summary, most of the student reports reflected comments on how much technology consumes life. This exercise clearly demonstrated that participants were more cognizant of their technology activity and the impact it had on their life and the lives of others. This recognition appears to be the impetus for their continued commitment to long-term change. To summarize the results of each theme in this study, Table 2 provides verbatim examples from the data accumulated from student reflective reports.

DISCUSSION

In this exploratory study, personal accounts from students about how they responded to a complete elimination of technology from their life was intriguing. Many of the accounts presented by the students began with an apprehension from participating in this assignment. A multitude of the reports commented that the student started the exercise on a holiday or weekend. Their expectation was that a standard day would be too difficult to pursue without the ability to use technology. It appeared that the phenomenon being investigated (i.e. going without technology for a 24-hour period) would prove more difficult to endure than students may have anticipated.

Table 2. Verbatim examples for each theme

Code short	Code long	Quotation	Quote Reference
SF	Start feelings	"I was very eager to go without my phone"	LUSU1759 p2
SF	Start feelings	"I thought to myself, there is no way that I am going to be able to do this"	LUSU1713 p2
SF	Start feelings	"I wasn't sure if I could fully commit to unplugging for a 24-hour timeframe."	LUSU1720 p2
TR	Technology Removed	"I was more productive that day as I ended up washing my car."	LUSU1768 p3
TR	Technology Removed	"instead I hunted up a book from my bookshelf"	LUSU1730 p3

The individuals that were able to complete the assignment successfully reported similar experiences such as having a strategy beforehand. Many reported that instead of using technology they would start the day by sleeping later, spending more time with loved ones and friends, as well as engaging in relaxing behavior like reading books. While many were anxious at the beginning of the exercise, there was a remarkable evolution seen in their reflections on how they felt about the exercise as they went throughout the day. Some identified that this assignment enabled them to realize that technology was indeed a pervasive force in their everyday life. For the majority that completed the exercise, many of them stated that they would make a concerted effort to minimize the use of technology in their future. It appears that this exercise may lead to a change in behavior that will have a positive effect in their life.

The invasive effect of technology on personal lives of students was well-documented by each of the of the student reports. Most of the students identified and documented what they may be missing if they allow their phone, television, and other electronics to hijack their attention. Although technology has proven to be very beneficial in providing entertainment, productivity, and social interactivity, if an individual does not remain cognizant of how often they rely on the use of technology it may overtake them. The ability to balance one's time and attention between the digital physical world is very important. As with any tool, self-regulation and determining how and when to use technology will govern its effectiveness and minimize its invasiveness.

FUTURE RESEARCH DIRECTIONS

The results of this research indicate that technology is a pervasive force that requires further investigation. While the scope of this assignment required students to eliminate the use of all technology, many of them reported the greatest apprehension in being unable to use their smart phone. There are many unique factors identified that can be explored to better understand the dynamics that encourage the use of smartphone technology. Additional research on how a student uses his or her phone would allow researchers to continue to further delineate the positive and negative consequences of technology usage. Furthermore, exploration of demographic characteristics may improve our understanding of this phenomenon.

Evaluation of both gender and age use of technology may help establish the underpinnings for more specific research. Gender differences have often been cited as differentiators in the adoption of technology. Meta-analysis studies have stated that males have generally shown a more favorable attitude about using technology than females. Two dimensions have been noted as differentiators between the sexes. Specifically, the components which reports the level of belief that technology is generally useful in society and how much self-confidence is expressed in the ability to learn and efficiently use new technology (See Cai, Fan, & Du, 2017). Similarly, generational differences are expected to play a key role in the use of mobile technology (Delello & McWhorter, 2017).

Generational differences are anticipated to become more apparent as technology is rapidly advancing and exposure to each new technology and when it becomes available may be a key differentiator (Ball, Francis, Huang, Kadylak, Cotten, & Rikard, 2019). Digital natives are common in society nowadays. Most of today's students have had access and have used different forms of technology in their everyday lives since they were young. Counterintuitively, empirical studies have not found that despite having easy access to technology, actual use varies among current generational cohorts. In short, just because technology is accessible, individual adoption is governed by a plethora of factors so one cannot assume that adoption and use of technology will always occur. In any event, technology is expected to continue as a fundamental part of the day to day life of most individuals. Therefore, continued research regarding the awareness of the impact of the use of technology should be pursued.

CONCLUSION

The exploratory study provided insight into the role of technology used by undergraduate students of a four-year university in the United States. It was shown that the pervasive use of technology in the day-to-day life of the student demonstrated a significant reliance on the same technology. Furthermore, when required to abstain from the use of technology, the student is capable of self-reflection and corrective behavior modification. Results of this study may provide a helpful foundation to design additional research investigating the impact of technology use on the well-being of individuals everywhere.

ACKNOWLEDGMENT

This research received no specific grant from any funding agency in the public, commercial, or not-for-profit sectors. Both authors wish to offer a special thanks to Dr. Rochell McWhorter for her encouragement and helpful guidance for the first book chapter project that has been undertaken by the authors.

REFERENCES

Adawi, M., Zerbetto, R., Re, T. S., Bisharat, B., Mahamid, M., Amital, H., ... Bragazzi, N. L. (2019). Psychometric properties of the brief symptom inventory in nomophobic subjects: Insights from preliminary confirmatory factor, exploratory factor, and clustering analyses in a sample of healthy Italian volunteers. *Psychology Research and Behavior Management*, *12*, 145–154. doi:10.2147/PRBM.S173282 PMID:30881158

Al-Zaidiyeen, N. J., Mei, L. L., & Fook, F. S. (2010). Teachers' attitudes and levels of technology use in classrooms: The case of Jordan schools. *International Education Studies*, *3*(2), 211–218. doi:10.5539/ies.v3n2p211

Ali-Hassan, H., Nevo, D., & Wade, M. (2015). Linking dimensions of social media use to job performance: The role of social capital. *The Journal of Strategic Information Systems*, *24*(2), 65–89. doi:10.1016/j.jsis.2015.03.001

Appold, K. (2015). Americans with chronic pain aren't getting sleep: The latest findings from the national sleep foundation (NSF) sleep in America poll report that pain is a key factor in the gap between the amount of sleep Americans say they need and the amount they are getting. *RT for Decision Makers in Respiratory Care*, *28*(5), 15.

Ball, C., Francis, J., Huang, K.-T., Kadylak, T., Cotten, S. R., & Rikard, R. V. (2019). The physical-digital divide: Exploring the social gap between digital natives and physical natives. *Journal of Applied Gerontology*, *38*(8), 1167–1184. doi:10.1177/0733464817732518 PMID:29165009

Bena, K., & James, M. (2001). *Information technology for schools, creating practical knowledge to improve students' performance.* San Fransico: Jossey–Bass.

Cai, Z., Fan, X., & Du, J. (2017). Gender and attitudes toward technology use: A meta-analysis. *Computers & Education, 105*, 1–13. doi:10.1016/j.compedu.2016.11.003

Cazzulino, F., Burke, R. V., Muller, V., Arbogast, H., & Upperman, J. S. (2014). Cell phones and young drivers: A systematic review regarding the association between psychological factors and prevention. *Traffic Injury Prevention, 15*(3), 234–242. doi:10.1080/15389588.2013.822075 PMID:24372495

Cheever, N. A., Rosen, L. D., Carrier, L. M., & Chavez, A. (2014). Out of sight is not out of mind: The impact of restricting wireless mobile device use on anxiety levels among low, moderate and high users. *Computers in Human Behavior, 37*, 290–297. doi:10.1016/j.chb.2014.05.002

Chen, X., & Wei, S. (2019). Enterprise social media use and overload: A curvilinear relationship. *Journal of Information Technology, 34*(1), 22–38. doi:10.1177/0268396218802728

Clark, S. C. (2000). Work/family border theory: A new theory of work/family balance. *Human Relations, 53*(6), 747–770. doi:10.1177/0018726700536001

Delello, J. A., Kiekel, J., Poyo, S. R., Williams, M., & Kerby, D. (2020in press). Educator perceptions of digital devices: Multitasking and distractions in the classroom. In J. A. Dellelo & R. R. McWhorter (Eds.), *Disruptive and Emerging Technology Trends Across Education and the Workplace.* Hershey, PA: IGI Global Inc.

Delello, J. A., & McWhorter, R. R. (2017). Reducing the digital divide: Connecting older adults to iPad technology. *Journal of Applied Gerontology, 36*(1), 3–28. doi:10.1177/0733464815589985 PMID:26084479

Dynarski, M., Agodini, R., Heaviside, S., Novak, T., Carey, N., Campuzano, L., . . . Javitz, H. (2007). *Effectiveness of reading and mathematics software products: Findings from the first student cohort.* Retrieved from https://ies.ed.gov/ncee/pdf/20074005.pdf

Eide, T. A., Aarestad, S. H., Andreassen, C. S., Bilder, R. M., & Pallesen, S. (2018). Smartphone restriction and its effect on subjective withdrawal related scores. *Frontiers in Psychology, 9*, 1444. doi:10.3389/fpsyg.2018.01444 PMID:30150959

Eyvazlou, M., Zarei, E., Rahimi, A., & Abazari, M. (2016). Association between overuse of mobile phones on quality of sleep and general health among occupational health and safety students. *Chronobiology International, 33*(3), 293–300. doi:10.3109/07420528.2015.1135933 PMID:26942630

Faulkner, S. L., & Trotter, S. P. (2017). *Data Saturation. In The International Encyclopedia of Communication Research Methods*. John Wiley & Sons. Retrieved from https://onlinelibrary.wiley.com/doi/book/10.1002/9781118901731

Gökçearslan, Ş., Mumcu, F. K., Haşlaman, T., & Çevik, Y. D. (2016). Modelling smartphone addiction: The role of smartphone usage, self-regulation, general self-efficacy and cyberloafing in university students. *Computers in Human Behavior, 63*, 639–649. doi:10.1016/j.chb.2016.05.091

Harwood, J., Dooley, J. J., Scott, A. J., & Joiner, R. (2014). Constantly connected; the effect of smart-devices on mental health. *Computers in Human Behavior, 34*, 267–272. doi:10.1016/j.chb.2014.02.006

Hilbrecht, M., Zuzanek, J., & Mannell, R. C. (2008). Time use, time pressures, and gendered behavior in early and late adolescence. *Sex Roles, 58*(5-6), 342–357. doi:10.100711199-007-9347-5

Hsieh, H. F., & Shannon, S. E. (2005). Three approaches to qualitative content analysis. *Qualitative Health Research, 15*(9), 1277–1288. doi:10.1177/1049732305276687 PMID:16204405

Ilango, A., & Lobo, M. (2015). Shining light on motivation, emotion, and memory processes. *Frontiers in Behavioral Neuroscience, 9*, 1–3. doi:10.3389/fnbeh.2015.00001 PMID:25653603

Israel, A. C., & Roderick, H. A. (2001). A measure of the stability of family activities: An initial examination. *Assessment, 8*(4), 417–424. doi:10.1177/107319110100800406 PMID:11785585

Kang, S., & Jung, J. (2014). Mobile communication for human needs: A comparison of smartphone use between the US and Korea. *Computers in Human Behavior, 35*, 376–387. doi:10.1016/j.chb.2014.03.024

Karr-Wisniewski, P., & Lu, Y. (2010). When more is too much: Operationalizing technology overload and exploring its impact on knowledge worker productivity. *Computers in Human Behavior, 26*(5), 1061–1072. doi:10.1016/j.chb.2010.03.008

Kuss, D. J., & Griffiths, M. D. (2017). Social networking sites and addiction: Ten lessons learned. *International Journal of Environmental Research and Public Health, 14*(3), E311. doi:10.3390/ijerph14030311 PMID:28304359

Lana, K., Johnson, R. E., & Barnes, C. M. (2014). Beginning the workday yet already depleted? Consequences of late-night smartphone use and sleep. *Organizational Behavior and Human Decision Processes, 124*(1), 11–23. doi:10.1016/j.obhdp.2014.01.001

Larson, R., & Richards, M. (1998). Waiting for the Weekend: Friday and Saturday Nights as the Emotional Climax of the Week. *New Directions for Child and Adolescent Development, 82*(82), 37–51. doi:10.1002/cd.23219988204 PMID:12794950

Larson, R. W., Richards, M. H., Sims, B., & Dworkin, J. (2001). How urban African American young adolescents spend their time: Time budgets for locations, activities, and companionship. *American Journal of Community Psychology, 29*(4), 565–597. doi:10.1023/A:1010422017731 PMID:11554153

Larson, R. W., & Verma, S. (1999). How children and adolescents spend time across the world: Work, play, and developmental opportunities. *Psychological Bulletin, 125*(6), 701–736. doi:10.1037/0033-2909.125.6.701 PMID:10589300

Lee, A. R., Son, S. M., & Kim, K. K. (2016). Information and communication technology overload and social networking service fatigue: A stress perspective. *Computers in Human Behavior, 55*, 51–61. doi:10.1016/j.chb.2015.08.011

Lei, J. (2010). Quantity versus quality: A new approach to examine the relationship between technology use and student outcomes. *British Journal of Educational Technology, 41*(3), 455–472. doi:10.1111/j.1467-8535.2009.00961.x

Lepp, A., Barkley, J., & Karpinski, A. (2014). The relationship between cell phone use, academic performance, anxiety, and satisfaction with life in college students. *Computers in Human Behavior, 31*, 343–350. doi:10.1016/j.chb.2013.10.049

Li, J., Lepp, A., Barkley, J. E., & Karpinski, A. C. (2014). The relationship between cell phone use, academic performance, anxiety, and satisfaction with life in college students. *Computers in Human Behavior, 31*, 343–350. doi:10.1016/j.chb.2013.10.049

Longstreet, P., & Brooks, S. (2017). Life satisfaction: A key to managing internet & social media addiction. *Technology in Society, 50*, 73–77. doi:10.1016/j.techsoc.2017.05.003

Mahmood, T., & Shahzad, M. (2016). Usage patterns and impact of smartphone on social behaviour of youth. *Journal of the Research Society of Pakistan, 53*(2).

Maier, C., Laumer, S., Eckhardt, A., & Weitzel, T. (2015). Giving too much social support: Social overload on social networking sites. *European Journal of Information Systems*, *24*(5), 447–464. doi:10.1057/ejis.2014.3

Mayring, P. (2014). *Qualitative content analysis: theoretical foundation, basic procedures and software solution.* Retrieved from https://www.ssoar.info/ssoar/handle/document/39517

McWhorter, R. R., & Delello, J. A. (2015). Green computing through virtual learning environments. In F. Nafukho (Ed.), *Integrating technology into higher education and in the workplace*. Hershey, PA: IGI Global. doi:10.4018/978-1-4666-8170-5.ch001

Mishra, M. (2017). Impact of smartphone on college students. *International Journal of Education and Management Studies*, *7*(3), 310–313.

Offer, S. (2013). Family time activities and adolescents' emotional well-being. *Journal of Marriage and the Family*, *75*(1), 26–41. doi:10.1111/j.1741-3737.2012.01025.x

Ong, T. (2018, January 9). Apple says it will introduce new features to help parents protect children. *The Wall Street Journal.* Retrieved from https://www.theverge.com/2018/1/9/16867330/apple-response-smartphone-addiction-youth

Oulasvirta, A., Rattenbury, T., Ma, L., & Raita, E. (2015). Habits make smartphone use more pervasive. *Personal and Ubiquitous Computing*, *16*(1), 105–114. doi:10.100700779-011-0412-2

Patton, M. Q. (2002). Qualitative interviewing. *Qualitative Research and Evaluation Methods*, *3*(1), 344–347.

Pearson.com. (2015). *Pearson student mobile device survey national report: College students.* Retrieved from https://www.pearsoned.com/wp-content/uploads/2015-Pearson-Student-Mobile-Device-Survey-College.pdf

Pew Research Center. (2015). *The smartphone difference.* Retrieved from http://www.pewinternet.org/2015/04/01/us-smartphone-use-in-2015/

Poushter, P., Bishop, C., & Chwe, H. (2018). *Global Attitudes and Trends; 2018: Social media use continues to rise in developing countries but plateaus across developed ones.* Retrieved from https://www.pewresearch.org/global/2018/06/19/social-media-use-continues-to-rise-in-developing-countries-but-plateaus-across-developed-ones/#table

Sánchez-Martínez, M., & Otero, A. (2009). Factors associated with cell phone use in adolescents in the community of Madrid (Spain). *Cyberpsychology & Behavior*, *12*(2), 131–137. doi:10.1089/cpb.2008.0164 PMID:19072078

Schlagwein, D., & Hu, M. (2017). How and why organisations use social media: Five use types and their relation to absorptive capacity. *Journal of Information Technology*, *32*(2), 194–209. doi:10.1057/jit.2016.7

Snyder, K. A. (2007). A vocabulary of motives: Understanding how parents define quality time. *Journal of Marriage and the Family*, *69*(2), 320–340. doi:10.1111/j.1741-3737.2007.00368.x

Statista.com. (2019). *Number of smartphone users worldwide from 2014 to 2020 (in billions).* Retrieved from: https://www.statista.com/statistics/330695/number-of-smartphoneusers-worldwide

Tarafdar, M., Gupta, A., & Turel, O. (2013). The dark side of information technology use. *Information Systems Journal*, *23*(3), 269–275. doi:10.1111/isj.12015

Turk, G. (2014). *Look Up Official Video*. Retrieved from https://www.youtube.com/watch?v=Z7dLU6fk9QY

Van Deursen, A. J., Bolle, C. L., Hegner, S. M., & Kommers, P. A. (2015). Modeling habitual and addictive smartphone behavior: The role of smartphone usage types, emotional intelligence, social stress, self-regulation, age, and gender. *Computers in Human Behavior*, *45*, 411–420. doi:10.1016/j.chb.2014.12.039

Wight, V. R., Price, J., Bianchi, S. M., & Hunt, B. R. (2009). The time use of teenagers. *Social Science Research*, *38*(4), 792–809. doi:10.1016/j.ssresearch.2009.05.009

Xie, Y., Szeto, G. P., Dai, J., & Madeleine, P. (2016). A comparison of muscle activity in using touchscreen smartphone among young people with and without chronic neck-shoulder pain. *Ergonomics*, *59*(1), 61–72. doi:10.1080/00140139.2015.1056237 PMID:26218600

Yu, X., Anaya, G. J., Miao, L., Lehto, X., & Wong, I. A. (2018). The impact of smartphones on the family vacation experience. *Journal of Travel Research*, *57*(5), 579–596. doi:10.1177/0047287517706263

Zhang, S., Zhao, L., Lu, Y., & Yang, J. (2016). Do you get tired of socializing? An empirical explanation of discontinuous usage behaviour in social network services. *Information & Management*, *53*(7), 904–914. doi:10.1016/j.im.2016.03.006

Zhitomirsky-Geffet, M., & Blau, M. (2016). Cross-generational analysis of predictive factors of addictive behavior in smartphone usage. *Computers in Human Behavior*, *64*, 682–693. doi:10.1016/j.chb.2016.07.061

ADDITIONAL READING

Delello, J. A., & McWhorter, R. R. (2016). New visual social media for the higher education classroom. In G. Mallia (Ed.), *The social classroom: Integrating social network use in education* (pp. 368–393). Hershey, PA: IGI Global.

Duke, É., & Montag, C. (2017). Smartphone addiction, Daily interruptions and self-reported Productivity. *Addictive Behaviors Reports*, *6*(1), 90–95. doi:10.1016/j.abrep.2017.07.002 PMID:29450241

Konok, V., Pogány, Á., & Miklósi, Á. (2017). Mobile attachment: Separation from the mobile phone induces physiological and behavioral stress and attentional bias to separation related stimuli. *Computers in Human Behavior*, *71*, 228–239. doi:10.1016/j.chb.2017.02.002

Kuss, D., Griffiths, M., Marila, L., & Billieux, J. (2014). Internet addiction: A systematic review of epidemiological research for the last decade. *Current Pharmaceutical Design*, *20*(25), 4026–4052. doi:10.2174/13816128113199990617 PMID:24001297

McWhorter, R. R. (2010). Exploring the emergence of Virtual Human Resource Development. *Advances in Developing Human Resources*, *12*(6), 623–631. doi:10.1177/1523422310395367

McWhorter, R. R., & Delello, J. A. (2016). Green technologies enabling virtual learning environments. *International Journal of Information Communication Technologies and Human Development*, *8*(4), 38–55. doi:10.4018/IJICTHD.2016100104

Miksch, L., & Schulz, C. (2018). Disconnect to reconnect: The phenomenon of digital detox as a reaction to technology overload. Lund University. Retrieved from http://lup.lub.lu.se/luur/download?func=downloadFile&recordOId=8944615&fileOId=8944623

KEY TERMS AND DEFINITIONS

Digital Detoxing: Separating oneself from technology devices to regain or strengthen connections with self, others, and activities.

Mandated Technological Abstinence: Absence of the use of technology, typically mobile devices and computer technology, required by parents, instructors, managers, or administrators.

Opportunity Cost of Digital Addiction: Trading real-life relationships and work achievements for technology dependence.

Technology Addiction: Obsessive behavior related to use of technology significantly affecting relationships with others, work tasks, and activities of daily living.

Technology Distraction: Not giving full attention to a person or activity due to engagement with a mobile or other technology device.

Technology Invasiveness: Devices or digital machines that intrude upon personal or business privacy.

Technology Lethargy: Absence of enthusiasm following sustained periods of technology usage.

Compilation of References

AARP.org. (2012). Framework for isolation in adults over 50. Retrieved from http://www.aarp.org/content/dam/aarp/aarp_foundation/2012_PDFs/AARP-Foundation-Isolation-Framework-Report.pdf

AARP.org. (2018a). *Loneliness and social connections*. Retrieved from https://www.aarp.org/content/dam/aarp/research/surveys_statistics/life-leisure/2018/loneliness-social-connections-2018.doi.10.26419-2Fres.00246.001.pdf

AARP.org. (2018b). *Tech training builds connections and confidence in older adults*. Retrieved from https://www.aarp.org/home-family/personal-technology/info-2018/technology-training-for-older-adults.html

Abbad, M. M., Morris, D., & De Nahlik, C. (2009). Looking under the bonnet: Factors affecting student adoption of e-learning systems in Jordan. *The International Review of Research in Open and Distributed Learning*, *10*(2). doi:10.19173/irrodl.v10i2.596

Abdullateef, B. N., Elias, N. F., Mohamed, H., Zaidan, A., & Zaidan, B. (2016). An evaluation and selection problems of OSS-LMS packages. *SpringerPlus*, *5*(1), 1. doi:10.118640064-016-1828-y PMID:27064567

Ackermann, E. (2001). Piaget's constructivism, Papert's constructionism: What's the difference. *Future of Learning Group Publication*, *5*(3), 438–448.

Adams, D. A., Nelson, R. R., & Todd, P. A. (1992). Perceived usefulness, ease of use, and usage of information technology: A replication. *Management Information Systems Quarterly*, *16*(2), 227–247. doi:10.2307/249577

Adawi, M., Zerbetto, R., Re, T. S., Bisharat, B., Mahamid, M., Amital, H., ... Bragazzi, N. L. (2019). Psychometric properties of the brief symptom inventory in nomophobic subjects: Insights from preliminary confirmatory factor, exploratory factor, and clustering analyses in a sample of healthy Italian volunteers. *Psychology Research and Behavior Management*, *12*, 145–154. doi:10.2147/PRBM.S173282 PMID:30881158

Ahmad, S. A., Chinade, U. B., Gambaki, A. M., Ibrahim, S., & Ala, N. A. (2012). The need for MOODLE as a learning management system in Nigerian Universities: Digesting University Utara Malaysia Learning Zone as a case study. *Academic Research International*, *2*(3), 444.

Ajana, B. (2017). Digital health and the biopolitics of the quantified self. *Digital Health*, *3*, 1–18. doi:10.1177/2055207616689509 PMID:29942580

Akyildiz, I. F., & Kak, A. (2019). The internet of space things/cubesats: A ubiquitous cyber-physical system for the connected world. *Computer Networks*, *150*(26), 134–149. doi:10.1016/j.comnet.2018.12.017

Al Amoush, A. B., & Sandhu, K. (2019). *A New Model for Acceptance of Analytics in Learning Management Systems at Jordanian Universities (JLMS). In Modern Technologies for Teaching and Learning in Socio-Humanitarian Disciplines* (pp. 138–161). IGI Global.

Al Musawi, A., Ambusaidi, A., Al-Balushi, S., & Al-Balushi, K. (2015). Effectiveness of E-Lab Use in Science Teaching at the Omani Schools. *TOJET: The Turkish Online Journal of Educational Technology*, *14*(1).

Alberici, A. (2002). *Imparare sempre nella società della conoscenza*. Milano: Mondadori.

Al-Dmour, R. (2014). *An integration model for identifying the determinants of the adoption and implementation level of HRIS applications and Its effectiveness in business organisations in Jordan* (Doctor of Philosophy). Brunel University. Retrieved from http://bura.brunel.ac.uk/handle/2438/9222

Al-Hujran, O., Al-Debei, M. M., Chatfield, A., & Migdadi, M. (2015). The imperative of influencing citizen attitude toward e-government adoption and use. *Computers in Human Behavior*, *53*, 189–203. doi:10.1016/j.chb.2015.06.025

Ali-Hassan, H., Nevo, D., & Wade, M. (2015). Linking dimensions of social media use to job performance: The role of social capital. *The Journal of Strategic Information Systems*, *24*(2), 65–89. doi:10.1016/j.jsis.2015.03.001

Alimisis, D. (2013). Educational robotics: Open questions and new challenges. *Themes in Science & Technology Education*, *6*(1), 63-71. Retrieved from: http://earthlab.uoi.gr/theste/index.php/theste/article/view/119/85

Allen, E. I., & Seaman, J. (2010). *Class differences: Online education in the United States, 2010*. Sloan Consortium. Retrieved from http://sloanconsortium.org/publications/survey/class_differences

Allen, I. E., Seaman, J., Poulin, R., & Straut, T. T. (2016). *Online report card: Tracking line education in the United States*. Babson Park, MA: Babson Survey Research Group and Quahog Research Group LLC.

Almarabeh, T. (2014). Students' Perceptions of E-learning at the University of Jordan. *iJET*, *9*(3), 31-35.

Alnsour, A., Muhsen, Z., Dababnah, M., Eljinini, M. A., & Barhoum, K. A. (2011). Implementing Moodle as a Tool to develop the Isra University e-learn system. *IJCSNS International Journal of Computer Science and Network Security*, *11*, 120–124.

AlQudah, A. A. (2014). Accepting Moodle by academic staff at the University of Jordan: Applying and extending TAM in technical support factors. *European Scientific Journal*, *10*(18).

Al-Shaikh, F. N. (2003). The practical reality theory and business ethics in non-Western context: Evidence from Jordan. *Journal of Management Development*, *22*(8), 679–693. doi:10.1108/02621710310487855

Al-Shboul, M. A., & Alsmadi, I. (2010). Challenges of Utilizing E-Learning Systems in Public Universities in Jordan. *iJET*, *5*(1), 4-10.

Al-Shboul, M. (2011). Potential Use of Course Management Systems in Higher Education Institutions in Jordan. *Online Submission*, *8*(2), 220–232.

Al-Shboul, M. (2013). The Level of E-Learning Integration at the University of Jordan: Challenges and Opportunities. *International Education Studies*, *6*(4). doi:10.5539/ies.v6n4p93

Al-Zaidiyeen, N. J., Mei, L. L., & Fook, F. S. (2010). Teachers' attitudes and levels of technology use in classrooms: The case of Jordan schools. *International Education Studies*, *3*(2), 211–218. doi:10.5539/ies.v3n2p211

Al-Zoubi, S. I., & Ali, M. (2019). *The Bearing of Culture upon Intention to Utilize D-learning Amongst Jordanian University Students: Modernizing with Emerging Technologies.* Paper presented at the International Conference for Emerging Technologies in Computing. 10.1007/978-3-030-23943-5_21

American Psychiatric Association. (2013). Diagnostic and statistical manual of mental disorders (5th ed.). Arlington, VA: American Psychiatric Publishing.

American Society of Addiction Medicine. (2011). *Public policy statement: Short definition of addiction.* Retrieved from https://www.asam.org/docs/default-source/public-policy-statements/1definition_of_addiction_short_4-11.pdf?sfvrsn=6e36cc2_0

Anderson, J. Q., Boyles, J. L., & Rainie, L. (2012, July 27). *The future impact of the Internet on higher education: Experts expect more efficient collaborative environments and new grading schemes; they worry about massive online courses, the shift away from on-campus life.* Pew Research Center's Internet & American Life Project. Retrieved from http://pewinternet.org/~/media//Files/Reports/2012/PIP_Future_of_Higher_Ed.pdf

Anderson, J., & Jiang, J. (2018). *Teens, social media & technology 2018.* Retrieved from http://www.pewinternet.org/2018/05/31/teens-social-media-technology-2018/

Anderson, J., & Rainie, L. (2014). *Internet of things.* Pew Research Internet Project. Retrieved from http://www.pewinternet.org/files/2014/05/PIP_Internet-of-things_0514142.pdf

Anderson, M., & Perrin, A. (2017, May). *Tech adoption climbs among older adults.* Pew Research Center. Retrieved from http://www.pewinternet.org/2017/05/17/tech-adoption-climbs-among-older-adults/

Anderson, T. (2004). Toward a theory of online learning. In T. Anderson & F. Elloumi (Eds.), *Theory and Practice of Online Learning* (pp. 33–60). Athabasca University.

Angelino, L. M., Williams, F. K., & Natvig, D. (2007). Strategies to engage online students and reduce attrition rates. *Journal of Educators Online*, *4*(2). doi:10.9743/JEO.2007.2.1

Anshari, M., Almunawar, M. N., Shahrill, M., Wicaksono, D. K., & Huda, M. (2017). Smartphones usage in the classrooms: Learning aid or interference? *Education and Information Technologies*, *22*(6), 3063–3079. doi:10.100710639-017-9572-7

Anslow, M. (2007). Creepy technology. *The Ecologist*, *37*(4), 16.

Apple, A. L., Billingsley, F., Schwartz, I. S., & Carr, E. G. (2005). Effects of video modeling alone and with self-management on compliment-giving behaviors of children with high-functioning ASD. *Journal of Positive Behavior Interventions*, *7*(1), 33–46. doi:10.1177/10983007050070010401

Appold, K. (2015). Americans with chronic pain aren't getting sleep: The latest findings from the national sleep foundation (NSF) sleep in America poll report that pain is a key factor in the gap between the amount of sleep Americans say they need and the amount they are getting. *RT for Decision Makers in Respiratory Care*, *28*(5), 15.

Aratani, L. (2018). Facial-recognition scanners at airports raise privacy concerns. *The Washington Post*. Retrieved from https://www.washingtonpost.com/local/trafficandcommuting,/facial-recognition-scanners-at-airports-raise-privacy-concerns/2018/09/15/a312f6d0-abce-11e8-a8d7-0f63ab8b1370_story.html

Arellano, I., & Jones, S. J. (2018). Exploration of University Faculty Perceptions and Experiences of Service-Learning as Engaged Scholarship of Teaching and Learning. *The Journal of Scholarship of Teaching and Learning*, *18*(4). doi:10.14434/josotl.v18i4.23178

Arguel, A., Lockyer, L., Kennedy, G., Lodge, J. M., & Pachman, M. (2019). Seeking optimal confusion: A review on epistemic emotion management in interactive digital learning environments. *Interactive Learning Environments*, *27*(2), 200–210. doi:10.1080/10494820.2018.1457544

Astin, A. W., Vogelgesang, L. J., Ikeda, E. K., & Yee, J. A. (2000). *How service learning affects students*. Academic Press.

Aten, J. (2019). *Instagram allowed an ad partner to track millions of users' data, and it's a major privacy problem.* Retrieved from https://www.inc.com/jason-aten/instagram-just-acknowledged-it-has-a-much-bigger-privacy-problem-than-we-thought.html

Atkinson, R. C., & Shiffrin, R. M. (1968). Human memory. A proposed system and its control process. In J. Spense & K. Spence (Eds.), *Psychology of learning and motivation* (Vol. 2, pp. 89–189). New York, NY: Academic Press.

Atmatzidou, S., & Demetriadis, S. (2014). *How to support students' computational thinking skills in educational robotics activities*. Paper presented at the 4th International Workshop Teaching Robotics, Teaching with Robotics & 5th International Conference Robotics in Education, Padova, Italy.

Atoum, I., Otoom, A., & Ali, A. A. (2017). Holistic Cyber Security Implementation Frameworks: A Case Study of Jordan. *International Journal of Information, Business and Management*, *9*(1), 108.

Ausebel, D. P., Novak, J. D., & Hanesian, H. (1978). *Educational psychology: A cognitive view* (2nd ed.). New York, NY: Holt, Rinehart and Winston.

Awwad, F., Ayesh, A., & Awwad, S. (2013). Are laptops distracting educational tools in classrooms. *Procedia: Social and Behavioral Sciences*, *103*, 154–160. doi:10.1016/j.sbspro.2013.10.320

Babić, S. (2012). Factors that influence academic teacher's acceptance of e-learning technology in blended learning environment. *E-Learning-Organizational Infrastructure and Tools for Specific Areas*, 3-18.

Babson Research Group. (2016). *WCET Distance Education Enrollment Report 2016*. Retrieved from http://wcet.wiche.edu/sites/default/files/WCETDistanceEducationEnrollmentRepo rt2016.pdf

Babson Survey Research Group. (2019). *International Reports*. Retrieved from https://www.onlinelearningsurvey.com/international.html

Baer, D. M., Wolf, M. M., & Risley, T. R. (1968). Some current dimensions of applied behavior analysis. *Journal of Applied Behavior Analysis*, *1*(1), 91–97. doi:10.1901/jaba.1968.1-91 PMID:16795165

Bailey, K. D. (1982). *Methods of social research*. New York: Free Press.

Baker, R. S., & Inventado, P. S. (2014). Educational data mining and learning analytics. In *Learning analytics* (pp. 61–75). Springer New York. doi:10.1007/978-1-4614-3305-7_4

Baldwin, R. (2019). *The globotics upheaval: Globalization, robotics, and the future of work*. New York: Oxford University Press.

Ball, C., Francis, J., Huang, K.-T., Kadylak, T., Cotten, S. R., & Rikard, R. V. (2019). The physical-digital divide: Exploring the social gap between digital natives and physical natives. *Journal of Applied Gerontology*, *38*(8), 1167–1184. doi:10.1177/0733464817732518 PMID:29165009

Bandura, A. (1991). Social cognitive theory of moral thought and action. In W. M. Kurtines & J. L. Gewirtz (Eds.), Handbook of moral behavior and development (Vol. 1, pp. 46-102). Florida International University.

Bandura, A. (1975). *Social Learning & Personality Development*. Holt, Rinehart & Winston.

Bandura, A. (1977). *Social Learning Theory*. Englewood Cliffs, NJ: Prentice Hall.

Bandura, A. (1986). *Social Foundations of Thought and Action*. Englewood Cliffs, NJ: Prentice Hall.

Bank My Cell. (2019). *Smartphone addiction facts & phone usage statistics: The Definitive Guide (2019 update)*. Retrieved from https://www.bankmycell.com/blog/smartphone-addiction

Barab, S.A., Scott, B., Siyahhan, S., Goldstone, R., Ingram-Goble, A., Zuiker, SJ., & Warren, S. (2009). Gioco trasformazionale come impalcatura curriculare: utilizzo dei videogiochi per sostenere l'educazione scientifica. *Journal of Science: educazione e tecnologia, 18*(4), 305-320.

Barker, B. S., Nugent, G., Grandgenett, N., & Adamchuk, V. I. (2012). *Robots in K-12 education: A new technology for learning*. Hershey, PA: Information Science Reference. doi:10.4018/978-1-4666-0182-6

Barnes, L. B., Christensen, C. R., & Hansen, A. J. (1994). *Teaching and the case method: Text, cases, and readings*. Harvard Business Press.

Barr, D., Harrison, J., & Conery, L. (2011). Computational thinking: A digital age skill for everyone. *Learning and Leading with Technology, 38*(6), 20–23.

Barr, V., & Stephenson, C. (2011). Bringing computational thinking to K-12: What is involved and what is the role of the computer science education community? *ACM Inroads, 2*(1), 48–54. doi:10.1145/1929887.1929905

Batchelder, T. H., & Root, S. (1994). Effects of an undergraduate program to integrate academic learning and service: Cognitive, prosocial cognitive, and identity outcomes. *Journal of Adolescence, 17*(4), 341–355. doi:10.1006/jado.1994.1031

Bates, T. (2016). *Teaching in a digital age: guidelines for designing teaching and learning*. SFU Document Solutions.

Bates, T. (2018). The 2017 national survey of online learning in Canadian post-secondary education: Methodology and results. *International Journal of Educational Technology in Higher Education, 15*(29). doi:10.118641239-018-0112-3

Bauman, Z. (2001). *Dentro La Globalizzazione. Conseguenze Sulle Persone*. Roma: Laterza.

Bauman, Z. (2005). *Vita liquida*. Roma, Bari: Laterza.

Bauman, Z. (2010). *La globalización. Consecuencias humanas*. México: Fondo de Cultura Económica.

Beach, L. R., & Mitchell, T. R. (1978). A contingency model for the selection of decision strategies. *Academy of Management Review, 3*(3), 439–449. doi:10.5465/amr.1978.4305717

Beattie, S., Woodley, C., & Souter, K. (2014). Creepy analytics and learner data rights. In B. Hegarty, J. McDonald, & S.-K. Loke (Eds.), Rhetoric and reality: Critical perspectives on educational technology. Proceedings ascilite Dunedin 2014 (pp. 421–425). Academic Press.

Becker, S., Brown, M., Dahlstrom, E., Davis, A., DePaul, K., Diaz, V., & Pomerantz, J. (2018). *NMC Horizon Report: 2018 Higher Education Edition*. Louisville, CO: Educause.

Bejan, A. (1997). Constructal-theory network of conducting paths for cooling a heat generating volume. *International Journal of Heat and Mass Transfer, 40*(4), 799–816. doi:10.1016/0017-9310(96)00175-5

Bellini, S., & Akullian, J. (2007). A meta-analysis of video modeling and video self-modeling interventions for children and adolescents with autism spectrum disorders. *Exceptional Children*, *73*(3), 264–287. doi:10.1177/001440290707300301

Bena, K., & James, M. (2001). *Information technology for schools, creating practical knowledge to improve students' performance.* San Fransico: Jossey–Bass.

Bennett, E. E., & McWhorter, R. R. (2020). Digital technologies for teaching and learning. In Handbook of Adult and Continuing Education. Academic Press.

Bennett, E. E. (2014). Introducing new perspectives on virtual human resource development. *Advances in Developing Human Resources*, *16*(3), 263–280. doi:10.1177/1523422314532091

Bennett, E. E. (2018). Intranets of people, things, and services: Exploring the role of virtual human resource development. In C. A. Simmers & M. Anandarajan (Eds.), *The Internet of people, things and services: Workplace transformations* (pp. 166–184). New York, NY: Routledge. doi:10.4324/9781315182407-10

Bennett, E. E., & McWhorter, R. R. (2016). Opening the black box and searching for smoking guns: Process causality in qualitative research. *European Journal of Training and Development*, *40*(8/9), 691–718. doi:10.1108/EJTD-07-2015-0049

Bennett, E. E., & McWhorter, R. R. (2017a). International HRD and Virtual HRD. In T. Garavan, A. McCarthy, & R. Carbery (Eds.), *A handbook of International HRD: Context, processes and people*. Edward Elgar Publishing.

Bennett, E. E., & McWhorter, R. R. (2017b). Reaction - Organizational learning, community, and virtual HRD: Advancing the discussion. *New Horizons in Adult Education & Human Resource Development*, *29*(3), 19–27. doi:10.1002/nha3.20188

Bennett, E. E., & McWhorter, R. R. (2019). Social movement learning and social innovation: Empathy, agency, and the design of solutions to unmet social needs. *Advances in Developing Human Resources*, *29*(2), 224–249. doi:10.1177/1523422319827939

Bennett, G., & Green, F. P. (2001). Promoting service learning via online instruction. *College Student Journal*, *35*(4), 491–497.

Benzoni, I. (2004). *Portfolio delle competenze e processi di personalizzazione. BG.* Junior.

Berry, S. (2019). Professional development for online faculty: Instructors' perspectives on cultivating technical, pedagogical and content knowledge in a distance program. *Journal of Computing in Higher Education*, *31*(1), 121–136. doi:10.100712528-018-9194-0

Bers, M. (2008). *Blocks to robots: Learning with technology in the early childhood classroom.* New York: Teachers College Press.

Bers, M. U. (2010). The tangible robotics program: Applied computational thinking for young children. *Early Childhood Research & Practice*, *12*(2).

Bers, M. U. (2012). *Designing digital experiences for positive youth development: From playpen to playground.* New York: Oxford University Press. doi:10.1093/acprof:oso/9780199757022.001.0001

Bers, M. U. (2018). *Coding as a playground: programming and computational thinking in the early childhood classroom.* New York, NY: Routledge.

Berthoz, A. (2011). *La semplessità.* Torino: Codice.

Biddle, S. (2018). Artificial intelligence experts issue urgent warning against facial scanning with a "dangerous history". *The Intercept.* Retrieved from https://theintercept.com/2018/12/06/artificial-intellgience-experts-issue-urgent-warning-against-facial-scanning-with-a-dangerous-history/

Billipp, S. H. (2001). The psychosocial impact of interactive computer use within a vulnerable elderly population: A report on a randomized prospective trial in a home health care setting. *Public Health Nursing (Boston, Mass.), 18*(2), 138–145. doi:10.1046/j.1525-1446.2001.00138.x PMID:11285108

Blikstein, P. (2013). Digital fabrication and "making" in education: The democratization of invention. In J. Walter-Herrmann & C. Büching (Eds.), FabLabs: Of machines, makers and inventors (pp. 1-21). Bielefeld, Germany: Transcript-Verlag Publishers.

Blikstein, P. (2018). *Seymour Papert's legacy: Thinking about learning and learning about thinking.* Retrieved from: https://tltl.stanford.edu/content/seymour-papert-s-legacy-thinking-about-learning-and-learning-about-thinking

Bloom, B. S. (1964). *Stability and change in human characteristics.* New York: John Wiley & Sons.

Bocci, F., De Angelis B., Fregola C., Olmetti Peja D., & Zona, U. (2016). *Rizodidattica. Teorie dell'apprendimento e modelli didattici inclusivi.* Lecce-Brescia: Pensa MultimeMedia Bologna: Il Mulino.

Bok, D. (2006). *Our underachieving colleges: A candid look at how much students learn and why they should be learning more.* Princeton University Press.

Bondì, A. (2013). *Percezione, semiosi e socialità del senso.* Milano: Mimesis.

Bouma, H., & Graafmans, J. A. M. (1992). *Gerontechnology.* Amsterdam: IOS Press.

Bourelle, T. (2014). Adapting service-learning into the online technical communication classroom: A framework and model. *Technical Communication Quarterly, 23*(4), 247–264. doi:10.1080/10572252.2014.941782

Bowman, L. L., Levine, L. E., Waite, B. M., & Gendron, M. (2010). Can students really multitask? An experimental study of instant messaging while reading? *Computers & Education, 54*(4), 927–931. doi:10.1016/j.compedu.2009.09.024

Boyd, C. (2019). *Location data leaks from family tracking app database.* Retrieved from https://blog.malwarebytes.com/cybercrime/2019/03/location-data-leaks-from-family-tracking-app-database/

Boyer, E. L. (1990). Scholarship reconsidered: Priorities ofthe professoriate. Princeton, NJ: The Carnegie Foundation for the Advancement of Teaching.

Boyer, E. L. (1997). The scholarship of engagement. *Journal of Public Service and Outreach, 1*(1), 11-20.

Boyer, E. (1996). The scholarship of engagement. *Journal of Public Service and Outreach, 1*(1), 11–20. Retrieved from http://openjournals.libs.uga.edu/index.php/jheoe/article/view/253/238

Bradley, T. (2017). Facebook AI creates its own language in creepy preview of our potential future. *Forbes*. Retrieved from https://www.forbes.com/sites/tonybradley/2017/07/31/facebook-ai-creates-its-own-language-in-creepy-preview-of-our-potential-future/#1fdc4a46292c

Brahms, L. J. (2014). *Making as a learning process: Identifying and supporting family learning in informal settings* (Order No. 3582510). Retrieved from ProQuest Dissertations & Theses Global. (1609201673)

Brahms, L. J., & Werner, J. (2014). Designing Makerspaces for Family Learning in Museums and Science Centers. In M. Honey & D. E. Kanter (Eds.), Design, make, play: Growing the next generation of STEM innovators (pp. 12-16). New York, NY: Routledge.

Brandstad, J. (2015). *A service-learning and transfer-oriented approach to teaching developmental reading and writing students*. Retrieved from https://bwe.ccny.cuny.edu/Brandstad_Service_Learning.htm

Bransford, J. D., Brown, A. L., & Cocking, R. R. (Eds.). (2000). *How People Learn: Brain, Mind, Experience, and School*. Washington, DC: National Academy Press.

Brennan, K., & Resnick, M. (2012). *New frameworks for studying and assessing the development of computational thinking*. Paper presented at the 2012 Annual Meeting at the American Educational Research Association, Vancouver, Canada.

Bringle, R. G., & Hatcher, J. A. (1996). Implementing service-learning in higher education. *The Journal of Higher Education, 67*(2), 221–239. doi:10.2307/2943981

Britt, P. (2015). When is speech technology too creepy? *Speech Technology*. Retrieved from http://www.speechtechmag.com/Articles/Editorial/Feature/When-Is-Speech-Technology-too-Creepy-107400.aspx

Bronfenbrenner, U. (1979). *Ecologia dello sviluppo umano*. Bologna: il Mulino.

Brotherton, D. (2019). *Beware: LinkedIn is a social network with privacy issues too*. Retrieved from https://theamericangenius.com/tech-news/linkedin-privacy-settings/

Bruner, J. (1960). The Process of Education. In Dopo Dewey: il processo di apprendimento nelle due culture. Roma: Armando.

Bruner, J. S. (1997). Celebrating divergence: Piaget and Vygotsky. *Human Development, 40*(2), 63–67. doi:10.1159/000278705

Bryman, A., & Bell, E. (2015). *Business research methods* (4th ed.). New York: Oxford University Press.

Buckinham Shum, S. (2012). *Learning Analytics Policy Brief.* UNESCO. Retrieved from http://iite.unesco.org/files/policy_briefs/pdf/en/learning_analytics.pdf

Buggey, T. (2005) Video self-modeling applications with students with autism spectrum disorder in a small private school setting. Sage Journals. doi:10.1177/10883576050200010501

Bullen, M. (1998). Participation and critical thinking in online university distance education. *Journal of Distance Education, 13*(2), 1–32.

Burge, L. (1988). Beyond andragogy: Some explorations for distance learning design. *International Journal of E-Learning & Distance Education/Revue internationale du e-learning et la formation à distance, 3*(1), 5-23.

Burns, A. J., Roberts, T. L., Posey, C., Bennett, R. J., & Courtney, J. F. (2018). Intentions to comply versus intentions to protect: A VIE theory approach to understanding the influence of insiders' awareness of organizational SETA efforts. *Decision Sciences, 49*(6), 1187–1228. doi:10.1111/deci.12304

Buzzetto-More, N., Johnson, R., & Elobaid, M. (2015). Communicating and sharing in the semantic web: An examination of social media risks, consequences, and attitudinal awareness. *Interdisciplinary Journal of e-Skills and Life Long Learning, 11*, 47-66. Retrieved from http://www.ijello.org/Volume11/IJELLv11p047-066Buzzetto1666.pdf

Cabero, J., Llorente, C., & Román, P. (2007). La tecnología cambió los escenarios: El efecto Pigmalión se hizo realidad [Technology changed scenarios: The Pygmalion effect became real]. *Comunicar, 28*, 167–175.

Cai, Z., Fan, X., & Du, J. (2017). Gender and attitudes toward technology use: A meta-analysis. *Computers & Education, 105*, 1–13. doi:10.1016/j.compedu.2016.11.003

Campus Technology. (2018, September). *Survey: 1 in 4 professors ban mobile phone use in class.* Retrieved from https://campustechnology.com/articles/2018/09/19/survey-1-in-4-professors-ban-mobile-phone-use-in-class.aspx

Cao, H. (2019). I'm developing AI that can read emotions. It's not as creepy as it sounds. *CNN Business Perspectives.* Retrieved from https://www.cnn.com/2019/07/15/perspectives/ai-reads-emotions/index.html

Caprara, G. V., & Cervone, D. (2003). *Personalità: Determinanti, Dinamiche, Potenzialità, 1ª.* Milano: Raffaello Cortina.

Cárdenas, A. (2015). *Lenguaje, analogía y sentido.* Doctorado Institucional en Educación. DIE: Universidad Pedagógica Nacional.

Carnevale, D. (2000). 'Boot camp' helps new online students at Boise State U. *The Chronicle of Higher Education, 46*(24), A48.

Carpenter, H. (2011). How we could measure community impact of nonprofit graduate students' service-learning projects: Lessons from the literature. *Journal of Public Affairs Education, 17*(1), 115–132. doi:10.1080/15236803.2011.12001630

Carr, S. (2000). As distance education comes of age, the challenge is keeping the students. *The Chronicle of Higher Education, 46*(23), A39–A41.

Castells, M. (1997). The Rise of the Network Society. In The Information Age: Economy, Society and Culture (Vol. 1). Cambridge, MA: Blackwell.

Cazzulino, F., Burke, R. V., Muller, V., Arbogast, H., & Upperman, J. S. (2014). Cell phones and young drivers: A systematic review regarding the association between psychological factors and prevention. *Traffic Injury Prevention, 15*(3), 234–242. doi:10.1080/15389588.2013.822075 PMID:24372495

Celio, C. I., Durlak, J., & Dymnicki, A. (2011). A meta-analysis of the impact of service-learning on students. *Journal of Experiential Education, 34*(2), 164–181. doi:10.1177/105382591103400205

Cellan-Jones, R. (2014). Stephen Hawking warns artificial intelligence could end mankind. *BBC News*. Retrieved from https://www.bbc.com/news/technology-30290540

Chalofsky, N. E. (2010). *Meaningful workplaces: Reframing how and where we work*. John Wiley & Sons.

Champagne, N. (2006). Service-learning: Its origin, evolution, and connection to health education. *American Journal of Health Education, 37*(2), 97–102. doi:10.1080/19325037.2006.10598885

Chan, M. Y., Haber, S., Drew, L. M., & Park, D. C. (2014). Training older adults to use tablet computers: Does it enhance cognitive function? *The Gerontologist, 56*(3), 475–484. doi:10.1093/geront/gnu057 PMID:24928557

Chan, P. E., Konrad, M., Gonzalez, V., Peters, M. T., & Ressa, V. A. (2014). The Critical Role of Feedback in Formative Instructional Practices. *Intervention in School and Clinic, 50*(2), 96–104. doi:10.1177/1053451214536044

Chaumon, M. B., Michel, C., Bernard, F. T., & Croisile, B. (2013). Can ICT improve the quality of life of elderly adults living in residential home care units? From actual impacts to hidden artefacts. *Behaviour & Information Technology, 33*(6), 574–590. doi:10.1080/0144929X.2013.832382

Cheever, N. A., Rosen, L. D., Carrier, L. M., & Chavez, A. (2014). Out of sight is not out of mind: The impact of restricting wireless mobile device use on anxiety levels among low, moderate and high users. *Computers in Human Behavior, 37*, 290–297. doi:10.1016/j.chb.2014.05.002

Chen, X., & Wei, S. (2019). Enterprise social media use and overload: A curvilinear relationship. *Journal of Information Technology, 34*(1), 22–38. doi:10.1177/0268396218802728

Chihuri, S., Mielenz, T. J., DiMaggio, C. J., Betz, M. E., DiGuiseppi, C., Jones, V. C., & Li, G. (2016). Driving cessation and health outcomes in older adults. *Journal of the American Geriatrics Society, 64*(2), 332–341. doi:10.1111/jgs.13931 PMID:26780879

Chomsky, N. (2002). *Syntactic structures*. Walter de Gruyter. doi:10.1515/9783110218329

Cho, S. K., & Berge, Z. L. (2002). Overcoming barriers to distance training and education. *USDLA Journal*, *16*(1), 16–34.

Christensson, P. (2006). *Input Definition*. Retrieved September 21, 2019 from https://techterms.com

Christensson, P. (2006). *Output Definition*. Retrieved September 21, 2019 from https://techterms.com

Cisco. (2013). *The Internet of everything: Global private sector economic analysis*. Retrieved from https://www.cisco.com/c/dam/en_us/about/business-insights/docs/ioe-economy-faq.pdf

Cisco. (2019). *Cisco visual networking index: Global mobile data traffic forecast update, 2017–2022 white paper.* Retrieved from https://www.cisco.com/c/en/us/solutions/collateral/service-provider/visual-networking-index-vni/mobile-white-paper-c11-520862.html

Clark, S. C. (2000). Work/family border theory: A new theory of work/family balance. *Human Relations*, *53*(6), 747–770. doi:10.1177/0018726700536001

Cloud based Web Filtering for Schools. (n.d.). Retrieved November 18, 2017, from https://www.securly.com/

Cole, B. (2019). Chinese university tests facial recognition system to monitor attendance and students' attention to lectures. *Newsweek*. Retrieved from https://www.newsweek.com/nanjing-china-facial-recognition-1457193

Common Core State Standards Initiative. (n.d.). English Language Arts Standards. Retrieved September 26, 2016, from http://www.corestandards.org/ELA-Literacy/RH/11-12/

Common Sense Media. (2012). *Children, teens, and entertainment media: The view from the classroom. A national survey of teachers about the role of entertainment media in students' academic and social development.* Retrieved from https://www.commonsensemedia.org/file/view-from-the-classroom-final-reportpdf-0/download

Common Sense Media. (2017). *The common sense census: Media use by kids age zero to eight*. Retrieved from https://www.commonsensemedia.org/sites/default/files/uploads/research/0-8_executivesummary_release_final_1.pdf

Common Sense Media. (2018). *Social media, social life*. Retrieved from https://www.commonsensemedia.org/sites/default/files/uploads/research/2018_cs_socialmediasociallife_executivesummary-final-release_3_lowres.pdf

Computational Thinking for Educators - - Unit 1 – Introducing Computational Thinking. (n.d.). Retrieved from https://computationalthinkingcourse.withgoogle.com/unit

Computational Thinking in K-12 Education. (2018). Retrieved June 1, 2019 from https://cspathshala.org/2017/10/25/computational-thinking-curriculum/

Condruz-Bacescu, M. (2015). e-Learning in companies. *Proceedings of the 10th international scientific conference eLearning and software for education*, 2, 101-107.

Constantinou, V. & Ioannou, A. (n.d.). Development of computational thinking skills through educational robotics. *Ceur-Ws*, 2193.

Corona, F. & De Giuseppe, T. (2017b). La Flipped Inclusion, tra impianto teoretico e didattica sperimentale di aula aumentata per una didattica inclusiva. *Pedagogia Più Didattica.*, (3), 2-7.

Costa, L. V., & Veloso, A. I. (2016). Factors influencing the adoption of video games in late adulthood: A survey of older adult gamers. *Technology and Human Interaction*, *12*(1), 35–50. doi:10.4018/IJTHI.2016010103

Cotten, S. R., Ford, G., Ford, S., & Hale, T. M. (2014). Internet use and depression among retired older adults in the United States: A longitudinal analysis. *The Journals of Gerontology. Series B, Psychological Sciences and Social Sciences*, *69*(5), 763–771. doi:10.1093/geronb/gbu018 PMID:24671896

Cottini, L. (2016). *L' autodeterminazione nelle persone con disabilità. Percorsi educativi per svilupparla.* Trento: Erikson.

Cottrell, G. J. (2017). *Thinking outside the box: A descriptive study focusing on the application of learning analytics on Instructional design* (Doctoral Dissertation). East Stroudsburg University.

Cowls, J., King, T. C., Taddeo, M., & Floridi, L. (2019). *Designing AI for social good: Seven essential factors*. Retrieved from https://ssrn.com/abstract=3388669

Cox, P., Geisen, T., & Green, R. (2011). *Qualitative re-search and social change. European context.* Palgrave Macmillan.

Creswell, J. W. (2007). Qualitative inquiry and research design: Choosing among five approaches. Thousand Oaks, CA: Sage publication Inc.

Creswell, J. (2007). *Qualitative Inquiry & Research Design: Choosing Among Five Approaches* (2nd ed.). Thousand Oaks, CA: Sage.

Creswell, J. W. (2003). *Research design: Qualitative, quantitative, and mixed method approaches* (2nd ed.). Thousand Oaks, CA: Sage.

Creswell, J. W. (2013). *Research design: Qualitative, quantitative, and mixed methods approaches.* Sage Publications.

Creswell, J. W. (2014). *A concise introduction to mixed methods research.* Sage Publications.

Creswell, J. W. (2015). *30 essential skills for the qualitative researcher*. Sage Publications.

CSTA. (n.d.). *About CSTA standards*. Retrieved May 13, 2019 from https://www.csteachers.org/page/about-csta-s-k-12-nbsp-standards

Cuador, C. (2017). From street photography to face recognition: Distinguishing between the right to be seen and the right to be recognized. *Nova Law Review*, *41*(2), 237–264.

Culhane, J., Niewolny, K., Clark, S., & Misyak, S. (2018). Exploring the Intersections of Interdisciplinary Teaching, Experiential Learning, and Community Engagement: A Case Study of Service Learning in Practice. *International Journal on Teaching and Learning in Higher Education*, *30*(3), 412–422.

Cumbley, R., & Church, P. (2013). Is 'big data' creepy? *Computer Law & Security Review*, *29*(5), 601–609. doi:10.1016/j.clsr.2013.07.007

Dahl, R. (2018). *Teaching tech skills to older adults*. Retrieved from https://edtech.worlded.org/teaching-tech-skills-older-adults/

Dailey-Hebert, A., Donelli-Salee, E., & DiPadova, L. N. (2008). *Service E-Learning: Educating for citizenship*. Charlotte, NC: Information Age Publishing.

Dalton, K. (2013). Their brains on Google: How digital technologies are altering the millennial generation's brain and impacting legal education. *SMU Science and Technology Law Review*, *16*, 409–501.

Dane, S. K., Mason, C. M., & O'Brien-McInally, B. A. (2013). *Household internet use in Australia: A study in regional communities*. CSIRO Report: EP1310907.

Dang, S. S. (2019). Artificial intelligence in humanoid robots. *Forbes*. Retrieved from https://www.forbes.com/sites/cognitiveworld/2019/02/25/artificial-intelligence-in-humanoid-robots/#6a7cd2b724c7

Datteri, E., Zecca, L., Laudisa, F., & Castiglioni, M. (2013). Learning to explain: The role of educational robots in science education. *Themes in Science and Technology Education*, *6*, 29–38.

Davies, T. (2018). *More privacy concerns hit Twitter*. Retrieved from https://gdpr.report/news/2018/12/19/more-privacy-concerns-hit-twitter/

Davis, F. D. (1989). Perceived usefulness, perceived ease of use, and user acceptance of information technology. *Management Information Systems Quarterly*, *13*(3), 319–340. doi:10.2307/249008

Day, M., Turner, G., & Drozdiak, N. (2019). *Amazon workers are learning to what you tell Alexa*. Retrieved from https://www.bloomberg.com/news/articles/2019-04-10/is-anyone-listening-to-you-on-alexa-a-global-team-reviews-audio

De Giuseppe T., Corona F. (2018). Il complex blended learning di spiral nella prospettiva sistemica della Flipped Inclusion. *Media Education – Studi, ricerche, buone pratiche*, *9*(2), 343-356.

De Giuseppe, T. (2018a). *Flipped Inclusion. L'impianto teoretico tra bisogni emergenti e prospettive epistemologiche. Collana Differenze*. Roma: Aracne.

De Giuseppe, T. (2018b). La media education nell'economia della formazione continua. *Formazione & Insegnamento*, *16*(2), 233–247.

De Giuseppe, T., & Corona, F. (2016). Flipped Inclusion per una didattica trasformativa. Convegno. In P. Magnoler, A. M. Notti, & L. Perla (Eds.), *La professionalità degli insegnanti. La ricerca e le pratiche* (pp. 551–570). Lecce: Pensa.

De Giuseppe, T., & Corona, F. (2017). Metodologia Flipped tra sistemica inclusione e prospettive didattico-assertive. *Formazione & Insegnamento, 15*(2), 409–420.

de Kerckhove, K. (1997). *Connected intelligence: the arrival of the web society*. Toronto: Somerville House Publ.

De Paola, A. (2014). An ontology-based autonomic system for ambient intelligence scenarios. In S. Gaglio & G. Lo Re (Eds.), *Advances onto The Internet of Things: How ontologies make the Internet of Things more meaningful* (pp. 1–17). Cham, Switzerland: Springer International Publishing. doi:10.1007/978-3-319-03992-3_1

de Saussure, F. (1970). *Corso di linguistica generale*. Bari: Laterza.

Delahunty, J., Verenikina, I., & Jones, P. (2014). Socio-emotional connections: Identity, belonging and learning in online interactions. A literature review. *Technology, Pedagogy and Education, 23*(2), 243–265. doi:10.1080/1475939X.2013.813405

Delello, J. A., & McWhorter, R. R. (2017). Reducing the digital divide: Connecting older adults to iPad technology. *Journal of Applied Gerontology, 36*(1), 3-28. Retrieved from https://www.ncbi.nlm.nih.gov/pubmed/26084479

Delello, J. A., Kiekel, J., Poyo, S. R., Williams, M., & Kerby, D. (2020in press). Educator perceptions of digital devices: Multitasking and distractions in the classroom. In J. A. Dellelo & R. R. McWhorter (Eds.), *Disruptive and Emerging Technology Trends Across Education and the Workplace*. Hershey, PA: IGI Global Inc.

Delello, J. A., & McWhorter, R. R. (2015). Reducing the digital divide: Connecting older adults to iPad technology. *Journal of Applied Gerontology, 36*(1), 3–28. doi:10.1177/0733464815589985 PMID:26084479

Delello, J. A., & McWhorter, R. R. (2016). New visual social media for the higher education classroom. In G. Mallia (Ed.), *The social classroom: Integrating social network use in education*. Hershey, PA: IGI Global.

Delello, J. A., Mokhtari, K., & Reichard, C. (2016). Multitasking among college students: Are freshmen more distracted? *International Journal of Cyber Behavior, Psychology and Learning, 6*(4), 1–12. doi:10.4018/IJCBPL.2016100101

De-Sola Gutiérrez, J., Rodríguez de Fonseca, F., & Rubio, G. (2016). Cell-Phone Addiction: A Review. *Frontiers in Psychiatry, 7*, 175. doi:10.3389/fpsyt.2016.00175 PMID:27822187

Devlin, M., & McKay, J. (2016). Teaching students using technology: Facilitating success for students from low socioeconomic status backgrounds in Australian universities. *Australasian Journal of Educational Technology, 32*(1). doi:10.14742/ajet.2053

Devlin, M., & McKay, J. (2018). Teaching inclusively online in a massified university system. *Widening Participation and Lifelong Learning: the Journal of the Institute for Access Studies and the European Access Network*, *20*(1), 146–166. doi:10.5456/WPLL.20.1.146

Dewey, J. (1961b). *Logica. Teoria dell'indagine*. Torino: Einaudi 1974.

Dewey, J. (1933). *How we think: A restatement of the relation of reflective thinking in the educative process*. Chicago: Henry Regnery.

Dewey, J. (1961a). *Come pensiamo. Una riformulazione del rapporto fra il pensiero riflessivo e l'educazione*. Firenze: La Nuova Italia.

Dewey, J. (1984). *Democrazia e educazione*. Firenze: La Nuova Italia.

Diaz, D. P., & Bontenbal, K. F. (2001). Learner preferences: Developing a learner-centered environment in the online or mediated classroom. *Ed at a Distance, 15*(8). Retrieved April 11, 2006, from http://www.usdla.org/html/journal/AUG01_Issue/article03.html

Dienhart, C., Maruyama, G., Snyder, M., Furco, A., McKay, M. S., Hirt, L., & Huesman, R. (2016). The impacts of mandatory service on students in service-learning classes. *The Journal of Social Psychology*, *156*(3), 305–309. doi:10.1080/00224545.2015.1111856 PMID:27064181

Dixon, E. (2018). *Does reading help with loneliness? A new report says it could be a solution to chronic isolation.* Retrieved from https://www.bustle.com/p/does-reading-help-with-loneliness-a-new-report-says-it-could-be-a-solution-to-chronic-isolation-13029607

Dougherty, D. (2005). *Maker faire founder Dale Dougherty on the past, present, and online future of the maker movement*. Retrieved from https://goo.gl/wRkKxF

Dougherty, D. (2013). The maker mindset. In M. Honey & D. E. Kanter (Eds.), *Design, make, play: Growing the next generation of STEM innovators*. New York, NY: Routledge.

Drachsler, H., & Greller, W. (2016, April). Privacy and learning analytics-It's a DELICATE issue. In *Proceedings of the Sixth International Conference on Learning Analytics & Knowledge* (pp.89-98). ACM. 10.1145/2883851.2883893

Draper, S. (2018). *Pentagon tells soldiers to leave wearable trackers at home when heading to war zones.* Retrieved from https://www.wearable-technologies.com/2018/08/pentagon-tells-soldiers-to-leave-wearable-trackers-at-home-when-heading-to-warzones/

Driscoll, M. P. (2000). *Psychology of learning for instructions* (2nd ed.). Needham Heights, MA: Allyn & Bacon.

Dwyer, S. C., & Buckle, J. L. (2009). The space between: On being an insider-outsider in qualitative research. *International Journal of Qualitative Methods*, *8*(1), 54–63. doi:10.1177/160940690900800105

Dynarski, M., Agodini, R., Heaviside, S., Novak, T., Carey, N., Campuzano, L., . . . Javitz, H. (2007). *Effectiveness of reading and mathematics software products: Findings from the first student cohort*. Retrieved from https://ies.ed.gov/ncee/pdf/20074005.pdf

Edwards, M. (2018). *The problem with oversharing on social media*. Retrieved from https://www.business2community.com/social-media/problem-oversharing-social-media-01983174

Eguchi, A. (2014). *Robotics as a learning tool for educational transformation*. Paper presented at the Teaching with Robotics & 5th International Conference Robotics in Education. Retrieved from http://www.terecop.eu/TRTWR-RIE2014/files/00_WFr1/00_WFr1_04.pdf

Eguchi, A. (2016). Computational thinking with educational robotics. *Site,* 79-84.

Eide, T. A., Aarestad, S. H., Andreassen, C. S., Bilder, R. M., & Pallesen, S. (2018). Smartphone restriction and its effect on subjective withdrawal related scores. *Frontiers in Psychology*, *9*, 1444. doi:10.3389/fpsyg.2018.01444 PMID:30150959

Emerson, K. G., & Jayawardhana, J. (2016). Risk factors for loneliness in elderly adults. *Journal of the American Geriatrics Society*, *64*(4), 886–887. doi:10.1111/jgs.14053 PMID:27100588

Ertmer, P. A., & Ottenbreit-Leftwich, A. T. (2010). Teacher technology change: How knowledge, confidence, beliefs, and culture intersect. *Journal of Research on Technology in Education*, *42*(3), 255–284. doi:10.1080/15391523.2010.10782551

Evans, D. (2011). *Internet of Things: How the next evolution of the Internet is changing everything*. Retrieved from https://www.cisco.com/c/dam/en_us/about/ac79/docs/innov/IoT_IBSG_0411FINAL.pdf

Eyler, J., Giles, D. E. Jr, & Braxton, J. (1997). The impact of service-learning on college students. *Michigan Journal of Community Service Learning*, *4*, 5–15.

Eyvazlou, M., Zarei, E., Rahimi, A., & Abazari, M. (2016). Association between overuse of mobile phones on quality of sleep and general health among occupational health and safety students. *Chronobiology International*, *33*(3), 293–300. doi:10.3109/07420528.2015.1135933 PMID:26942630

Falloon, G. (2011). Usare avatar e ambienti virtuali nell'apprendimento: Cosa offrono? *Journal of Educational Technology*, *41*(2), 108–122.

Family Educational Rights and Privacy Act (FERPA). (2015, June 26). [Guides]. Retrieved January 25, 2016, from http://www2.ed.gov/policy/gen/guid/fpco/ferpa/index.html

Faulkner, S. L., & Trotter, S. P. (2017). *Data Saturation. In The International Encyclopedia of Communication Research Methods*. John Wiley & Sons. Retrieved from https://onlinelibrary.wiley.com/doi/book/10.1002/9781118901731

Ford, M., & Palmer, W. (2019). Alexa, are you listening to me? An analysis of Alexa voice service network traffic. *Personal and Ubiquitous Computing*, *23*(1), 67–79. doi:10.100700779-018-1174-x

Forrest, C., & Bayern, M. (2019). 20 terrifying uses of artificial intelligence. *TechRepublic*. Retrieved from https://www.techrepublic.com/pictures/10-terrifying-uses-of-artificial-intelligence/5/

Foucault, M. (1967). *Le parole e le cose*. Milano: BUR.

Fox, S., & Rainie, L. (2014). The web at 25 in the US: The overall verdict: The internet has been a plus for society and an especially good thing for individual users. *Pew Research Center's Internet and American Life Project*. Retrieved from http://www.pewinternet.org/2014/02/25/the-web-at-25-in-the-u-s

Franceschi-Bicchierai, L. (2018). *Spyware company leaves 'terabytes' of selfies, text messages, and location data exposed online*. Retrieved https://www.vice.com/en_us/article/9kmj4v/spyware-company-spyfone-terabytes-data-exposed-online-leak

Franceschi-Bicchierai, L. (2019). *This spyware data leak is so bad we can't even tell you about it*. Retrieved from https://www.vice.com/en_us/article/j573k3/spyware-data-leak-pictures-audio-recordings

Francis, K., & Davis, B. (2018). *Coding Robots as a Source of Instantiations for Arithmetic*. Digital Experiences in Mathematics Education. doi:10.100740751-018-0042-7

Frankenfield, J. (2019). *General data protection regulation (GDPR)*. Retrieved from https://www.investopedia.com/terms/g/general-data-protection-regulation-gdpr.asp

Friedman, J. (2016). *Blended Learning Options Rise at Medical Schools*. Retrieved from https://www.usnews.com/education/best-graduate-schools/top-medical-schools/articles/2 016-03-23/blended-learning-options-rise-at-medical-schools

Fromm, E. (1994). *Io difendo l'uomo*. Milano: Rusconi.

Fusilli, A. (2016). *Effetto borderline. Soggettivazione e movimenti del desiderio*. Milano: Franco Angeli.

Gagné, R. (1977). *The conditions of learning*. Boston, MA: Thomson Learning.

Gagné, R. M., & Medsker, K. L. (1995). *Conditions of learning: Training applications*. Belmont, CA: Wadsworth Publishing.

Garb, H. N., Lilienfeld, S. O., & Fowler, K. A. (2008). Psychological assessment and clinical judgement. In B. A. Winstead (Ed.), *Psychopathology: Foundations for a contemporary understanding* (2nd ed.; pp. 103–124). New York, NY: Routledge/Taylor & Francis Group.

García-Romero, D., Sànchez-Busqués, S., & Lalueza-Sazatornil, J. L. (2018). *Exploring the Value of Service Learning: Students' assessments of Personal*. Procedural and Content Learning.

Garrison, D. R. (1993). A cognitive constructivist view of distance education: An analysis of teaching-learning assumptions. *Distance Education, 14*(2), 199–211. doi:10.1080/0158791930140204

Gates Annual Letter. (2015). *Our Big Bet for the Future*. Retrieved from https://www.appropriations.senate.gov/imo/media/doc/hearings/032615%20Gates%20Foundation%20Annual%20Letter%202015%20-%20SFOPS.pdf

Gazzaley, A., & Rosen, L. D. (2016). *The distracted mind: Ancient brains in a high-tech world*. Cambridge, MA: MIT Press.

Gee, J. P. (1990). *Linguistica e alfabetizzazione sociale: Ideologia nei discorsi*. Falmer Press.

Gee, J. P. (2005). Learning by Design: Buoni videogiochi come macchine per l'apprendimento. *E-learning*, *2*(1), 5–1.

Gee, J. P. (2008). *Quali videogiochi devono insegnarci sull'apprendimento e l'alfabetizzazione, rivisti e aggiornati*. Basingstoke, UK: Palgrave Macmillan.

Gefen, D., & Straub, D. W. (1997). Gender differences in the perception and use of e-mail: An extension to the technology acceptance model. *Management Information Systems Quarterly*, *21*(4), 389–400. doi:10.2307/249720

Gelinas, J. (2019). *Alex IS listening to your conversations*. Retrieved from https://www.komando.com/security-privacy/alexa-is-listening-to-your-conversations/561016/

Georgetown Law Library. (2019). *International and foreign law research guide*. Retrieved from https://guides.ll.georgetown.edu/cyberspace

Germain, M.-L. (2019). *Integrating Service-Learning and Consulting in Distance Education*. London, UK: Emerald Group Publishing. doi:10.1108/9781787694095

Gerstein, J. (2015). *The Educator as a Maker Educator*. Kindle Edition. Amazon Digital Services LLC. ASIN: B016Z5NZ6O

Gerst-Emerson, J. K., & Jayawardhana, J. (2015). Loneliness as a public health issue: The impact of loneliness on health care utilization among older adults. *American Journal of Public Health*, *105*(5), 1013–1019. doi:10.2105/AJPH.2014.302427 PMID:25790413

Gibson, W. (1994). *Luce Virtuale*. Milano: Mondadori.

Giddens, A. (2000). *Il mondo che cambia. Come la globalizzazione ridisegna la nostra vita*. Bologna: Il Mulino.

Giummarra, M. J., Haralambous, B., Moore, K., & Nankervis, J. (2007). The concept of health in older age: Views of older people and health professionals. *Australian Health Review*, *31*(4), 642–650. doi:10.1071/AH070642 PMID:17973624

Glaser, A. (2016). *Biometrics are coning, along with serious security concerns*. Retrieved from https://www.wired.com/2016/03/biometrics-coming-along-serious-security-concerns/

Glass, A. L., & Kang, M. (2018). Dividing attention in the classroom reduces exam performance. *Educational Psychology*, *39*(3), 395–408. doi:10.1080/01443410.2018.1489046

Glazatov, T. R. (2012). Applying Instructional Design System Theory to Mobile Learning Environments. *Journal of Applied Learning Technology*, *2*(2), 29–35.

GNU.org. (2018). *What is free software?* The Free Software Definition. The GNU Project.

Godfrey, P. C., Illes, L. M., & Berry, G. R. (2005). Creating breadth in business education through service-learning. *Academy of Management Learning & Education*, *4*(3), 309–323. doi:10.5465/amle.2005.18122420

Goertzel, Mossbridge, Monroe, Hanson, & Yu. (2017). *Loving AI: Humanoid robots as agents of human consciousness expansion*. Retrieved from https://arxiv.org/pdf/1709.07791.pdf

Goffman, E. (1974). *Frame analysis. L'organizzazione dell'esperienza*. Roma: Armando.

Gökçearslan, Ş., Mumcu, F. K., Haşlaman, T., & Çevik, Y. D. (2016). Modelling smartphone addiction: The role of smartphone usage, self-regulation, general self-efficacy and cyberloafing in university students. *Computers in Human Behavior*, *63*, 639–649. doi:10.1016/j.chb.2016.05.091

Goldacre, B. (2014). When data gets creepy: The secrets we don't realise we're giving away. *The Guardian*. Retrieved from https://www.theguardian.com/technology/2014/dec/05/when-data-gets-creepy-secrets-were-giving-away

Goldbeck, J. (2014). Why we overshare online. *Psychology Online*. Retrieved from https://www.psychologytoday.com/us/blog/your-online-secrets/201410/why-we-overshare-online

Goldberg, I. (1996). *Internet addiction disorder*. Retrieved from http://www.urz.uni-heidel-berg.de/Netzdienste/anleitung/ wwwtips/8/addict.htm

Goldy-Brown, S. (2017, December 6). *11 Benefits of MOOCs (Massive Open Online Courses)*. Retrieved from https://plexuss.com/news/article/benefits-of-moocs

Goleman, D. (2013). *Focus: The hidden driver of excellence*. New York: HarperCollins.

Goll, J. C., Charlesworth, G., Scior, K., & Stott, J. (2015). Barriers to social participation among lonely older adults: The influence of social fears and identity. *PLoS One*, *10*(2), 1–17. doi:10.1371/journal.pone.0116664 PMID:25706933

González, A., Ramírez, M. P., & Viadel, V. (2012). Attitudes of the elderly toward information and communications technologies. *Educational Gerontology*, *38*(9), 585–594. doi:10.1080/03601277.2011.595314

Graham, J. (2019). *Hey Siri, Google and Alexa — enough with the snooping*. Retrieved from https://www.usatoday.com/story/tech/2019/08/24/hey-siri-google-alexa-and-cortana-stop-snooping-now/2083899001/

Greenland, S., & Moore, C. (2014). Patterns of online student enrolment and attrition in Australian open access online education: A preliminary case study. *Open Praxis*, *6*(1), 45–54. doi:10.5944/openpraxis.6.1.95

Greenleaf, R. K. (1977). *Servant Leadership: A Journey into the Nature of Legitimate Power and Greatness*. New York, NY: Paulist Press.

Grover, S. (2018). *The 5th 'C' of 21st century skills? try computational thinking (not coding)*. Retrieved on March 1, 2018 from: https://www.edsurge.com/news/2018-02-25-the-5th-c-of-21st-century-skills-try-computational-thinking-not-coding?

Grover, S., & Pea, R. (2013). Computational thinking in K–12: A review of the state of the field. *Educational Researcher*, *42*(1), 38–43.

Gupta, S. K., & Rao, D. (2018). A case study on contemporary water management innovation (CWMI) in sustainable agriculture: Farmers joined hands for mutual survival and prosperity: The case of Kummarvandlapalli. *IJCS*, *6*(6), 2283–2287.

Guthrie, K. L., & McCracken, H. (2010). Teaching and learning social justice through online service-learning courses. *The International Review of Research in Open and Distributed Learning*, *11*(3), 78–94. doi:10.19173/irrodl.v11i3.894

Hacihasanoğlu, R., Yildirim, A., & Karakurt, P. (2011). Loneliness in elderly individuals, level of dependence in activities of daily living (ADL) and influential factors. *Archives of Gerontology and Geriatrics*, *54*(1), 61–66. doi:10.1016/j.archger.2011.03.011 PMID:21514680

Hackerspace. (2015). *Hackerspace*. Retrieved from http://hackerspaces.org/

Harasim, L. (1999). A framework for online learning: The Virtual-U. *Computer*, *32*(9), 44–49. doi:10.1109/2.789750

Hardigan, S. (2007). *The Learning Revolution has Begun*. Retrieved from http://www.stevehargadon.com/2007/04/web-20-and-school-20-connection.html

Harwood, J., Dooley, J. J., Scott, A. J., & Joiner, R. (2014). Constantly connected; the effect of smart-devices on mental health. *Computers in Human Behavior*, *34*, 267–272. doi:10.1016/j.chb.2014.02.006

Hastings, O. P. (2016). Not a lonely crowd? Social connectedness, religious service attendance, and the spiritual but not religious. *Social Science Research*, *57*, 63–79. doi:10.1016/j.ssresearch.2016.01.006 PMID:26973032

Hauge, S., & Kirkevold, M. (2012). Variations in older persons' descriptions of the burden of loneliness. *Scandinavian Journal of Caring Sciences*, *26*(3), 553–560. doi:10.1111/j.1471-6712.2011.00965.x PMID:22233299

Heidari, M., Ghodusi, M., & Shirvani, M. (2016). Loneliness in elderly and non-elderly residents of nursing homes. *International Journal of Nursing Education*, *8*(2), 71–76. doi:10.5958/0974-9357.2016.00051.9

Hendrickson, K. (n.d.). *Computational Thinking*. Retrieved May 11, 2019 from https://k12cs.org/computational-thinking

Herold, B. (2019, April). Ed-Tech supporters promise innovations that can transform schools: Teachers not seeing impact. *Education Week*. Retrieved from https://www.edweek.org/ew/articles/2019/04/24/ed-tech-supporters-promise-innovations-that-can-transform.html

Hesser, G. (1995). Faculty Assessment of Student Learning: Outcomes Attributed to Service-Learning and Evidence of Changes in Faculty Attitudes About Experiential Education. *Michigan Journal of Community Service Learning*, *2*, 33–42.

Hicks, T. J. (2000). What is your life like now? Loneliness and elderly individuals residing in nursing homes. *Journal of Gerontological Nursing*, *26*(8), 15–19. doi:10.3928/0098-9134-20000801-05 PMID:11276607

Hilbrecht, M., Zuzanek, J., & Mannell, R. C. (2008). Time use, time pressures, and gendered behavior in early and late adolescence. *Sex Roles*, *58*(5-6), 342–357. doi:10.100711199-007-9347-5

Hilhorst, S., Lockey, A., & Seight, T. (2018). *A society of readers*. London: Demos Publishing.

Hill, T., & Westbrook, R. (1997). SWOT Analysis: It's time for a product recall. *Long Range Planning*, *30*(1), 46–52. doi:10.1016/S0024-6301(96)00095-7

Hofstede, G. (1983). National cultures in four dimensions: A research-based theory of cultural differences among nations. *International Studies of Management & Organization*, *13*(1-2), 46–74. doi:10.1080/00208825.1983.11656358

Holland, B. (1997). Analyzing institutional commitment to service: A model of key organizational factors. *Michigan Journal of Community Service Learning*, *4*(1), 30–41.

Honey, M., & Kanter, D. E. (Eds.). (2013). *Design, make, play: Growing the next generation of STEM innovators*. New York, NY: Routledge. doi:10.4324/9780203108352

Howard, J. (2017). *When kids get their first cell phones around the world*. Retrieved from https://www.cnn.com/2017/12/11/health/cell-phones-for-kids-parenting-without-borders-explainer-intl/index.html

Hsbollah, H. M., & Idris, K. M. (2009). E-learning adoption: The role of relative advantages, trialability and academic specialisation. *Campus-Wide Information Systems*, *26*(1), 54–70. doi:10.1108/10650740910921564

Hsieh, H. F., & Shannon, S. E. (2005). Three approaches to qualitative content analysis. *Qualitative Health Research*, *15*(9), 1277–1288. doi:10.1177/1049732305276687 PMID:16204405

HUD.gov. (2016). *Housing choice vouchers sheet*. U.S. Department of Housing and Urban Development. Retrieved from http://portal.hud.gov/hudportal/HUD?src=/topics/housing_choice_voucher_program_section_8

Hughes, J. A. (2004). Supporting the online learner. In T. Anderson & F. Elloumi (Eds.), *Theory and Practice of Online Learning* (pp. 367–384). Athabasca University.

Hui, W., Hu, P. H., Clark, T. H., Tam, K. Y., & Milton, J. (2008). Technology-assisted learning: A longitudinal field study of knowledge category, learning effectiveness and satisfaction in language learning. *Journal of Computer Assisted Learning*, *24*(3), 245–259. doi:10.1111/j.1365-2729.2007.00257.x

Hunter, D. (2007). The virtual student/client experience. *The Journal of American Academy of Business, Cambridge*, *12*(1), 88–92.

Igbaria, M., Zinatelli, N., Cragg, P., & Cavaye, A. L. (1997). Personal computing acceptance factors in small firms: A structural equation model. *Management Information Systems Quarterly*, *21*(3), 279–305. doi:10.2307/249498

Ilango, A., & Lobo, M. (2015). Shining light on motivation, emotion, and memory processes. *Frontiers in Behavioral Neuroscience*, *9*, 1–3. doi:10.3389/fnbeh.2015.00001 PMID:25653603

Ilgaz, H., & Gülbahar, Y. (2015). A snapshot of online learners: e-Readiness, e-Satisfaction and expectations. *The International Review of Research in Open and Distributed Learning*, *16*(2). doi:10.19173/irrodl.v16i2.2117

Imperial, M. T., Perry, J. L., & Katula, M. C. (2007). Incorporating service learning into public affairs programs: Lessons from the literature. *Journal of Public Affairs Education*, *13*(2), 243–264.

Isaac, M. (2019). *Facebook's new tool lets you see which apps and websites tracked you*. Retrieved from https://www.nytimes.com/2019/08/20/technology/facebook-tool-privacy-apps-websites.html

Israel, A. C., & Roderick, H. A. (2001). A measure of the stability of family activities: An initial examination. *Assessment*, *8*(4), 417–424. doi:10.1177/107319110100800406 PMID:11785585

ISTAT. (2014). *Internet@Italia 2014. L'uso di Internet da parte di cittadini e imprese*. Roma: Fondazione Ugo Bordoni. Retrieved from https://www.istat.it/it/files//2014/12/Cittadini_e_nuove_tecnologie_anno-2014.pdf

Jacelon, C., LeBlanc, R., Alkhawaldeh, M., Ridgeway, J., Marquard, J., & Choi, J. M. (2018). Developing ASSISTwell, a tablet application to support older adult's self-management of symptoms of chronic conditions. *Gerontechnology (Valkenswaard)*, *17*(1), 1–18. doi:10.4017/gt.2018.17.1.002.00

Jaeger, A. J., Shipley, T. F., & Reynolds, S. J. (2017). The roles of working memory and cognitive load in geoscience learning. *Journal of Geoscience Education*, *65*(4), 506–518. doi:10.5408/16-209.1

Jerald, J., & Block, M. D. (2008). Issues for DSM-V: Internet Addiction. *The American Journal of Psychiatry*, *3*, 306–307.

Jiang, J. (2018). *How teens and parents navigate screen time and device distractions*. Pew Research Center. Retrieved from http://www.pewinternet.org/2018/08/22/how-teens-and-parents-navigate-screen-time-and-device-distractions/

Jiménez-Murcia, S., Fernández-Aranda, F., Granero, R., Chóliz, M., La Verde, M., & Aguglia, E. (2014). Video game addiction in gambling disorder: Clinical, psychopathological, and personality correlates. *Biomed Reserach International, 2014*, 1–11. doi:10.1155/2014/31506

Johnson, A. (2017). Facial recognition hearing looks to give lawmakers clearer picture on creepy technology. *NBC News*. Retrieved from https://www.nbcnews.com/news/us-news/facial-recognition-hearing-looks-give-lawmakers-clearer-picture-creepy-technology-n736771

Johnson, D. (2019). The 11 creepiest technologies that exist today. *Business Insider.* Retrieved from https://www.businessinsider.com/creepiest-technologies-tech-examples-ai-2019-8

Johnson, N. (2012). *The Institutional Costs of Student Attrition*. Delta Cost Project at American Institutes for Research. Retrieved from https://deltacostproject.org/sites/default/files/products/Delta-Cost-Attrition-Research-Paper.pdf

Johnson, A. M., Jacovina, M. E., Russell, D. E., & Soto, C. M. (2016). Challenges and solutions when using technologies in the classroom. In S. A. Crossley & D. S. McNamara (Eds.), *Adaptive educational technologies for literacy instruction* (pp. 13–29). New York: Taylor & Francis. Published with acknowledgment of federal support. doi:10.4324/9781315647500-2

Jonassen, D. H. (1995). Operationalizing mental models: Strategies for assessing mental models to support meaningful learning and design supportive learning environments. *Proceeding of the first international Conference Computer Support for Collaborative Learning '95*, 182-186.

Jones, T., Kay, D., Upton, P., & Upton, D. (2013). An evaluation of older adults' use of iPads in eleven UK care-homes. *International Journal of Mobile Human Computer Interaction, 5*(3), 62–76. doi:10.4018/jmhci.2013070104

Judd, T. (2015). Task selection, task switching and multitasking during computer-based independent study. *Australasian Journal of Educational Technology, 31*(2), 193–207. doi:10.14742/ajet.1992

Junco, R. (2012). In class multitasking and academic performance. *Computers in Human Behavior, 28*(6), 2236–2243. doi:10.1016/j.chb.2012.06.031

Jung, C. G. (1999). *Simboli della trasformazione*. Torino: Bollati Boringieri.

Kafai, Y. B., & Resnick, M. (1996). *Constructionism in practice: Designing, thinking, and learning in a digital world.* Mahwah, NJ: Erlbaum.

Kaiser Family Foundation. (2013). *Generation M2: Media in the lives of 8-to-18-year-olds.* Retrieved from https://kaiserfamilyfoundation.files.wordpress.com/2013/04/8010.pdf

Kalil, T. (2013). Have fun-learn something, do something, make something. In M. Honey & D. E. Kanter (Eds.), Design, make, play: Growing the next generation of STEM innovators (pp. 12-16). New York, NY: Routledge.

Kalyuga, S., & Singh, A.-M. (2015). Rethinking the boundaries of cognitive load theory in complex learning. *Educational Psychology Review, 28*(4), 831–852. doi:10.100710648-015-9352-0

Kambil, A. (2008). What is your Web 5.0 strategy? *The Journal of Business Strategy, 29*(6), 56–58. doi:10.1108/02756660810917255

Kang, S., & Jung, J. (2014). Mobile communication for human needs: A comparison of smartphone use between the US and Korea. *Computers in Human Behavior, 35*, 376–387. doi:10.1016/j.chb.2014.03.024

Kaplan, J. (2016). *Artificial intelligence: What everyone needs to know*. New York: Oxford University Press.

Karr-Wisniewski, P., & Lu, Y. (2010). When more is too much: Operationalizing technology overload and exploring its impact on knowledge worker productivity. *Computers in Human Behavior, 26*(5), 1061–1072. doi:10.1016/j.chb.2010.03.008

Kats, Y. (2010). *Learning Management System Technologies and Software Solutions for Online Teaching: Tools and Applications: Tools and Applications*. IGI Global.

Kay, A. (2003). Interview with Alan Kay. *Computers in Entertainment, 1*(1), 8. doi:10.1145/950566.950580

Kaye, S., & Farrell, M. (2014). Disorders associated with excessive use of Internet, computers, smartphones and similar electronic devices. In *Proceedings from the First Act of the World Health Organization Congress: Public health implications of behavioral addictions associated with excessive use of Internet, computers,smart phones and similar electronic devices*. Geneva, Switzerland: WHO Document Production Services.

Kay, R., Benzimra, D., & Li, J. (2017, January 26). Exploring factors that influence technology-based distractions in bring your own device classrooms. *Journal of Educational Computing Research, 55*(7), 974–995. doi:10.1177/0735633117690004

Keats, D., & Schmidt, J. P. (2007). The genesis and emergence of Education 3.0 in higher education and its potential for Africa. *First Monday, 12*(3). doi:10.5210/fm.v12i3.1625

Keh-Wen, C., & Kuan-Chou, C. (2013). Designing service-learning: Project in systems analysis and design course. *Academy of Educational Leadership Journal, 17*(2), 47–60.

Keller, J. M. (2010). *Motivational design for learning and performance: The ARCS model approach*. New York, NY: Springer. doi:10.1007/978-1-4419-1250-3

Khalil, M., & Ebner, M. (2016). De-identification in learning analytics. *Journal of Learning Analytics, 3*(1), 129–138. doi:10.18608/jla.2016.31.8

Khosravi, P., Rezvani, A., & Wiewiora, A. (2016). The impact of technology on older adults' social isolation. *Computers in Human Behavior, 63*, 594–603. doi:10.1016/j.chb.2016.05.092

Khwaldeh, S. M., Al-Hadid, I., Masa'deh, R. e., & Alrowwad, A. (2017). The Association between E-Services Web Portals Information Quality and ICT Competence in the Jordanian Universities. *Asian Social Science, 13*(3), 156. doi:10.5539/ass.v13n3p156

Killian, J. (2004). Pedagogical experimentation: Combining traditional, distance, and service learning techniques. *Journal of Public Affairs Education, 10*(3), 209–224. doi:10.1080/15236803.2004.12001360

Kim, B., Park, H., & Baek, Y. (2009). Non solo per divertimento, ma strategie serie: utilizzo di strategie meta-cognitive nell'apprendimento basato sul gioco. *Computer e istruzione, 52,* 800-810.

Koh, K., Repenning, A., Nickerson, H., & Motter, P. (2013). Will it stick? Exploring the sustainability of computational thinking education through game design conference. *Proceeding of the 44th ACM technical symposium on computer science education*, 597-602. doi:10.1145/2445196.2445372

Kolb, D. A. (2014). *Experiential learning: Experience as the source of learning and development.* Pearson FT Press.

Korbach, A., Brünken, R., & Park, B. (2018). Differentiating different types of cognitive load: A comparison of different measures. *Educational Psychology Review, 30*(2), 503–529. doi:10.100710648-017-9404-8

Kordaki, M., & Kakavas, P. (2017). *Digital storytelling as an effective framework for the development of computational thinking skills.* Paper presented at the International Conference on Education and New Learning Technologies, Greece. 10.21125/edulearn.2017.2435

Korkmaz, O. (2016). The effect of scratch-and Lego Mindstorms Ev3-based programming activities on academic achievement, problem-solving skills and Logical Mathematical thinking skills of students. *Malaysian Online Journal of Educational Sciences, 4*(3).

Koster, R. (2004). *A theory of fun for game design* (1st ed.). Phoenix, AZ: Paraglyph Press.

Kraushaar, J. M., & Novak, D. C. (2010). Examining the effects of student multitasking with laptops during the lecture. *Journal of Information Systems Education, 21*(2), 241–251.

Kuhn, D., Cheney, R., & Weinstock, M. (2000). The development of epistemological understanding. *Cognitive Development, 15*(3), 309–328. doi:10.1016/S0885-2014(00)00030-7

Kuss, D. J., & Griffiths, M. D. (2017). Social networking sites and addiction: Ten lessons learned. *International Journal of Environmental Research and Public Health, 14*(3), E311. doi:10.3390/ijerph14030311 PMID:28304359

Kuznekoff, J. H., & Titsworth, S. (2013). The impact of mobile phone usage on student learning. *Communication Education, 62*(3), 233–252. doi:10.1080/03634523.2013.767917

Kwenani, D. F., & Yu, X. (2019). Maximizing international students' service-learning and community engagement experience: A case study of student voices on the benefits and barriers. *Journal of Higher Education Outreach & Engagement, 22,* 29–52. Retrieved from http://openjournals.libs.uga.edu/index.php/jheoe/article/view/2153

L'Amoreaux, C. (2019). *Who is Seymour Papert (and why does it matter)?* Retrieved June 7, 2019 from https://www.claudialamoreaux.com/who-is-seymour-papert-and-why-does-it-matter/

Lana, K., Johnson, R. E., & Barnes, C. M. (2014). Beginning the workday yet already depleted? Consequences of late-night smartphone use and sleep. *Organizational Behavior and Human Decision Processes, 124*(1), 11–23. doi:10.1016/j.obhdp.2014.01.001

Larrey, P. (2017). Would super-human machine intelligence really be super human? In G. Dodig-Crnkovic & R. Giovagnoi (Eds.), *Representation and Reality in Humans, Other Living Organisms and Intelligent Machines* (pp. 365–378). Cham, Switzerland: Springer. doi:10.1007/978-3-319-43784-2_19

Larson, P.J., & Maag, J. W. (1999). L'assessment funzionale dei comportamenti problema in classe. *Difficoltà di apprendimento, 4*(4), 539 – 556.

Larson, R. W., Richards, M. H., Sims, B., & Dworkin, J. (2001). How urban African American young adolescents spend their time: Time budgets for locations, activities, and companionship. *American Journal of Community Psychology, 29*(4), 565–597. doi:10.1023/A:1010422017731 PMID:11554153

Larson, R. W., & Verma, S. (1999). How children and adolescents spend time across the world: Work, play, and developmental opportunities. *Psychological Bulletin, 125*(6), 701–736. doi:10.1037/0033-2909.125.6.701 PMID:10589300

Larson, R., & Richards, M. (1998). Waiting for the Weekend: Friday and Saturday Nights as the Emotional Climax of the Week. *New Directions for Child and Adolescent Development, 82*(82), 37–51. doi:10.1002/cd.23219988204 PMID:12794950

Lauby, S. (2019). *Thinking about an encore career? Here's why you should.* Retrieved from http://worklife.coloniallife.com/2019/06/thinking-encore-career-heres/

Laurillard, D. (1993). *Rethinking University Teaching: Rethinking University Teaching: a Framework for the Effective Use of Educational Technology*. Routledge.

Lauzon, A. C. (1992). Integrating computer-based instruction with computer conferencing: An evaluation of a model for designing online education. *American Journal of Distance Education, 6*(2), 32–46. doi:10.1080/08923649209526784

Lawson, D., & Henderson, B. B. (2015). The costs of texting in the classroom. *College Teaching, 63*(3), 119–124. doi:10.1080/87567555.2015.1019826

Lazar, J., & Preece, J. (1999). Implementing service learning in an online communities course. In *14th Annual Conference International Academy for Information Management* (pp. 22–27). Charlotte, NC: Academic Press.

Le Breton, D. (2016). *Fuggire da sé. Una tentazione contemporanea*. Milano: Cortina Raffaello.

Lederman, D. (2018). Who is studying online (and where). *Inside Higher Ed.* Retrieved from https://www.insidehighered.com/digital-learning/article/2018/01/05/new-us-data-show-continued-growth-college-students-studying

Lee, A. R., Son, S. M., & Kim, K. K. (2016). Information and communication technology overload and social networking service fatigue: A stress perspective. *Computers in Human Behavior*, *55*, 51–61. doi:10.1016/j.chb.2015.08.011

Leedahl, S. N., Brasher, M. S., Estus, E., Breck, B. M., Dennis, C. B., & Clark, S. C. (2019). Implementing an interdisciplinary intergenerational program using the Cyber Seniors® reverse mentoring model within higher education. *Gerontology & Geriatrics Education*, *40*(1), 71–89. doi:10.1080/02701960.2018.1428574 PMID:29364782

Lee, I., Martin, F., & Apone, K. (2014). Integrating computational thinking across the K–8 curriculum. *ACM Inroads*, *5*(4), 64–71. doi:10.1145/2684721.2684736

Lee, R. M., & Robbins, S. B. (1998). The relationship between social connectedness and anxiety, self-esteem, and social identity. *Journal of Counseling Psychology*, *45*(3), 338–345. doi:10.1037/0022-0167.45.3.338

Lehman, R., & Conceição, S. (2010). *Creating a sense of presence in online teaching: How to "be there" for distance learners*. San Francisco, CA: John Wiley & Sons.

Lei, J. (2010). Quantity versus quality: A new approach to examine the relationship between technology use and student outcomes. *British Journal of Educational Technology*, *41*(3), 455–472. doi:10.1111/j.1467-8535.2009.00961.x

Lenhart, A., Ling, R., Campbell, S., & Purcell, K. (2010). *Chapter Four: How parents and schools regulate teens' mobile phones*. Retrieved from https://www.pewinternet.org/2010/04/20/chapter-four-how-parents-and-schools-regulate-teens-mobile-phones/

Leonardi, P. M. (2012). Materiality, Sociomateriality, and Socio-Technical Systems: What Do These Terms Mean? How are They Related? Do We Need Them? In P. M. Leonardi, B. A. Nardi, & J. Kallinikos (Eds.), *Materiality and Organizing: Social Interaction in a Technological World* (pp. 25–48). Oxford, UK: Oxford University Press. doi:10.1093/acprof:oso/9780199664054.003.0002

Leonard, J., Mitchell, M., Barnes-Johnson, J., Unertl, A., Outka-Hill, J., Robinson, R., & Hester-Croff, C. (2018). Preparing teachers to engage rural students in computational thinking through robotics, game design, and culturally responsive teaching. *Journal of Teacher Education*, *69*(4), 386–407. doi:10.1177/0022487117732317

Lepp, A., Barkley, J., & Karpinski, A. (2014). The relationship between cell phone use, academic performance, anxiety, and satisfaction with life in college students. *Computers in Human Behavior*, *31*, 343–350. doi:10.1016/j.chb.2013.10.049

Lewandowski, J. (2018). *Back to school without mobile phones: France imposes nationwide ban*. Retrieved from https://www.newsweek.com/back-school-without-mobile-phones-france-imposes-nationwide-ban-1104060

Lieberman, M. (2018). *I know how you felt this semester*. Retrieved from https://www.insidehighered.com/digital-learning/article/2018/02/20/sentiment-analysis-allows-instructors-shape-course-content

Lin, V. (2015). *Computational thinking and technology toys* (Honors Computer Science Thesis Collection). Available from Wellesley College. (307). Retrieved on June 28, 2019 from https://repository.wellesley.edu/thesiscollection/307

Lincoln, Y. S., & Guba, E. G. (1989). Ethics: The failure of positivist science. *The Review of Higher Education*, *12*(3), 221–240. doi:10.1353/rhe.1989.0017

Lindland, E., Volmert, A., Haydon, A., & Ford, A. (2014). *Talkin' bout my generation: Mapping the gaps between expert and public understandings of demographic change in the U.S.* Frameworks Institute Research Report. Retrieved from http://frameworksinstitute.org/pubs/mtg/demographicchange/index.html

Lipovetsky, G., & Charles, S. (2006). *Los tiempos hipermodernos*. Barcelona: Anagrama.

Littlefield, J. (2017, August 31). *Earn an Ivy League Degree Online*. Retrieved from https://www.thoughtco.com/earn-an-ivy-league-degree-online-1098183

Liu, S. (2019). *Number of digital voice assistants in use worldwide 2019-2023*. Retrieved from https://www.statista.com/statistics/973815/worldwide-digital-voice-assistant-in-use/

Liu, F., Ritzhaupt, A., Dawson, K., & Barron, A. (2017). Explaining technology integration in K-12 classrooms: A multilevel path analysis model. *Educational Technology Research and Development*, *65*(4), 795–813. doi:10.100711423-016-9487-9

LoBuono, D. L., Leedahl, S. N., & Maiocco, E. (2019). *Older adults learning technology in intergenerational program: Qualitative analysis of areas of technology requested for assistance*. Academic Press.

Lockyer, L., Heathcote, E., & Dawson, S. (2013). Informing Pedagogical Action: Aligning Learning Analytics with Learning Design. *The American Behavioral Scientist*, *57*(10), 1439–1459. doi:10.1177/0002764213479367

Loe, M. (2011). Doing it my way: Old women, technology and wellbeing. *Sociology of Health & Illness*, *32*(2), 319–334. doi:10.1111/j.1467-9566.2009.01220.x PMID:20149150

Logo Foundation. (n.d.). Retrieved June 7, 2019 from https://el.media.mit.edu/logo-foundation/

Long, P., & Siemens, G. (2011, September/October). Penetrating the fog: Analytics in learning and education. *EDUCAUSE Review*, 31–40. Retrieved from http://er.educause.edu/

Longstreet, P., & Brooks, S. (2017). Life satisfaction: A key to managing internet & social media addiction. *Technology in Society*, *50*, 73–77. doi:10.1016/j.techsoc.2017.05.003

Longworth, N. (2003). *El aprendizaje a lo largo de toda la vida*. Barcelona: Lifelong Learning.

Loomis, L. (2017). *Alexa told me where the bodies are buried*. Retrieved from https://thoughtcatalog.com/lane-loomis/2017/01/alexa-told-me-where-the-bodies-are-buried/

López-Vargas, O., Ibáñez-Ibáñez, J., & Racines-Prada, O. (2017). Students' metacognition and cognitive style and their effect on cognitive load and learning achievement. *Journal of Educational Technology & Society*, *20*(3), 145–157.

Loxterkamp, D. (2014). Life after work: A doctor's lonely hearts club. *BMJ: British Medical Journal*, *348*(7962), 26. doi:10.1136/bmj.g3715 PMID:24904030

Lu, D., & Duan, J. (2018, July). Systematic Thinking on University Service-learning. In *2018 3rd International Conference on Education, Sports, Arts and Management Engineering (ICESAME 2018)*. Atlantis Press. 10.2991/icesame-18.2018.62

Luhman, M., & Hawkley, L. (2016). Age differences in loneliness from late adolescence to oldest old age. *Developmental Psychology*, *50*(6), 943–959. doi:10.1037/dev0000117 PMID:27148782

Lum, C. M. K. (2006). *Perspectives on Culture, Technology and Communication: The Media Ecology Tradition*. Cresskill, NJ: Hampton Press.

Mack, E. (2017). *Elon Musk called out by creepy artificial intelligence robot*. Retrieved from https://www.cnet.com/news/elon-musk-artificial-intelligence-ai-robot-sophia-hanson-robotics/

Madsen, C. (2019). *Global privacy concerns: The world's top five cities using invasive technology*. Retrieved from https://www.webroot.com/blog/2019/05/29/global-privacy-concerns-the-worlds-top-five-cities-using-invasive-technology/

Magid, L. (2018). *The good and bad of location sharing*. Retrieved from https://www.connectsafely.org/the-good-and-bad-of-location-sharing/

Mahmood, T., & Shahzad, M. (2016). Usage patterns and impact of smartphone on social behaviour of youth. *Journal of the Research Society of Pakistan*, *53*(2).

Maier, C., Laumer, S., Eckhardt, A., & Weitzel, T. (2015). Giving too much social support: Social overload on social networking sites. *European Journal of Information Systems*, *24*(5), 447–464. doi:10.1057/ejis.2014.3

Malkin, N., Deatrick, J., Tong, A., Wijeskera, P., Egelman, S., & Wagner, D. (2019). Privacy attitudes of smart speaker users. *Proceedings on Privacy Enhancing Technologies*, *2019*(4), 250-271. doi:10.2478/popets-2019-0068

Malvey, D. M., Hamby, E. F., & Fottler, M. D. (2006). E-service learning: A pedagogic innovation for healthcare management education. *Journal of Health Administration Education*, *33*(2), 181-198.

Mannheim, I., Schwartz, E., Xi, W., Buttigieg, S. C., McDonnell-Naughton, M., Wouters, E. J. M., & van Zaalen, Y. (2019). Inclusion of older adults in the research and design of digital technology. *International Journal of Environmental Research and Public Health*, *16*(19), 1–17. doi:10.3390/ijerph16193718 PMID:31581632

Marchesini, R. (2009). *Il tramonto dell'uomo: la prospettiva post-umanista*. Bari: Dedalo.

Marks, J. (2012). *Who's Afraid of the Big Bad Disruption?* Retrieved from https://www.insidehighered.com/views/2012/10/05/why-moocs-wont-replace-traditional-i

Markus, G. B., Howard, J. P. F., & King, D. C. (1993). Integrating community service and classroom instruction enhances learning: Results from an experiment. *Educational Evaluation and Policy Analysis, 15*, 410–419.

Marr, B. (2016). 21 scary things big data knows about you. *Forbes.* Retrieved from https://www.forbes.com/sites/bernardmarr/2016/03/08/21-scary-things-big-data-knows-about-you/#7db6abaf6e7d

Marr, B. (2018). *What is big data?* Retrieved from https://www.bernardmarr.com/default.asp?contentID=766

Marr, B. (2019). 5 Internet of things trends everyone should know about. *Forbes.* Retrieved from https://www.forbes.com/sites/bernardmarr/2019/02/04/5-internet-of-things-trends-everyone-should-know-about/#45e991824b1f

Martin, N. (2019). The major concerns around facial recognition technology. *Forbes.* Retrieved from https://www.forbes.com/sites/nicolemartin1/2019/09/25/the-major-concerns-around-facial-recognition-technology/#2062df884fe3

Martin, S. A. (2016). *A descriptive and explorative case study of a scratch programming experience involving the creation of a lunar Simulation/Model with grade six learners.* Retrieved from https://prism.ucalgary.ca/bitstream/handle/11023/2851/ucalgary_2016_martin_stephen.pdf?sequence=7&isAllowed=y

Martinez, S. L., & Stager, G. (2013). *Invent to Learn: Making, tinkering, and engineering in the classroom.* Torrance, CA: Constructing Modern Knowledge Press.

Marzano, R. J. (2010). *The art and science of teaching: a comprehensive framework for effective instruction.* Alexandria, VA: Association for Supervision and Curriculum Development.

Marzilli, C., Delello, J. A., Marmion, S., McWhorter, R. R., Brown, P., & Marzilli, T. S. (2014). Faculty attitudes towards integrating technology and innovation. *International Journal on Integrating Technology in Education, 3*(1), 1–20. doi:10.5121/ijite.2014.3101

Masika, R., & Jones, J. (2016). Building student belonging and engagement: Insights into higher education students' experiences of participating and learning together. *Teaching in Higher Education, 21*(2), 138–150. doi:10.1080/13562517.2015.1122585

Mastel-Smith, B., Duke, G., & He, Z. (2018). A randomized controlled trial examining the effects of Tai Chi and electronic tablet use on Older Adults' cognition and health. *Journal of Holistic Nursing, 37*(2), 163–174. doi:10.1177/0898010118792961 PMID:30111233

Mayring, P. (2014). *Qualitative content analysis: theoretical foundation, basic procedures and software solution.* Retrieved from https://www.ssoar.info/ssoar/handle/document/39517

McCoy, B. R. (2016). Digital distractions in the classroom phase II: Student classroom use of digital devices for non-class related purposes. *Faculty Publications, College of Journalism & Mass Communications,* Paper 90. Retrieved from http://digitalcommons.unl.edu/ journalismfacpub/90

McFadden, C. (2018, September 24). *The origin of the term 'computer bug'*. Retrieved September 21, 2019 from https://interestingengineering.com/the-origin-of-the-term-computer-bug

McIntosh, J., & Torres, M. I. (2014). *Learning Management Systems: Revisiting Learning Management Systems at the University Level.* The University of Tampa.

McLaughlin, E. (2010). The" real-world" experience: Students' perspectives on service-learning projects. *American Journal of Business Education, 3*(7), 109–118. doi:10.19030/ajbe.v3i7.463

McLean, G., & Osei-Frimpong, K. (2019). Hey Alexa…examine the variables influencing the use of artificial intelligent in-home voice assistants. *Computers in Human Behavior, 99*, 28–37. doi:10.1016/j.chb.2019.05.009

McLeod, S. A. (2018, June 6). *Jean Piaget's theory of cognitive development.* Retrieved October 9, 2019 from https://www.simplypsychology.org/piaget.html

McWhorter, R. R. (2014b). *What's the 'Internet of Everything'?* Retrieved from https://virtualhrd.wordpress.com/2014/04/12/whats-the-internet-of-everything/

McWhorter, R., Delello, J. A., & Roberts, P. B. (2016). *Giving back: Exploring service-learning in an online learning environment.* Academic Press.

McWhorter, R. R. (2014a). A synthesis of new perspectives on Virtual HRD. *Advances in Developing Human Resources, 16*(3), 391–401. doi:10.1177/1523422314532126

McWhorter, R. R., & Delello, J. A. (2015). Green computing through virtual learning environments. In F. Nafukho (Ed.), *Integrating technology into higher education and in the workplace.* Hershey, PA: IGI Global. doi:10.4018/978-1-4666-8170-5.ch001

McWhorter, R. R., & Delello, J. A. (2015). The use of iPads to reduce loneliness and foster a sense of community in older adults. *Mixed Methods International Research Association Conference*, San Antonio, TX.

McWhorter, R. R., & Ellinger, E. E. (2018). Qualitative case study research: An initial primer. In V. C. X. Wang (Ed.), *Handbook of Research on Innovative Techniques, Trends, and Analysis for Optimized Research Methods*. Hershey, PA: IGI Global. doi:10.4018/978-1-5225-5164-5.ch012

McWhorter, R. R., Lynham, S. A., & Porter, D. E. (2008). Scenario planning as the development of leadership capability and capacity. *Advances in Developing Human Resources, 10*(2), 259–284. doi:10.1177/1523422307313332

Merleau-Ponty, M. (1945). *Phénoménologie de la perception.* Milano: Bompiani 2003.

Merleau–Ponty, M. (1942). La struttura del comportamento. *Mimesis.*

Merriam, S. B., & Tisdell, E. J. (2016). Designing your study and selecting a sample. *Qualitative research: A guide to design and implementation*, 73-104.

Merriam, S. B. (1998). *Qualitative research and case study applications in education*. San Francisco: Jossey-Bass.

Merriam, S. B. (1998). *Qualitative research: A guide to design and implementation*. Hoboken, NJ: John Wiley & Sons.

Merriam, S. B., & Bierema, L. L. (2013). *Adult learning: Linking theory and practice*. San Francisco, CA: Jossey-Bass Higher and Adult Education.

Meshi, D., Cotton, S. R., & Bender, A. R. (2019). Problematic social media use and perceived social isolation in older adults: A cross-sectional study. *Gerontology*, 1–9. doi:10.1159/000502577 PMID:31522178

Mezirow, J. (1981). A critical theory of adult learning and education. *Adult Education, 32*(1), 3–24. doi:10.1177/074171368103200101

Micera, S., Bonato, P., & Tamura, T. (2008). Gerontechnology. *IEEE Engineering in Medicine and Biology Magazine: The Quarterly Magazine of the Engineering in Medicine & Biology Society*, *27*(4), 10–14. doi:10.1109/MEMB.2008.925213 PMID:18650144

Mills, J., Harrison, H., Franklin, R., & Birks, M. (2017). *Case study research: Foundations and methodological orientations.* Paper presented at the Forum Qualitative Sozialforschung/Forum: Qualitative Social Research.

Mills, S. (2001). Electronic journaling: Using the web-based, group journal for service-learning reflection. *Michigan Journal of Community Service Learning*, *8*(1), 27–35.

Milton, L. E., & Otty, R. (2018). Innovations in occupational therapy education: The centralized service-learning model. *Journal of Occupational Therapy Education*, *2*(1), 8. doi:10.26681/jote.2018.020108

Minich, E. (1996). Using student feedback to improve distance education. Florida Community College, Jacksonville, FL. (ERIC No. ED 397893)

Mishra, M. (2017). Impact of smartphone on college students. *International Journal of Education and Management Studies*, *7*(3), 310–313.

Mishra, P., & Koehler, M. J. (2006, June). Technological pedagogical content knowledge: A framework for teacher knowledge. *Teachers College Record*, *108*(6), 1017–1054. doi:10.1111/j.1467-9620.2006.00684.x

Mitzner, T., Rogers, W., Fisk, A., Boot, W., Charness, N., Czaja, S., & Sharit, J. (2016). Predicting older adults' perceptions about a computer system designed for seniors. *Universal Access in the Information Society*, *15*(2), 271–280. doi:10.100710209-014-0383-y PMID:31186624

Mkhize, P., Mtsweni, E. S., & Buthelezi, P. (2016). Diffusion of Innovations Approach to the Evaluation of Learning Management System Usage in an Open Distance Learning Institution. *The International Review of Research in Open and Distributed Learning*, *17*(3). doi:10.19173/irrodl.v17i3.2191

Mokhtari, K., Delello, J. A., & Reichard, C. (2015). Millennial multitasking: Constantly Connected yet Distracted. *Journal of College Reading and Learning*, *45*(2), 164–180. doi:10.1080/10790195.2015.1021880

Moody, J. (2004). Distance education: Why are the attrition rates so high? *The Quarterly Review of Distance Education*, *5*(3), 205–210.

Moore, D. R., Cheng, M. I., & Dainty, A. R. J. (2002). Competence, competency and competencies. Performance assessment in organizations. *Work Study*, *6*(6), 314–319. doi:10.1108/00438020210441876

Moreau, E. (2019). *The 3 best location sharing apps*. Retrieved from https://www.lifewire.com/top-best-location-sharing-apps-3485922

Morgan, D. L. (1997). *Qualitative research methods: Focus groups as qualitative research*. Thousand Oaks, CA: SAGE Publications, Inc.; doi:10.4135/9781412984287

Morgan, D. L. (1999). *The focus group guidebook*. Thousand Oaks, CA: SAGE.

Morris, M., Adair, B., Ozanne, E., Kurowski, W., Miller, K., Pearce, A., ... Said, C. (2014). Smart technologies to enhance social connectedness in older people who live at home. *Australasian Journal on Ageing*, *33*(3), 142–152. doi:10.1111/ajag.12154 PMID:24730370

Muchado, L. R., Silva da Silva Mendes, J., Sampaio, D. C. F., Grande, T. P. F., & Behar, P. A. (2017). M-learning and the elderly: Construction of inclusive pedagogies. In V. L. Uskov, R. J. Howlett, & L. C. Jain (Eds.), *Smart Education and e-Learning* (pp. 391–399). Cham, Switzerland: Springer International Publishing.

Mupinga, D. M. (2017). School-wide and classroom policies on the use of mobile technologies: An exploratory study. *Journal of Technology Studies*, *43*(2), 70–79. doi:10.21061/jots.v43i2.a.2

Murnane, K. (2017). *Creepy privacy invasions using Macs, Roombas and RFID chips*. Retrieved from https://www.forbes.com/sites/kevinmurnane/2017/07/27/3-creepy-ways-technology-is-being-used-to-invade-your-privacy/#691911b53858

Mutlag, A. A., Ghani, M. K. A., Arunkumar, N., Mohamed, M. A., & Mohd, O. (2019). Enabling technologies for fog computing in healthcare IoT systems. *Future Generation Computer Systems*, *90*, 62–78. doi:10.1016/j.future.2018.07.049

Nash, R. (2005, Winter). Course completion rates among distance learners: Identifying possible methods to improve retention. *Online Journal of Distance Learning Administration*, *8*(4). Retrieved from http://www.westga.edu/%7Edistance/ojdla/winter84

National Digital Inclusion Alliance. (2019). *Definitions*. Retrieved from https://www.digitalinclusion.org/definitions/

National Research Council. (2011). *Successful K-12 STEM education: Identifying effective approaches in science, technology, engineering, and mathematics*. Washington, DC: National Academies Press.

Navarro, F. (2018). *5 smartphone apps that might be spying on you*. Retrieved from https://www.komando.com/smartphones-gadgets/5-smartphone-apps-that-might-be-spying-on-you/47578/

Neeves, B. B., Franz, R., Judges, R., Beermann, C., & Baecker, R. (2019). Can digital technology enhance social connectedness among older adults? A feasibility study. *Journal of Applied Gerontology*, *38*(1), 49–72. doi:10.1177/0733464817741369 PMID:29166818

Neulight, N., Kafai, Y. B., Kao, L., Foley, B., & Galas, C. (2007). Children's participation in a virtual epidemic in the science classroom: Making connections to natural infectious diseases. *Journal of Science Education and Technology*, *16*(1), 47–58. doi:10.100710956-006-9029-z

Neville, S., Adams, J., Montayre, J., Larmer, P., Garrett, N., Stephens, C., & Alpass, F. (2018). Loneliness in men 60 years and over: The association with purpose in life. *American Journal of Men's Health*, *12*(4), 730–739. doi:10.1177/1557988318758807 PMID:29458295

Ngai, G., Chan, S. C., & Kwan, K. P. (2018). Challenge, Meaning, Interest, and Preparation: Critical Success Factors Influencing Student Learning Outcomes from Service- Learning. *Journal of Higher Education Outreach & Engagement*, *22*(4), 55–80.

Norman, D. A., & Spohrer, J. C. (1996). Learner-centered education. *Communications of the ACM*, *39*(4), 24–27. doi:10.1145/227210.227215

Novotney, A. (2019). Social isolation: It can kill you. *American Psychological Association Monitor*, *50*(5). Retrieved from https://www.apa.org/monitor/2019/05/ce-corner-isolation

Nystrom, C. (1973). *Towards a Science of Media Ecology* (PhD dissertation). New York: New York University.

Offer, S. (2013). Family time activities and adolescents' emotional well-being. *Journal of Marriage and the Family*, *75*(1), 26–41. doi:10.1111/j.1741-3737.2012.01025.x

Ong, T. (2018, January 9). Apple says it will introduce new features to help parents protect children. *The Wall Street Journal*. Retrieved from https://www.theverge.com/2018/1/9/16867330/apple-response-smartphone-addiction-youth

Ophir, E., Nass, C., & Wagner, A. D. (2009). Cognitive control in media multitaskers. *Proceedings of the National Academy of Sciences of the United States of America*, *106*(37), 15583–15587. doi:10.1073/pnas.0903620106 PMID:19706386

Organizzazione Mondiale della Sanità. (2001). *ICF Classificazione Internazionale del Funzionamento, della Disabilità e della Salute*. Trento: Erikson2004.

Orlikowski, W. J. (1992). The duality of technology: Rethinking the concept of technology in organizations. *Organization Science*, *3*(3), 398–427. doi:10.1287/orsc.3.3.398

Ortiz, L. (2016). *Configuración Sociosemiótica del chat: Una visión multimodal* (Unpublished doctoral dissertation). UPTC.

Oulasvirta, A., Rattenbury, T., Ma, L., & Raita, E. (2015). Habits make smartphone use more pervasive. *Personal and Ubiquitous Computing*, *16*(1), 105–114. doi:10.100700779-011-0412-2

Pahomov, L. (2015). *In pursuit of a cell phone policy*. ASCD. Retrieved from http://www.ascd.org/publications/educational-leadership/may15/vol72/num08/In-Pursuit-of-a-Cell-Phone-Policy.aspx

Pak, C. S. (2018). Linking service-learning with sense of belonging: A culturally relevant pedagogy for heritage students of Spanish. *Journal of Hispanic Higher Education*, *17*(1), 76–95. doi:10.1177/1538192716630028

Palladino, V. (2018). *Unprompted, creepy laughter from Alexa is freaking out Echo users*. Retrieved from https://arstechnica.com/gadgets/2018/03/unprompted-creepy-laughing-from-alexa-is-freaking-out-echo-users/?comments=1&start=200

Pan, M. L., Lawrence, L., Creswell, J. W., Hacker, D., Merriam, S., Cone, J. D., & Creswell, J. W. (2007). *Preparing literature reviews*. Los Angeles, CA: Pyrczak Publishing.

Papamitsiou, Z., & Economides, A. A. (2014). Learning analytics and educational data mining in practice: A systematic literature review of empirical evidence. *Journal of Educational Technology & Society*, *17*(4), 49.

Papert, S. (1993a, February 1). Obsolete skill set: The 3Rs--Literacy and literacy in the media ages. *Wired Magazine*. Retrieved from http://www.wired.com/1993/02/1-2-papert/

Papert, S. (1980). *Mindstorms: Children, computers, and powerful ideas* (1st ed.). New York: Basic Books.

Papert, S. (1993b). *The children's machines: Rethinking schools in the age of computers*. New York, NY: Basic Books.

Papert, S. (1996). An exploration in the space of mathematics educations. *International Journal of Computers for Mathematical Learning*, 95–123.

Papert, S. (2000). What's the big idea? Toward a pedagogy of idea power. *IBM Systems Journal*, *29*(3 & 4), 720–729. doi:10.1147j.393.0720

Parag, K. (2016). *Connectography: Le mappe del futuro ordine mondiale*. Roma: Fazi.

Partnership for 21st Century Learning (P21). (2019). *Framework for 21st century learner*. Retrieved from http://www.p21.org/

Pashler, H. (2000). Task switching and multitask performance. In S. Monsell & J. Driver (Eds.), *Control of Cognitive Processes: Attention and Performance XVIII* (pp. 275–423). Cambridge, MA: The MIT Press.

Pate, A. (2014). *Social isolation: Its impact on the mental health and wellbeing of older Victorians.* Melbourne: COTA Victoria. Retrieved from http://www.cotavic.org.au/wp-content/uploads/2014/02/Working-Paper_Social-Isolation.pdf

Patton, M. Q. (2002). Qualitative interviewing. *Qualitative Research and Evaluation Methods, 3*(1), 344–347.

Patton, M. Q. (2015). *Qualitative research & evaluation methods: Integrating theory and practice.* Los Angeles, CA: SAGE.

Paulré, B. (2000). De la New economy au capitalisme cognitif. *Multitudes, 2*(2), 25–42. doi:10.3917/mult.002.0025

Pearson. (2015). *Pearson student mobile device survey 2015 national report: Students in grades 4-12.* Retrieved from https://www.pearsoned.com/wp-content/uploads/2015-Pearson-Student-Mobile-Device-Survey-Grades-4-12.pdf

Pearson.com. (2015). *Pearson student mobile device survey national report: College students.* Retrieved from https://www.pearsoned.com/wp-content/uploads/2015-Pearson-Student-Mobile-Device-Survey-College.pdf

Perez, S. (2000). *Assessing service learning using pragmatic principles of education: A Texas charter school case study.* Academic Press.

Perkins-Eastman. (2019). *The state of senior living: An industry grappling with autonomy.* Retrieved from http://perkinseastman.com/white_papers

Perla, L. (2010). *La didattica dell'implicito. Ciò che l'insegnante non sa.* Brescia: La Scuola.

Petersen, J., Austin, D., Mattek, N., & Kaye, J. (2015). Time out-of-home and cognitive, physical, and emotional wellbeing of older adults: A longitudinal mixed effects model. *PLoS One, 10*(10), 1–16. doi:10.1371/journal.pone.0139643 PMID:26437228

Petrich, M., Wilkinson, K., & Bevan, B. (2013). It looks like fun, but are they learning? In M. Honey & D. Kanter (Eds.), *Design, make, play: Growing the next generation of STEM innovators* (pp. 50–70). New York, NY: Routledge.

Pew Research Center. (2015). *The smartphone difference.* Retrieved from http:// www.pewinternet.org/2015/04/01/us-smartphone-use-in-2015/

Pham, C. M. (2019). Building a maturity framework for big data cybersecurity analytics. In S. Miah & W. Yeoh (Eds.), *Applying Business Intelligence Initiatives in Healthcare and Organizational Settings* (pp. 164–183). Hershey, PA: IGI Global Inc. doi:10.4018/978-1-5225-5718-0.ch009

Phillipo, J., & Krongard, S. (2012). *Learning Management System.* LMS.

Piaget, J. (1954). *The construction of reality in the child.* New York, NY: Basic Books. doi:10.1037/11168-000

Poon, P., Chan, T. S., & Zhou, L. (2011). Implementation of service-learning in business education: Issues and challenges. *Journal of Teaching in International Business*, *22*(3), 185–192. doi:10.1080/08975930.2011.653746

Popham, W. J. (2008). *Transformative assessment*. Alexandria, VA: Association for Supervision and Curriculum Development.

Postman, N. (1968). Ecologia dei media. L'insegnamento come attività conservatrice. Roma: Armando, 1983.

Postman, N. (1970). The Reformed English Curriculum. In High School 1980. The Shape of Future in American Secondary Education (pp. 160-168). New York: Pitman.

Poushter, P., Bishop, C., & Chwe, H. (2018). *Global Attitudes and Trends; 2018: Social media use continues to rise in developing countries but plateaus across developed ones*. Retrieved from https://www.pewresearch.org/global/2018/06/19/social-media-use-continues-to-rise-in-developing-countries-but-plateaus-across-developed-ones/#table

PrivacyPolicies.com (2019). *Privacy policies are legally required*. Retrieved from privacypolicies.com/blog/privacy-policies-legally-required/

Purshouse, J., & Campbell, L. (2019). Privacy, crime control and police use of automated facial recognition technology. *Criminal Law Review (London, England)*, *3*, 188–204.

Qian, Y., & Li, L. (2017). Student off-task electronic multitasking predictors: Scale development and validation. *The Journal of Scholarship of Teaching and Learning*, *17*(2), 53–73. doi:10.14434/josotl.v17i2.20682

Radu, I. (2017). Emergent technologies for learning analytics. *ELearning & Software for Education*, 2509-524.

Rama, D. V., Ravencroft, S., Wolcott, S. K., & Zlotkowski, E. (2000). Service- Learning Outcomes: Guidelines for Educators and Researchers. *Issues in Accounting Education*, *15*(4), 656–689.

Rank, U. U. (2019). *Jordanian University Ranking 2019*. Retrieved from https://www.4icu.org/jo/

Reader, R. (2019). *Nine technologies creeping us out in 2019*. Retrieved from https://www.fastcompany.com/90285664/technology-creeping-us-out-in-2019

Reidenberg, J. R. (2014). Privacy in public. *University of Miami Law Review*, *69*, 141-159. Retrieved from Available at: http://repository.law.miami.edu/umlr/vol69/iss1/6

Reigeluth, C. M., & Carr-Chellman, A. (1999). *Instructional-design theories and models. A new paradigm of instructional theory*. Academic Press.

Reiser, R. A., & Dempsey, J. Y. (2012). *Trends and issues in instructional design and technology* (3rd ed.). Saddlebrook, NJ: Pearson.

Rekart, J. L. (2011). Taking on multitasking. *Phi Delta Kappan*, *93*(4), 60–63. doi:10.1177/003172171109300415

Report, O. C. (2019). *The Best Resource for Open Source, Paid, and Career-centered Online Education.* Retrieved from http://www.onlinecoursereport.com/

Resende, I. C. (2018). *An action research study of teachers' use of learning analytics as a formative practice* (Doctoral Dissertation). East Stroudsburg University.

Resnick, M. (2018). *Computational Fluency – Mitchel Resnick – Medium.* Retrieved from https://medium.com/@mres/computational-fluency-776143c8d725

Resnick, M., & Robinson, K. (2018). *Lifelong kindergarten: Cultivating creativity through projects, passion, peers, and play.* Cambridge, MA: The MIT Press.

Retailwire.com. (2019). *What makes voice assistants creepy?* Retrieved from https://retailwire.com/discussion/what-makes-voice-assistants-creepy/

Reyes, J. A. (2015). The Skinny on Big Data in Education: Learning Analytics Simplified. *TechTrends, 59*(2), 75–80. doi:10.100711528-015-0842-1

Reyns, B. W. (2019). Online pursuit in the Twighlight Zone: Cyberstalking perpetration by college students. *Victims & Offenders, 14*(2), 183–198. doi:10.1080/15564886.2018.1557092

Richmond County School System. (2018). *BYOT (Bring your own technology).* Retrieved from https://www.rcboe.org/domain/7853

Ridder, H.-G. (2017). The theory contribution of case study research designs. *Business Research, 10*(2), 281–305. doi:10.100740685-017-0045-z

Rideout, V. J., Foehr, U. G., & Roberts, D. F. (2010). *Generation M2: Media in the Lives of 8-to 18-Year-Olds.* Menlo Park, CA: Henry J. Kaiser Family Foundation.

Rideout, V., & Robb, M. B. (2018). *Social media, social life: Teens reveal their experiences.* San Francisco, CA: Common Sense Media.

Rincón, B., & Castillo-Montoya, M. (2018). Learning Assessment in Student Affairs through Service-Learning. *Research & Practice in Assessment, 13*, 39–50.

Roche, R. (1995). *La condotta pro-sociale. Terapia del comportamento.* Roma: Bulzoni.

Rose, E. (2010). Continuous partial attention: Reconsidering the role of online learning in the age of interruption. *Educational Technology, 50*(4), 41–46.

Rosen, L. D., Cheever, N. A., & Carrier, L. M. (2012). iDisorder: Understanding our obsession with technology and overcoming its hold on us. New York, NY: Palgrave Macmillan

Rosen, L. D., Carrier, M. L., & Cheever, N. A. (2013). Facebook and texting made me do it: Media-induced task-switching while studying. *Computers in Human Behavior, 29*(3), 948–958. doi:10.1016/j.chb.2012.12.001

Rosen, L. D., Whaling, K., Rab, S., Carrier, L. M., & Cheever, N. A. (2013). Is Facebook creating "iDisorders"? The link between clinical symptoms of psychiatric disorders and technology use, attitudes and anxiety. *Computers in Human Behavior, 29*(3), 1243–1254. doi:10.1016/j.chb.2012.11.012

Rovai, A. (2002). Building sense of community at a distance. *International Review of Research in Open and Distance Learning, 4*(1), 1–9.

Rowley, J. (2002). Using case studies in research. *Management Research News, 25*(1), 16–27. doi:10.1108/01409170210782990

Russell, D. (1996). UCLA Loneliness Scale (Version 3): Reliability, validity, and factor structure. *Journal of Personality Assessment, 66*(1), 20–40. doi:10.120715327752jpa6601_2 PMID:8576833

Ruston, D. (2017). *Screenagers' national survey reveals that the majority of U.S. middle schools allow students to carry cell phones throughout the school day.* Screenagers: Away for the Day. Retrieved from https://awayfortheday.squarespace.com/survey

Ryan. (2018, July 31). Tinkering about thinking about Seymour [Blog post]. Retrieved from https://www.exploratorium.edu/tinkering/blog/2017/02/03/tinkering-about-thinking-about-seymour

Saadé, R., & Bahli, B. (2005). The impact of cognitive absorption on perceived usefulness and perceived ease of use in on-line learning: An extension of the technology acceptance model. *Information & Management, 42*(2), 317–327. doi:10.1016/j.im.2003.12.013

Salam, M., Iskandar, D. N. A., Ibrahim, D. H. A., & Farooq, M. S. (2019). Service learning in higher education: A systematic literature review. *Asia Pacific Education Review*, 1–21.

Sale, P. (2018). Gerontechnology, domotics, and robotics. In S. Masiero & U. Carraro (Eds.), *Rehabilitation medicine for elderly patients. Practical Issues in Geriatrics.* Cham, Switzerland: Springer. doi:10.1007/978-3-319-57406-6_19

Samak, A., & Tawfik, Z. (2017). *Exploration of Jordanian English Language Teachers' Attitudes.* Skills, and Access as Indicator of Information and Communication Technology Integration in Jordan.

Sánchez-Martínez, M., & Otero, A. (2009). Factors associated with cell phone use in adolescents in the community of Madrid (Spain). *Cyberpsychology & Behavior, 12*(2), 131–137. doi:10.1089/cpb.2008.0164 PMID:19072078

Sánchez-Valle, M., Vinarás Abad, M., & Llorente-Barroso, C. (2017). Empowering the elderly and promoting active ageing through the Internet: The Benefit of e-inclusion Programmes. In Safe at Home with Assistive Technology (pp. 95-108). Cham, Switzerland: Springer. doi:10.1007/978-3-319-42890-1_7

Satici, S. A., Uysal, R., & Deniz, M. E. (2016). Linking social connectedness to loneliness: The mediating role of subjective happiness. *Personality and Individual Differences, 97*, 306–310. doi:10.1016/j.paid.2015.11.035

Sax, L. J., & Astin, A. W. (1997). The benefits of service: Evidence from undergraduates. *The Educational Record, 78*(3/4), 25–32.

Scaradozzi, D., Sorbi, L., Pedale, A., Valzano, M., & Vergine, C. (2015). Teaching robotics at the primary school: An innovative approach. *Procedia: Social and Behavioral Sciences, 174*, 3838–3846. doi:10.1016/j.sbspro.2015.01.1122

Schein, E. H. (2010). *Organizational cultural and leadership* (4th ed.). San Francisco, CA: John Wiley & Sons, Inc.

Schlagwein, D., & Hu, M. (2017). How and why organisations use social media: Five use types and their relation to absorptive capacity. *Journal of Information Technology, 32*(2), 194–209. doi:10.1057/jit.2016.7

Schneider, M. (2010). *Finishing the first lap: The cost of first year student attrition in America's four year colleges and universities*. American Institutes for Research. Retrieved from https://www.air.org/sites/default/files/AIR_Schneider_Finishing_the_First_Lap_Oct101.pdf

Schuelke-Leech, B.-A. (2018). A model for understanding the orders of magnitude of disruptive technologies. *Technological Forecasting and Social Change, 129*, 261–274. doi:10.1016/j.techfore.2017.09.033

Selingo, J. J. (2014). *MOOC U: Who is getting the most out of online education and why*. New York, NY: Simon and Schuster.

Selingo, J. J. (2014a). *Demystifying the MOOC*. Retrieved from https://www.nytimes.com/2014/11/02/education/edlife/demystifying-the-mooc.html

Senior Health, N. I. H. (2013). *Helping older adults search for health information online: A toolkit for trainers*. Retrieved from http://nihseniorhealth.gov/toolkit/toolkit.html

Shah, D. (2016). *By the numbers: MOOCs in 2016*. Retrieved from https://www.class-central.com/report/mooc-stats-2016/

Shochat, T. (2012). Impact of lifestyle and technology developments on sleep. *Nature and Science of Sleep, 4*, 19–31. doi:10.2147/NSS.S18891 PMID:23616726

Shute, V., Sun, C., & Asbell-Clarke, J. (2017). Demystifying computational thinking. *Educational Research Review, 22*, 1–18. doi:10.1016/j.edurev.2017.09.003

Sibilio, M. (2014). *La didattica semplessa*. Napoli: Liguori.

Sigmon, R. L. (1975). *Service-learning: A new path for the transition from adolescence to adulthood*. Western Carolina University Journal of Education.

Sigmon, R. L. (1979). Service-learning: Three principles. *Synergist, 8*(1), 9–11.

Simon, M., & Brown, G. (2013). Raft: A maker palace for educators. In M. Honey & D. E. Kanter (Eds.), Design, make, play: Growing the next generation of STEM innovators (pp. 138-162). New York, NY: Routledge.

Simondon, G. (2001). *El modo de existencia de los objetos técnicos*. Buenos Aires: Prometeo libros.

Simondon, G. (2008). *Imagination et Invention (1965-1966)*. Chatou: Éditions de La Transparence

Smith, A. (2015). U.S. smartphone use in 2015. *Pew Research Center's Internet & American Life Project*. Retrieved from http://www.pewinternet.org/2015/04/01/us-smartphone-use-in-2015/

Smith, J., Fisher, G., Ryan, L., Clarke, P., House, J., & Weir, D. (2013). Psychosocial and lifestyle questionnaire (2006-2010): Documentation Report, Core Section LB. *Survey Research Center, Institute for Social Research*. Retrieved from http://hrsonline.isr.umich.edu/sitedocs/userg/HRS2006-2010SAQdoc.pdf

Smith, B. L., MacGregor, J., Mathews, R., & Gabelnick, F. (2004). *Learning Communities: Reforming Undergraduate Education*. Jossey-Bass.

Smolensky, P. (1988). On the Proper Treatment of Connectionism. *Behavioral and Brain Sciences*, *11*(1), 1–61. doi:10.1017/S0140525X00052432

Snyder, K. A. (2007). A vocabulary of motives: Understanding how parents define quality time. *Journal of Marriage and the Family*, *69*(2), 320–340. doi:10.1111/j.1741-3737.2007.00368.x

Sokolova, S. (2014, December 28). *What are the advantages of MOOCs and how can you benefit from them?* Retrieved from https://www.linkedin.com/pulse/what-advantages-moocs-benefits-from-them-siyana-sokolova

Solon, O. (2016). *The rise of the robots: Forget evil AI – the real risk is far more insidious*. Retrieved from https://www.theguardian.com/technology/2016/aug/30/rise-of-robots-evil-artificial-intelligence-uc-berkeley

Sorzio, P. (2005). *La ricerca qualitativa in educazione*. Roma: Carocci.

Spiezio, C. (2019). Building a data privacy program in-house. BNY Mellon's Anju Khurana weighs in. *Law.com*. Retrieved from https://www.law.com/corpcounsel/2019/08/15/building-a-data-privacy-program-in-house-bny-mellons-anju-khurana-weighs-in/

Spoonauer, M. (2016). *This PC's pop-up privacy cam is brilliant*. Retrieved from https://www.tomsguide.com/us/hp-all-in-one-privacycam,news-22618.html

Spreng, R.N., McKinnon, M.C., Mar, R.A., & Levine, B.J. (2009). The Toronto Empathy Questionnaire. Scale development and initial validation of a factor-analytic solution to multiple empathy measures. *NCBI Resources, 91*(1), 62.

St. John, A. (2018). How Facebook tracks you, even when you are not on Facebook. *Consumer Reports*. Retrieved from https://www.consumerreports.org/privacy/how-facebook-tracks-you-even-when-youre-not-on-facebook/

Stafford, P. (2019). *The dangers of in-game data collection*. Retrieved from https://www.polygon.com/features/2019/5/9/18522937/video-game-privacy-player-data-collection

Stake, R. E. (1995). *The art of case study research*. Thousand Oaks, CA: Sage Publications.

Statista. (2019). *Number of mobile phone users worldwide from 2015 to 2020 (in billions).* Retrieved from https://www.google.com/url?q=https://www.statista.com/statistics/274774/forecast-of-mobile-phone-users-worldwide/&sa=D&ust=1559138725843000&usg=AFQjCNEnu_FZ6yA9rbWeaZeuqrg5GOrIQQ

Statista.com. (2019). *Number of smartphone users worldwide from 2014 to 2020 (in billions).* Retrieved from: https://www.statista.com/statistics/330695/number-of-smartphoneusers-worldwide

Steers, R. M., Meyer, A. D., & Sanchez-Runde, C. J. (2008). National culture and the adoption of new technologies. *Journal of World Business*, *43*(3), 255–260. doi:10.1016/j.jwb.2008.03.007

Steptoe, A., Shankar, A., Demakakos, P., & Wardle, J. (2013). Social isolation, loneliness, and all-cause mortality in older men and women. *Proceedings of the National Academy of Sciences of the United States of America*, *110*(15), 5797–5801. doi:10.1073/pnas.1219686110 PMID:23530191

Stern, J. (2019). You should definitely track your loved ones' phones. Actually maybe not. *Wall Street Journal.* Retrieved from https://www.wsj.com/articles/you-should-definitely-track-your-loved-ones-phones-actually-maybe-not-11565883576

Stillman, J. (2018). 7 creepy things robots have already done that shocked their makers. *Inc.* Retrieved from https://www.inc.com/jessica-stillman/7-creepy-things-robots-have-already-done-that-shocked-their-makers.html

Stoecker, R., Hilgendorf, A., & Tryon, E. (2008). Incorporating technology in service-learning: A case study of appropriate use. In A. Dailey-Hebert, E. Donnelli-Sallee, & L. DiPadova-Stocks (Eds.), *Service-eLearning: Educating for citizenship*, (pp. 31-44). Charlotte, NC: Information Age Publishing.

Stokes, T. F., & Baer, D. M. (1977). On Implicit Technology of Generalization. *Journal of Applied Behavior Analysis*, *2*(10), 349–367. doi:10.1901/jaba.1977.10-349 PMID:16795561

Stone, C., O'Shea, S., May, J., Delahunty, J., & Partington, Z. (2016). Opportunity through online learning: Experiences of first-in-family students in online open-entry higher education. Academic Press.

Stone, Z. (2017). Everything you need to know about Sophia, the world's first robot citizen. *Forbes.* Retrieved from https://www.forbes.com/sites/zarastone/2017/11/07/everything-you-need-to-know-about-sophia-the-worlds-first-robot-citizen/#5590748646fa

Strait, J., & Sauer, T. (2004). *Constructing experiential learning for online courses: The birth of e-service.* Retrieved from http://er.educause.edu/articles/2004/1/constructing-experiential-learning-for-

Strait, J., & Sauer, T. (2004). Constructing experiential learning for online courses: The birth of e-service. *EDUCAUSE Quarterly*, *1*, 62–65.

Strate, L. (1999). Understanding MEA. *Medias Res*, *1*(1), 1–9.

Strauss, V. A. (2018). Schools are banning smartphones. Here's an argument for why they shouldn't--and what they should do instead. *The Washington Post*. Retrieved from https://www.washingtonpost.com/education/2018/09/21/schools-are-banning-smartphones-heres-an-argument-why-they-shouldnt-what-they-should-do-instead/?noredirect=on&utm_term=.322c4563af0e

Street, B. V., Bloome, D., & Pahl, K. (2012). Lettera: Apprendere l'empowerment attraverso la formazione nella ricerca in stile etnografico. In M. Grenfell, C. Hardy, & J. Rowsell (Eds.), Lingua, etnografia e istruzione: colmare nuovi studi di alfabetizzazione e Bourdieu (pp. 73-88). New York: Routledge.

Street, B.V., & Lefstein, A. (2007). *Literacy: An Advanced Resource Book. Applied Linguistics.* London: Routledge Applied Linguistics.

Su, J. (2019). Confirmed: Apple caught in Siri privacy scandal, let contractors listen to private voice recordings. *Forbes*. Retrieved from https://www.forbes.com/sites/jeanbaptiste/2019/07/30/confirmed-apple-caught-in-siri-privacy-scandal-let-contractors-listen-to-private-voice-recordings/#113cbb877314

Sulleyman, A. (2017). *Google's AI future: So impressive it's scary*. Retrieved from http://www.independent.co.uk/life-style/gadgets-and-tech/features/google-lens-ai-preview-features-so-impressive-its-scary-a7745686.html

Sum, S., Saboor, M., & Sahaf, R. (2015). Older people, loneliness and depression. *Medicinski Glasnik / Medical Gazette, 20*(56), 31-42.

Sweller, J. (1998). Cognitive load during problem solving: Effects on learning. *Cognitive Science, 12*(2), 257–285. doi:10.120715516709cog1202_4

Symanovich, S. (2018). *How does facial recognition work?* Retrieved from https://us.norton.com/internetsecurity-iot-how-facial-recognition-software-works.html

Szajna, B. (1996). Empirical evaluation of the revised technology acceptance model. *Management Science, 42*(1), 85–92. doi:10.1287/mnsc.42.1.85

Tabor, S. W. (2007). Narrowing the distance: Implementing a hybrid learning model for information security education. *Quarterly Review of Distance Education, 8*(1), 47-59.

Tarafdar, M., Gupta, A., & Turel, O. (2013). The dark side of information technology use. *Information Systems Journal, 23*(3), 269–275. doi:10.1111/isj.12015

Tashea, J. (2019). *As facial recognition software become more ubiquitous, some governments slam on the breaks.* Retrieved from http://www.abajournal.com/web/article/facial-recog-bans

Teich, A. G. (2019). *Repurposing school libraries as vibrant hubs and centers of change.* Retrieved September 21, 2019 from https://www.techlearning.com/resources/repurposing-school-libraries-as-vibrant-hubs-and-centers-of-change

Tene, O., & Polontsky, J. (2013). A theory of creepy: Technology, privacy and shifting social norms. *Yale Journal of Law and Technology, 16*(1). Retrieved from http://digitalcommons.law.yale.edu/cgi/viewcontent.cgi?article=1098&context=yjolt

Terry, N. (2001). Assessing enrollment and attrition rates for the online MBA. *T.H.E. Journal, 28*(7), 64–68.

The Alberta Teachers' Association. (2015). *Technology promise and peril: Growing up digital.* Retrieved from https://www.teachers.ab.ca/Public%20Education/EducationResearch/Pages/ResearchPublications.aspx

The Economist. (2019). *How creepy is your smart speaker?* Retrieved from https://www.economist.com/leaders/2019/05/11/how-creepy-is-your-smart-speaker

The Guardian. (2017). *Do Online Courses Spell the End for the Traditional University?* Retrieved from https://www.theguardian.com/education/2012/nov/11/online-free-learning-end-of-university

The White House. (2019, March). *Emerging technologies to support an aging population.* A report by the Task Force on Research and Development for Technology to Support Aging Adults. Committee on Technology of the National Science & Technology Council. Retrieved from https://www.whitehouse.gov/wp-content/uploads/2019/03/Emerging-Tech-to-Support-Aging-2019.pdf

Thomas, L., Herbert, J., & Teras, M. (2014). A sense of belonging to enhance participation, success and retention in online programs. *The International Journal of the First Year in Higher Education, 5*(2), 69–80. doi:10.5204/intjfyhe.v5i2.233

Thurston, A. (2005). Building online learning communities. *Technology, Pedagogy and Education, 14*(3), 353-369.

Time.com. (2019). *Amazon is making it easier to delete your Alexa recordings.* Retrieved from https://time.com/5686352/delete-amazon-alexa-recordings/

Tinto, V. (1975). Dropout from higher education: A theoretical synthesis of recent research. *Review of Educational Research, 45*(1), 89–125. doi:10.3102/00346543045001089

Tobenkin, D. (2019). *HR needs to stay ahead of automation.* Retrieved from https://www.shrm.org/hr-today/news/hr-magazine/spring2019/pages/hr-needs-to-stay-ahead-of-automation.aspx

Towles, D. E., Ellis, J. R., & Spencer, J. (1993, May). *Student persistence in a distance education program: The effect of faculty-initiated contact.* Paper at the Annual Forum of the Association for Institutional Research, Chicago, IL. (ERIC No. ED 360931)

Trinchero, R. (2004). *I metodi della ricerca educativa.* Bari: Laterza.

Trines, S. (2018). *Educating the Masses: The Rise of Online Education in Sub-Saharan Africa and South Asia.* Retrieved from https://wenr.wes.org/2018/08/educating-the-masses-the-rise-of-online-education

Tsai, H. S., Shillair, R., & Cotten, S. R. (2015). Social support and 'playing around': An examination of how older adults acquire digital literacy with tablet computers. *Journal of Applied Gerontology*, *36*(1), 29–55. doi:10.1177/0733464815609440 PMID:26491029

Turk, G. (2014). *Look Up Official Video*. Retrieved from https://www.youtube.com/watch?v=Z7dLU6fk9QY

U.S. Department of Education. (2012). *Enhancing Teaching and Learning Through Educational Data Mining and Learning Analytics: An Issue Brief*. Retrieved from http://tech.ed.gov/wp-content/uploads/2014/03/edm-la-brief.pdf

U.S. Department of Education. (2017). *Reimagining the Role of Technology in Education: 2017 National Education Technology Plan Update*. Retrieved from https://tech.ed.gov/files/2017/01/NETP17.pdf

Uncapher, M. R., Lin, L., Rosen, L. D., Kirkorian, H. L., Baron, N. S., Bailey, K., & Wagner, A. D. (2017, November). Media multitasking and cognitive, psychological, neural, and learning differences. *Pediatrics*, *140*(Supplement 2), S62–S66. doi:10.1542/peds.2016-1758D PMID:29093034

UNESCO. (2005). *Literacy for life. EFA Global Monitoring Report*. Paris, France: Author. Retrieved from http://www.unesco.org/education/gmr_download/chapter1.pdf

UNESCO-Institute for Lifelong Learning. (2009). *Harnessing the power and potential of adult learning and education for a viable future: Belém framework for action*. Hamburg, Germany: Author.

UNFPA.org. (2012). *Ageing in the twenty-first century*. New York: United Nations Population Fund. Retrieved from http://www.unfpa.org/pds/ageing.html

United States Census Bureau. (2018). *Older People Projected to Outnumber Children for first Time in U.S. History*. Retrieved from https://www.census.gov/newsroom/press-releases/2018/cb18-41-population-projections.html

United States Senate's Special Committee on Aging. (2017). *America's aging workforce: Opportunities and challenges*. Retrieved from https://www.congress.gov/115/crpt/srpt191/CRPT-115srpt191.pdf

University of Michigan. (1996). *Definitions of Instructional Design*. Retrieved from http://www.umich.edu/~ed626/define.html

Unwin, T., Kleessen, B., Hollow, D., Williams, J. B., Oloo, L. M., Alwala, J., . . . Muianga, X. (2010). Digital learning management systems in Africa: myths and realities. *Open Learning: The Journal of Open, Distance and e-Learning*, *25*(1), 5-23.

Valentino-DeVries, J., Singer, N., Keller, M. H., & Krolik, A. (2018). *Your apps know where you were sleeping night and they are not keeping it a secret*. Retrieved from https://www.nytimes.com/interactive/2018/12/10/business/location-data-privacy-apps.html

Van Deursen, A. J., Bolle, C. L., Hegner, S. M., & Kommers, P. A. (2015). Modeling habitual and addictive smartphone behavior: The role of smartphone usage types, emotional intelligence, social stress, self-regulation, age, and gender. *Computers in Human Behavior*, *45*, 411–420. doi:10.1016/j.chb.2014.12.039

Varis, T., & Al-Agtash, S. (2008). *Ubiquitous ICT for sustainable education and cultural literacy*. Citeseer.

Venkatesh, V. (1999). Creation of favorable user perceptions: Exploring the role of intrinsic motivation. *Management Information Systems Quarterly*, *23*(2), 239–260. doi:10.2307/249753

Venkatesh, V., & Davis, F. D. (2000). A theoretical extension of the technology acceptance model: Four longitudinal field studies. *Management Science*, *46*(2), 186–204. doi:10.1287/mnsc.46.2.186.11926

Victor, C., Scambler, S., Bond, J., & Bowling, A. (2000). Being alone in later life: Loneliness, social isolation and living alone. *Reviews in Clinical Gerontology*, *10*(4), 407–417. doi:10.1017/S0959259800104101

Vinez, K. E. (2017). *The admissibility of data colleted from wearable devices*. Retrieved from https://www2.stetson.edu/advocacy-journal/the-admissibility-of-data-collected-from-wearable-devices/

Vladimir, S. (2018). Machine learning approach for student engagement automatic recognition from facial expressions. *Scientific Publications of the State University of Novi Pazar Series A: Applied Mathematics, Infomatics and Mechanics, 10*(2), 79-96. Retrieved from https://scindeks.ceon.rs/article.aspx?artid=2217-55391802079S

Volery, T., & Lord, D. (2000). Critical success factors in online education. *International Journal of Educational Management*, *14*(5), 216–223. doi:10.1108/09513540010344731

Vygotsky, L. (1978). Interaction between learning and development. In M. A. Gauvin (Ed.), *Reading on development of children* (pp. 34–40). New York, NY: Scientific American Books.

Vygotsky, L. S. (1978). *Mind in society, the development of higher psychological processes*. Cambridge, MA: Harvard University Press.

Vygotsky, L. S., & Cole, M. (1978). *Mind in society: The development of higher psychological processes*. Cambridge, MA: Harvard University Press.

Wagner, D. A., & Srivastava, A. B. L. (1989). *Measuring literacy through household surveys*. New York, NY: United Nations Statistical Office.

Wagner, E. D., & McCombs, B. L. (1995). Learner centered psychological principles in practice: Designs for distance education. *Educational Technology*, *35*(2), 32–35.

Wagner, T. (2008). *The global achievement gap: Why even our best schools don't teach the new survival skills our children need--and what we can do about it*. New York, NY: Basic Books.

Wagner, W., & Pigza, J. M. (2016). The intersectionality of leadership and service-learning: A 21st-century perspective. *New Directions for Student Leadership, 2016*(150), 11–22. doi:10.1002/yd.20167 PMID:27150901

Waldner, L. S., McGorry, S.Y., & Widener, M. C. (2010). Extreme e-service learning (XE-SL): E-service learning in the 100% online course. *MERLOT Journal of Online Learning & Teaching, 6*(4), 839–851.

Waldner, L. S., McGorry, S. Y., & Widener, M. C. (2012). E-Service-Learning: The Evolution of Service-Learning to Engage a Growing Online Student Population. *Journal of Higher Education Outreach & Engagement, 16*(2), 123–150.

Waldner, L. S., Widener, M. C., & McGorry, S. Y. (2012). E-service learning: The evolution of service-learning to engage a growing online student population. *Journal of Higher Education Outreach & Engagement, 16*(2), 123–150.

Walker, W., & Walker, C. (2018). Do Course-Based Service Learning Programs Offer More Benefits to Students than Non-Course-Based Programs?" The Research Says' Yes'. *Journal of Education and Learning, 7*(6), 197–202. doi:10.5539/jel.v7n6p197

Wang, Y., & Kosinski, M. (2018). Deep neural networks are more accurate than humans at detecting sexual orientation from facial images. *Journal of Personality and Social Psychology, 114*(2), 246–257. doi:10.1037/pspa0000098 PMID:29389215

Ward, A. F., Duke, K., Gneezy, A., & Bos, M. W. (2017). Brain drain: The mere presence of one's own smartphone reduces available cognitive capacity. *Journal of the Association for Consumer Research, 2*(2), 140–154. doi:10.1086/691462

Warner, C. B., Roberts, A. R., Jeanblanc, A. B., & Adams, K. B. (2017). Coping resources, loneliness, and depressive symptoms of older women with chronic illness. *Journal of Applied Gerontology: The Official Journal of The Southern Gerontological Society*. doi:10.1177/0733464816687218 PMID:28380713

Warren-Gordon, K., & Graff, C. S. (2018). Critical Service-Learning as a Vehicle for Change in Higher Education Courses. Change. *The Magazine of Higher Learning, 50*(6), 20–23. doi:10.1080/00091383.2018.1540817

Washor, E., & Mojkowski, C. (2013). *Leaving to learn: How out-of-school learning increases student engagement and reduces dropout rates*. Portsmouth, NH: Heinemann.

Watson, M. M. (2019). Encore! A Boomer's guide to rocking your retirement. Littleton, CO: Bird Mountain Press.

Wehlage, G. G., Rutter, R. A., & Smith, G. A. (1989). *Reducing the risk: Schools as communities of support*. New York: The Falmer Press.

Weick, K. (2001). *Making sense of the organization*. Malden, MA: Blackwell Publishing.

Weick, K. E. (1995). *Sensemaking in organizations*. Thousand Oaks, CA: Sage.

Weintrop, D., Beheshti, E., Horn, M., Orton, K., Jona, K., Trouille, L., & Wilensky, U. (2016). Defining computational thinking for mathematics and science classrooms. *Journal of Science Education and Technology, 25*(1), 127–147. doi:10.100710956-015-9581-5

Wenger, E. (1998). Communities of practice: Learning as a social system. *The Systems Thinker, 9*(5), 2–3.

Wiggins, G. P. (1993). *Assessing student performance*. San Francisco: Jossey-Bass Publishers.

Wight, V. R., Price, J., Bianchi, S. M., & Hunt, B. R. (2009). The time use of teenagers. *Social Science Research, 38*(4), 792–809. doi:10.1016/j.ssresearch.2009.05.009

Wilhite, S., & Silver, P. (2005). Educating citizens vs. educating technicians: A false dichotomy for higher education. *National Civic Review, 94*(2), 46–54.

Wilkins, J. (2018). *Improving productivity with facial recognition*. Retrieved from https://www.automation.com/automation-news/industry/improving-productivity-with-facial-recognition

Williams, P. (2002). The learning web: The development, implementation and evaluation of internet-based undergraduate materials for the teaching of key skills. *Active Learning in Higher Education, 3*(1), 40–53. doi:10.1177/1469787402003001004

Wilson, M., Scalise, K., & Gochyyev, P. (2016). Assessment of learning in digital interactive social networks: A learning analytics approach. *Online Learning, 20*(2).

Wilson, C., & Moulton, B. (2010). *Loneliness among older adults: A national survey of adults 45+. Prepared by Knowledge Networks and Insight Policy Research*. Washington, DC: AARP.

Wilson, F. R. (1999). *The hand: How its use shapes the brain, language and human culture*. New York, NY: Vintage.

Wimpenny, K., Adefila, A., & DeWinter, A. (2018). *Jordan Opportunity for Virtual Innovative Teaching and Learning: A Needs Analysis Report Contextualising the State of the Art in International Online Teaching and Learning, with Particular Attention to the Jordanian Case*. Academic Press.

Wing, J. M. (2006). Computational thinking. *Communications of the ACM, 49*(3), 33–35. doi:10.1145/1118178.1118215

Wing, J. M. (2017). Computational thinking's influence on research and education for all. *Italian Journal of Educational Technology, 25*(2), 7–14. doi:10.17471/2499-4324/922

Wissinger, E. (2018). Blood, sweat, and tears: Navigating creepy versus cool in wearable biotech. *Information Communication and Society, 21*(5), 779–785. doi:10.1080/1369118X.2018.1428657

With, A. C. (n.d.). *Mining the Computational Universe*. Retrieved from https://www.edge.org/conversation/stephen_wolfram-mining-the-computational-universe

Witherspoon, E. B., Hugashi, R., Schunn, C., Baehr, E., & Shoop, R. (2017). Developing computational thinking through a virtual robotics programming curriculum. *ACM Transactions on Computing Education, 32*(1), 1–20. doi:10.1145/3104982

Wojciechowski, A., & Palmer, L. B. (2005). Individual student characteristics: Can any be predictors of success in online classes? *Online Journal of Distance Learning Administration, 8*(2). Retrieved from http://www.westga.edu/~distance/ojdla/summer82/wojciechowski82.htm

World Health Organization (WHO). (2015). *Public Health Implications of Excessive Use of the Internet, Computers, Smartphones and Similar Electronic Devices- Meeting report.* Geneva, Switzerland: WHO Press, World Health Organization. Retrieved from https://apps.who.int/iris/bitstream/handle/10665/184264/9789241509367_eng.pdf;sequence=1

Worldbank.org. (2018). *Technology landscape for digital identification.* Retrieved from http://documents.worldbank.org/curated/en/199411519691370495/Technology-Landscape-for-Digital-Identification.pdf

Xie, Y., Szeto, G. P., Dai, J., & Madeleine, P. (2016). A comparison of muscle activity in using touchscreen smartphone among young people with and without chronic neck-shoulder pain. *Ergonomics, 59*(1), 61–72. doi:10.1080/00140139.2015.1056237 PMID:26218600

Yániz, C., & Villardón, L. (2006). *Planificar desde competencias para promover el aprendizaje. El reto de la sociedad del conocimiento para el profesorado universitario* [Planning from the competence perspective to promote learning. Challenge of the knowledge society for university teachers]. Bilbao, España: Deusto University.

Yao, M. (2017). *Why building social robots is much harder than you think.* Retrieved from https://www.topbots.com/building-social-robots-jibo-anki-cozmo-much-harder-think/

Yen, J. Y., Ko, C. H., Yen, C. F., Wu, H. Y., & Yang, M. J. (2007). The comorbid psychiatric symptoms of Internet addiction: Attention deficit and hyperactivity disorder (ADHD), depression, social phobia, and hostility. *The Journal of Adolescent Health, 41*(1), 93–98. doi:10.1016/j.jadohealth.2007.02.002 PMID:17577539

Yin, R. K. (2011). *Applications of case study research.* Sage.

Yin, R. K. (2012). A (very) brief refresher on the case study method. *Applications of Case Study Research*, 3-20.

Yin, R. K. (1981). The case study crisis: Some answers. *Administrative Science Quarterly, 26*(1), 58–65. doi:10.2307/2392599

Yin, R. K. (2014). *Case study research: Design and methods* (5th ed.). Thousand Oaks, CA: Sage.

Yin, R. K. (2015). *Qualitative research from start to finish.* Guilford Publications.

Yoo, S. J., & Huang, W. D. (2013). Engaging online adult learners in higher education: Motivational factors impacted by gender, age, and prior experiences. *The Journal of Continuing Higher Education, 61*(3), 151–164. doi:10.1080/07377363.2013.836823

Young, J. E. (1982). Loneliness, depression, and cognitive therapy: Theory and applications. In L. A. Peplau & D. Perlman (Eds.), *Loneliness. A sourcebook of current theory, research and therapy* (pp. 379–405). New York: Wiley.

Yu, X., Anaya, G. J., Miao, L., Lehto, X., & Wong, I. A. (2018). The impact of smartphones on the family vacation experience. *Journal of Travel Research*, *57*(5), 579–596. doi:10.1177/0047287517706263

Zaharias, P. (2002). On-line Learning Courses: A review and usability attributes. In M. Driscoll & T. Reeves (Eds.), *Proceedings of E-Learn 2002--World Conference on E-Learning in Corporate, Government, Healthcare, and Higher Education* (pp. 1056-1062). Montreal, Canada: Association for the Advancement of Computing in Education (AACE). Retrieved from https://www.learntechlib.org/primary/p/15341/

Zammuner, V. L. (1998). *Tecniche dell'intervista e del questionario*. Bologna: Il Mulino.

Zhang, B. (2018). *Why creepy ad is the new black?* Radbound University. Retrieved from https://www.ru.nl/publish/pages/769526/zb_michael_zhang.pdf

Zhang, S., Zhao, L., Lu, Y., & Yang, J. (2016). Do you get tired of socializing? An empirical explanation of discontinuous usage behaviour in social network services. *Information & Management*, *53*(7), 904–914. doi:10.1016/j.im.2016.03.006

Zhitomirsky-Geffet, M., & Blau, M. (2016). Cross-generational analysis of predictive factors of addictive behavior in smartphone usage. *Computers in Human Behavior*, *64*, 682–693. doi:10.1016/j.chb.2016.07.061

About the Contributors

Julie Anne Delello is an Associate Professor in the College of Education and Psychology at The University of Texas at Tyler. She also serves as the Director of the Center for Excellence in Teaching and Learning at the university. She received her Ph.D. in Curriculum and Instruction with a specialization in science and technology from Texas A&M University. She has extensive experience in curriculum and technology implementation and faculty training and development. She currently serves on the leadership board for The International Society of Technology Education's Teacher Education Network. She recently served as an expert panelist for the EDUCAUSE 2019 Higher Education Horizon Report and she is a Microsoft Innovative Educator Expert. She has authored numerous publications and her professional interests focus on teaching pedagogies, academic innovations, visual media technologies, artificial intelligence and STEM explorations, gerontechnology, and social media platforms for authentic learning.

Rochell R. McWhorter is an Associate Professor of Human Resource Development in the Soules College of Business at The University of Texas at Tyler. She received her Ph.D. degree from Texas A&M University. Dr. McWhorter currently serves as an Associate Editor of the Advances in Developing Human Resource Development journal. She has authored numerous journal articles and book chapters around technology-facilitated learning in higher education and in the workplace. Her publications include topics such as Virtual Human Resource Development (VHRD), professional virtual conferences, virtual scenario planning, visual social media, real-time group meetings, virtual leadership development, and eService-Learning.

* * *

Abdeleh Bassam Al Amoush is a Ph.D. candidate in the Business School at The University of New England, Australia. Dr. Bassam has a M.S. degree in Computer Information System from AABFS and a BSc degree in Computer Science from the Princess Sumaya University for Technology.

Elisabeth E. Bennett, PhD, is Associate Teaching Professor of Organizational Leadership Studies at Northeastern University, Boston, MA. Dr. Bennett is a leading scholar in virtual human resource development and her research includes technology, organizational culture, and informal learning. She co-edited two Advances in Developing Human Resources issues on virtual HRD as well as writing book chapters and articles on the subject. Dr Bennett previously was director of education research and development at the Western Campus of Tufts University School of Medicine, and has worked in financial services, manufacturing, and university-based continuing education. She served on the board of the Academy of Human Resource Development and presently is a member of the editorial boards for Human Resource Development Quarterly, Adult Education Quarterly, Advances in Developing Human Resources, and New Horizons in Adult Education and Human Resource Development.

Kleanthe Caruso, MSN, is a Clinical Instructor in the College of Nursing and Health Sciences. She received her MSN in Administration from The University of Texas Medical Branch School of Biomedical Sciences, Galveston, Texas and has BSN and ANCC certifications. She teaches courses in nursing competencies, gerontological nursing, as well as leadership and management.

Felice Corona is a Full Professor of Education and Special Pedagogy of the University of Salerno, Italy. Dr. Corona is the President of the Teaching Council in Education Sciences for Inclusion and Wellness (SFIB) and Director of the specialization course in Education and Psychopedagogy for students with autistic disorders in the Department of Human Sciences, Philosophy and Education at the University of Salerno. Dr. Corona is a member of the Editorial Board of Autism Insight, Journal of Experimental Neuroscience, Rehabilitation Process and Outcome, and Annals of Neurosciences. His recent scientific contributions deal with the issue of special educational needs with a transdisciplinary approach, which integrates the contributions of medical sciences, neurosciences and technoscience to the heritage of educational and didactic research.

Gregory Cottrell has eight years of experience as a technology education teacher and five years as an assistant principal. In 2017, Dr. Cottrell became the Principal of North Hunterdon High School in Annandale, NJ. In 2017, he also earned his doctorate from East Stroudsburg University.

Tonia De Giuseppe has a PhD in Language, Society, Politics and Education: Corporality, Technology and Inclusion, at the University of Salerno, Italy. She works in the Department of Political, Social and Communication Sciences (DISPSC).

Dr. Giuseppe is a member of the international editorial board for the magazine IJCRSE. Lecturer of the Ministry, University, and Research (MIUR) since 1991, she has worked indefinitely on adult education and the teaching of Italian. Since 1998, she has been involved in intercultural and inclusive projects for which she has received both national and international awards. She is the author of scientific papers on magazines, texts, and monographs on topics of didactic innovation of inclusive design.

Greg Duckworth is currently a Ph.D. student at the University of Texas at Tyler. Mr. Duckworth is an experienced Human Resources Director with a demonstrated history of working in the information technology and services industry. Skilled in HR consulting, executive coaching, HR Policies, and management. Mr. Duckworth is a strong human resource professional with a M.S. HR and Training and M.B.A. focused in Human Resources Development from Amberton University.

Marie-Line Germain is an internationally recognized scholar who holds a tenured position as Associate Professor of Human Resources and Leadership and serves as the Director of the HR Consulting Initiative at Western Carolina University (a campus of the University of North Carolina system). Professor Germain founded the WCU HR Consulting Initiative, which has provided pro bono HR support to close to 200 organizations since 2011. Professor Germain was selected as a Fulbright Specialist in 2017 in Ghana, Africa, where she provided guidance to a major Ghanian university on furthering research initiatives. She has provided expertise and support to small and large for-profit (Fortune 50) and nonprofit organizations on psychosocial support of employees, among many others. Professor Germain has been a visiting professor in Finland, South Korea, China, and Africa. In addition to her scholarship, Dr. Germain is an experienced consultant specialized in psychosocial risks, mental health in the workplace, and personality disorders or corporate leaders. She has served as a consultant to CAC 40 Companies, emerging businesses, and nonprofits. She has published nearly 50 peer-reviewed research articles, book chapters, and editorials and has given approximately 150 presentations throughout the country and around the world. She is the author of two books: Narcissism at Work: Personality Disorders of Corporate Leaders and Integrating Service-Learning and Consulting in Distance Education. She holds a Ph.D. in Leadership and HRD. She is a member of the Eastern Academy of Management and the Academy of Human Resource Development.

Christine S. Gipson is an Assistant Professor in the College of Nursing and Health Sciences at The University of Texas at Tyler where she earned her PhD in 2016. She is also a licensed RN and CNE and teaches community health nursing and health policy for population health. Her research interests include health promotion, community health, wellness and education at the university.

Catherine Glaser received her degree from The University of Northern Colorado. Dr. Otieno has background in K-12, higher education and academic libraries. She has taught many educational technology courses geared towards integration in K-12 environment. Dr. Otieno has worked closely with practicing preservice teachers and instructors interested in integrating technology in their curriculum. Her research interests range from effective technology integration in pedagogy, mobile learning, Makerspaces, libraries and information literacy.

Deborah Kerby teaches K-6 Computer Skills and Computer Science at a public school in Northeast, PA. Before teaching elementary she taught 16 years of Business and Computer classes at Pocono Mountain East High School. Deborah earned a Doctorate in Education Leadership, as well as a Certificate in Instructional Technology, from Wilkes University. Before a career in teaching she worked in Business doing Computer and Networking Support, Database Design, Programming and Accounting. She has presented on Adaptive Learning with Google Forms in a Webinar and a face-to-face session through ISTE.

Jean Kiekel is a professor at The University of St. Thomas, a small Catholic, liberal arts college in Houston, Texas where she teaches in the School of Education and Human Services.

Andrew Krouse is a Ph.D. student at The University of Texas at Tyler. Krouse has been in higher education where he has served as Director of Facilities Management and as an adjunct professor in finance and international business. His experience includes planning and coordinating facilities; participating on an oversight team for all campus capital projects, coordinating projects with various departments such as environmental health and safety and campus police and parking, fiscal administration, human resource actions, work flow management and operational improvement.

Mandi Laurie is a Clinical Researcher at Hanger Prosthetics and Orthotics. She is currently a doctoral student at The University of Texas at Tyler pursuing her PhD in Human Resource Development focused on organizational development and change. Her research interests are nonprofit organizations, community engagement, and eService-Learning.

Beth Mastel-Smith is an Associate Professor of Nursing at The University of Texas at Tyler. She holds a PhD in Nursing from Texas Woman's University and her research interests include quality of life for older adults and caregivers and transcultural nursing. Her teaching has included courses in Gerontological Nursing, Transcultural and Mixed Methods Research, and Scholarship in Nursing.

Carol Munn is a tech-savvy, dedicated technology teacher with over fifteen years of career experience in computer science education. She is highly adept in providing professional development on integrating technology into the existing curriculum and creating new lessons to encompass digital technology in all subjects. Carol is skilled in implementing a methodology in project-based learning, responsive classroom, cooperative learning groups, performance assessments, and differentiated instruction. She designs and compliments lessons with high-level engaging learning activities integrating STEM curricula, ISTE and NGSS standards. Ms. Munn is currently working on a dissertation titled A Qualitative Study Exploring Robots as a Potential Teaching Tool for Computational Thinking in a Sixth-Grade Classroom.

Susan R. Poyo has been a Catholic educator for over 25 years. Her experiences extend from K-12 to Higher Education and Adult Education. Dr. Poyo earned her doctoral degree in Instructional Technology from Duquesne University and is currently the Director of Undergraduate Education and Associate Professor at Franciscan University of Steubenville. Preparing teachers to educate the 21st century learner is her passion, engaging her students in conversation, inquiry and practical application of effective integration of technology in planning and assessment for learning. Dr. Poyo is a leader for the ISTE Teacher Education Network and has presented both nationally and internationally on TPACK, Curriculum Design and Assessment, and Online and Blended Learning.

Isabel Resende is the Assistant Superintendent of the Nazareth Area School District. With 19 years of experience, Dr. Resende has served as a high school biology teacher, K-12 Director of Math, Science, and Technology. Dr. Resende earned her M.A. in Educational Leadership from Lehigh University, and completed her doctorate at East Stroudsburg University. Dr. Resende's study described the influence of learning analytics on teacher instructional practices at the student level.

Kamaljeet Sandhu is an active scholar, experienced and passionate about research in digital innovation and strategy, business data analytics, digital health, AI, IoT, cryptocurrency, blockchain, cloud computing, IT startups, FinTech, cybersecurity, accounting, corporate governance & CSR, supply chain, ERP, SMEs, and entrepreneurship. He is the editor of three books and has published over 180 articles in peer-reviewed journals, book chapters, and conference presentations. He has held multiple senior leadership positions at universities in Australia, Canada, Fiji, and

India. In addition, Dr. Sandhu serves on several committees at The University of New England Business School and in professional associations, on editorial boards, and he also works with industry experts.

Mia Kim Williams is known for her work in innovative and critical pedagogy. She supports students' learning of innovation and development of creativity, design, critical thinking, and community engagement. She currently coordinates the Learning, Design, and Technology programs at the University of Wyoming. She works with practicing and preservice teachers as well as doctoral students interested in promoting change in a variety of educational contexts. Her current research projects include using game design as a critical pedagogy, designing transformative learning spaces, and a collaboration with Microsoft to study the Minecraft Global Mentoring Program. She earned her PhD in Curriculum Studies from Arizona State University and has a history in teaching high school English and Earth Science in rural Arizona.

Index

F

G

H

I

J

K

L

M

O

P

Q

R

S

T

U

V

W

www.ingramcontent.com/pod-product-compliance
Lightning Source LLC
LaVergne TN
LVHW081348300926
849161LV00008B/2096
* 9 7 8 1 7 9 9 8 2 9 1 4 0 *